Encyclopedia of Milk Production: Animal Health and Nutrition

Volume I

Encyclopedia of Milk Production: Animal Health and Nutrition
Volume I

Edited by **Christian Snider**

R CALLISTO
REFERENCE

New York

Published by Callisto Reference,
106 Park Avenue, Suite 200,
New York, NY 10016, USA
www.callistoreference.com

Encyclopedia of Milk Production: Animal Health and Nutrition
Volume I
Edited by Christian Snider

International Standard Book Number: 978-1-63239-272-5 (Hardback)

Printed in the United States of America.

Contents

Preface

The purpose of the book is to provide a glimpse into the dynamics and to present opinions and studies of some of the scientists engaged in the development of new ideas in the field from very different standpoints. This book will prove useful to students and researchers owing to its high content quality.

This book is aimed at presenting new developments in knowledge related to milk production. It is dedicated to detailed analysis of various fields with aspects of genetics factors and the molecular and cellular mechanisms, animal supervision, nutrition and husbandry. This book will prove to be beneficial for students, researchers, teaching staff and practicing professionals associated with dairy science, animal science, food science, nutrition, physiology, biochemistry, veterinary medicine and related fields. There are extensive references in this book which will help readers in future endeavors regarding the developments in this field.

At the end, I would like to appreciate all the efforts made by the authors in completing their chapters professionally. I express my deepest gratitude to all of them for contributing to this book by sharing their valuable works. A special thanks to my family and friends for their constant support in this journey.

Editor

Genetics Factors and Cellular Mechanism

Genetic Improvement of Livestock for Milk Production

Sammy K. Kiplagat, Moses K. Limo and Isaac S. Kosgey

Additional information is available at the end of the chapter

1. Introduction

This chapter presents issues pertaining to genetic improvement of livestock for production. It covers aspects from basic population to quantitative genetics to molecular genetics, and their application in animal breeding. Genetics is the science of heredity which is concerned with physical and chemical properties of the hereditary material, how the material is transmitted from one generation to the next and how the information it contains is expressed in the development of an individual. Genetic make-up of animals control their structural configuration and productive abilities either via single genes or by multiple genes situated in different loci. Genes are compost of nucleotide sequences packaged into chromosomes in the nucleus. Milk production is largely affected by a combination of factors namely; genetic make-up in terms of the use of improved breeds selected for milk production, a favourable nutritional environment and improved managerial practices. Consequently, genetic make-up of dairy animals plays a great role in the variation of milk yield and composition. Milk production is, therefore, a factor of genotype-environment interactions. It is important to balance selection for both production (e.g., milk yield and composition) and functional (e.g., fertility, disease resistance, feed intake and body weight) traits. Techniques applied in molecular genetics in conjunction with conventional animal breeding techniques could be used to optimize animal breeding programmes, resulting in higher yields (i.e., greater genetic gains), as it is possible to determine the potential of an animal, even before the trait is expressed phenotypically. A genetic marker serves to favourably relate alleles for quantitative characteristics with information about the individual mode of action and their interaction of genes, helping to understand the quantitative variations and their practical use in animal husbandry. DNA markers present two possible future applications in animal selection; the combination of the best alleles of two or more breeds, and the selection of the best alleles within a breed or lineage. Commonly used genetic markers are the DNA-based markers; RFLPs and minisatellites, and PCR-based markers like microsatellites, and SNP. DNA-based markers are more direct molecular markers that survey DNA variation itself rather than rely on variations in the electrophoresis mobility of protein that the DNA encodes. They allow the number of mutations between different alleles to be quantified. DNA containing genetic information identified to influence milk production traits can be artificially introduced into a dairy animal, using recombinant DNA technology, and then it must be transmitted through the germ line so that every cell, including germ cells, of the animal contain the same modified genetic material. Such techniques enable dairy animals to acquire desired milk production characteristics. However, use of such transgenic techniques attracts ethical questions. Generally, genetic marker approaches is a promising tool for milk production improvement. It is imperative that genetic improvement for milk production is approached holistically, taking into consideration all the factors that may affect a breeding programme.

2. Functional traits genetics

Functional traits are those traits that increase biological and economic efficiency not by higher outputs of products, but by reduced costs of production. Functional traits have become important for efficient breeding schemes in the dairy industries, due to increased costs of production relative to milk prices and consumers demand for safe, quality food and attention to animal welfare. Since dairy animals are bred in a wide range of local production conditions, the list of candidate functional traits may be large, including components of milk feed efficiency (body weight, feed intake and body reserves), reproduction traits (sexual precocity, out-of-season calving/lambing/kidding ability, female fertility), calf/lamb/kid meat production (suckling ability, prolificacy), milking ability (udder morphology, milking speed), resistance to disease (mastitis, scrapie, internal parasites), adaptation to local breeding conditions (fitness, wool, longevity), and others (Barillet et al., 2007). The relationships between milk production and functional traits are often null or antagonistic, illustrating the importance of knowledge of the genetic correlations between milk production and the functional traits of interest (Barillet et al., 2007).

Functional traits that increase efficiency not by higher output of products but by reduced costs of input, might have a greater impact on the profit of dairy farmers and should, therefore, be included in breeding programmes. Apart from economic reasons for including functional traits in the breeding programmes, there are several non-economic reasons, for example ethical and consumer concerns, which are becoming increasingly important (Olesen et al., 1999). The inclusion of functional traits in breeding programmes will likely have a major impact on the expected selection response of the functional traits, and will result in only small losses of the expected selection response of the production traits. Depending on the number of functional traits included in a breeding scheme, the relative importance of production versus functional traits varies from 70:30 to 30:70, sometimes even more. Relative weightings for the two groups of traits, production and functional, for the Estonian Holstein population were 79:21. In any of breeding scenarios tested, selection response in financial terms will come largely from production traits, because genetic parameters favour fat and protein yields (high heritability, high positive genetic correlation) (Pärnal et al., 2003).

Smallholders, pastoralists and their animals often live in harsh environments which may be hot and dry, hot and humid, or high in altitude and cold. Moreover, these environments can be characterized by scarce feed and water resources and high disease pressure with large seasonal and annual variation. Adaptation to these factors is largely based on genetics, but animals can "learn" to live under such stressful conditions. To match genotype with the environment, breeders can follow two alternative strategies: adapt the environment to the needs of the animals as is the case in industrial animal production systems or keep animals that are adapted to the respective environment as is the case in low input smallholder and pastoral systems. Because of this, smallholders and pastoralists need different and diverse animal genotypes, species mix and types to enable animal husbandry in their specific physical environment locations and production systems (Mirkena et al., 2010).

3. Milk production traits genetic determinants

The phenotypic expression of milk production traits (e.g., milk yield and composition) are controlled by genes, which may or may not be transferred to the offspring. The genetic value of a trait indicates the likelihood that the genes responsible for that trait will be transferred to any offspring. Consequently, when dairy producers are selecting animals for breeding stock, they are typically more concerned with an animal's genetic value rather than its phenotypic value of a particular trait. The difference is that while the phenotypic value refers to the presence or absence of particular traits, the genetic value indicates the potential (or probability) that this animal, if bred, will give birth to calves with certain desired traits. The challenge of the dairy breeder is, therefore, to determine which cows and bulls to breed in order to obtain progeny with high quality milk production traits, as well as any other desirable attributes.

Two main reasons for the decline in fitness traits of cows associated with increased genetic merit for milk yield are: (i) fitness traits are ignored in the construction of selection indices because they are considered to have lower heritability or are not easy to record and (ii) use of inappropriate breeding programmes while the underlying genetic process (selection and inbreeding depression) is not well understood (Goddard, 2009). However, the low heritability of some fitness traits does not imply negligible genetic variance; often heritability is low because the phenotypic variance is rather larger than the genetic variance as evidenced by as high genetic coefficient of variation for fitness traits as for some production traits (Goddard, 2009).

The appropriate strategy for any breeding programme would, therefore, be to set suitable selection goals that match the production system rather than ambitious performance objectives that cannot be reached under the prevailing environment. Area-specific approach utilizing the existing resources and taking into account the prevailing constraints appears to be the only reasonable sustainable solution. Such approach would also enable *in situ* conservation of farm animal genetic resources, the only viable and practical conservation method in less developed countries compared to *ex situ* or cryopreservation approaches. This would support the importance of identifying the most adapted genotype capable of coping with the environmental challenges posed by any particular production systems (Mirkena et al., 2010).

Karugia et al. (2000) analyzed the impact of crossbreeding zebu with temperate cattle breeds for dairy improvement in Kenya using sector- and farm-level approaches. The agricultural sector model showed that a dairy technology that involved crossbreeding and complementary improvements in nutrition and management has had a positive impact on Kenyan economy and welfare but this approach ignored important social cost components of crossbreeding. The farm-level approach, however, indicated that farm performance was little improved by replacing the indigenous zebu with exotic breeds. Conversely, this analysis indicated that a breeding programme that concentrates on improving the local zebu breeds would improve the financial performance of the farm level with important implications for the conservation of farm animal biodiversity.

4. Genotype by environment interactions

The external environmental stimuli (physical, chemical, climatic and biological) to which animals respond interact with their genotypes to determine level of performance. In the absence of genotype by environment interaction, the expected genetic correlation across environments is one. However, all species respond to changing natural environments through altering phenotype and physiology; in livestock production the situations become more complex since human intervention influences both genotype and external environment (King, 2006).

When genotypes have significant differences between the quantitative measures of the phenotypic plasticity, then there is a genotype by environment interaction. Plastic genotypes are known by highly variable phenotypes across environments, whereas robust or stable genotypes are known by relatively constant phenotypes across environments. Differences in the phenotypic plasticity could be explained by the fact that some alleles may only be expressed in some specific environment due to change in some gene regulations depending on the environment; favorable genes in some environments may become unfavorable under other environmental conditions. Developing countries in the tropics often rely on exotic germplasm for breeding purposes. They, however, have climatic conditions, production systems and markets that are different from those where animals were evaluated. Consequently, the genotype-environment interactions can cause reduced efficiency of their genetic improvement programmes. When genotype by environment interactions exists and the environment is under the control of the breeders (i.e., genotype by ration or genetic by housing interaction), it would be easier for breeders to modify the environment to allow optimum expression of the genotype. However, when environments are beyond the breeders' control, they have to choose the genotypes able to adapt to those environments (Hammami et al., 2008). In low input systems, the best alternative to circumvent the consequences of genotype by environment interactions is to select for adaptive traits.

According to Mirkena et al. (2010), imported improved temperate breeds produce more than indigenous tropical breeds if supplied with high quality feed; however, they lose weight and fail to survive when fed poor quality grass or straw, whereas adapted indigenous animals still grow, give some milk and reproduce. Adapted tropical animals recycle nutrients more efficiently than do improved temperate breeds and can also reduce their basic metabolism during periods of weight loss (Bayer and Feldmann, 2003). Leitóna et al. (2008) observed that the average genetic variance for 305-d milk yield in Costa Rican populations of Holstein and Jersey cows was near 20%. The environmental and genetic trends for milk production in both breeds were positive, although the proportion attributable to genetic improvement was low compared to the phenotypic increase. The genotype by environment interaction had a significant effect on milk production in both breeds, but was particularly marked for the Holstein breed. The study concluded that these were probably caused by the lack of control over import of genetic material into Costa Rica and how it was used, which implied that dairy producers needed to reconsider the genetic improvement strategy based almost exclusively on importing genetic material.

In Zebu cattle, Freitas et al. (2010), in a preliminary study involving Gyr dairy cattle, observed the effect of herd on milk production of daughters of sires with different breeding values, pointing out for the possibility of G-E interaction. In another study, Ayalew et al. (2003) compared productivity of indigenous breeds of goats (Hararghe Highland and Somali) with that of crossbred (Anglo-Nubian X Somali) goats in Ethiopia and concluded that the crossbreds did not improve households' income in the mixed crop–livestock production system. The study indicated that there were increased net benefits per unit of land or labor from mixed flocks (i.e. both indigenous goats and Anglo-Nubian crosses) under improved management compared with indigenous goats under traditional management. In flocks using an improved management package, the crossbreds did not produce more net benefits than indigenous goats either in mixed or separate flocks. The improved management package, however, increased net benefits of farmers keeping indigenous goats; these findings that explained the low adoption rate of exotic crosses by smallholder farmers and superior adaptability of indigenous goats to the prevailing production system (Mirkena et al., 2010). It is, therefore, imperative that the genetic improvement of locally adapted breeds will be important to realizing sustainable production systems.

5. Genetic variations

Allele frequencies and, therefore, genotypic frequencies do not change on their own accord. They will tend to remain the same generation after generation, and each progeny generation will tend to resemble its parental generation. Counteracting this tendency is a number of processes that can change allele frequency and thereby lead to the genetic modification of progeny. In the long term, the most important of the modifying processes is natural selection, the process in which the most adapted to survive and reproduce in their environment and, subsequently, contribute more than an equal share of alleles to the next generation; when repeated over a course of many generations, a disproportionate contribution of alleles, even if small, will significantly increase the frequency of alleles responsible for the superior adaptation. Other causes of change in the frequency of alleles are as follows: mutation that occurs when a DNA gene is damaged or changed in such a way as to alter the genetic message carried by that gene and/ or changes in sequences of introns and/ or promoter regions; migration, the movement of individuals among subpopulations within a larger population; random genetic drift that results from random undirected changes in all population and, especially, occurring in small populations. These four processes account for most or all of the changes in allele frequencies that occur in populations. They form the bases for cumulative change in the genetic characteristics of populations, leading to descent with modification.

Interestingly, mutation can result due to replication errors caused by both endogenous and exogenous factors. Endogenous factors consist of transversion, spontaneous depurination of bases, deamination of cytosine and sometimes adenine residues, yielding uracil and hypoxanthine, respectively. Exogenous reactions for mutations include dimerization of pyrimidine bases induced by ultra violet light, various chemicals such as alkylating agents forming adducts with DNA bases, reactive oxygen species damaging pyrimidine and purine rings and ionizing radiation causing DNA strand nicking and breakage. The majority of these modifications are generally recognized and corrected by the DNA repair system.

6. Molecular genetic technologies

Molecular genetics is the study of the genetic makeup of individuals at the DNA level; it is the identification and mapping of genes and genetic polymorphisms. There are opportunities for using molecular genetics to identify genes that influence milk production traits. Armed with this information, it would be possible to select improved livestock on the basis of their genetic makeup. If applied with care, the use of molecular information in selection programmes has the potential to increase productivity, enhance environmental adaptation and maintain genetic diversity (Naqvi, 2007). The first task is to understand the genetic control of the trait of interest and then to identify the genes and genotypes involved.

Molecular genetic technologies have been used to identify loci or chromosomal regions that affect single-gene traits and quantitative traits. Single-gene traits include genetic defects, genetic disorders, and appearance. For the purposes of quantitative traits loci (QTL) detection and application, quantitative traits can be categorized into (a) routinely recorded traits; (b) difficult to record traits (e.g., feed intake and product quality); and (c) unrecorded traits (disease resistance). Each of these can be further subdivided into traits that are (i) recorded on both sexes; (ii) sex-limited traits; and (iii) traits that are recorded late in life. The ability to detect QTL depends on the availability of phenotypic data and decreases in the order a, b, c and within each of those in the order i, ii and iii. For related reasons, genome scans, which require more phenotypic data than candidate gene analyses, are often used to detect QTL for traits in category a, whereas candidate gene approaches are more often used to identify QTL for traits that are not routinely recorded (b and c) (Meuwissen and Goddard, 1996).

The use of molecular genetic technologies potentially offer a way to select breeding animals at an early age (even embryos); to select for a wide range of traits and to enhance reliability in predicting the mature phenotype of the individual. The broad categories of existing gene technologies based options include; molecular analysis of genetic diversity, animal identification and traceability, reproductive enhancement; transgenic livestock; germ line manipulation and; marker/ gene based trait selection; animal health: diagnosis, protection and treatment; ruminant and non-ruminant nutrition and metabolism (Naqvi, 2007).

7. Genetic markers

Recent developments in molecular biology and statistics have opened the possibility of identifying and using genomic variation and major genes for the genetic improvement of livestock. Molecular techniques allow detection of the existence of variation or polymorphisms among individuals in the population for specific regions of the DNA. These polymorphisms can be used to build up genetic maps and to evaluate differences between markers in the expression of particular traits in a family that might indicate a direct effect of these differences in terms of genetic determination on the trait (Montaldo et al., 1998).

Application of molecular genetics for genetic improvement relies on the ability to genotype individuals for specific genetic loci. Genetic markers can be used to identify specific regions of chromosomes where genes affecting quantitative traits are located, i.e., QTL (Davis and DeNise, 1998). These techniques can directly confirm the potential parent-to-offspring transfer of those genes associated with a desired trait (Akhimienmhonan and Vercammen, 2007). For these purposes, three types of observable polymorphic genetic loci can be distinguished: (i) direct markers: loci that code for the functional mutation; (ii) LD markers: loci that are in population- wide linkage disequilibrium with the functional mutation; and (iii) LE markers: loci that are in population wide linkage equilibrium with the functional mutation in outbred populations; linked markers can be used within families segregating marker and QTL alleles following the establishment of the phase relationship (Davis and DeNise, 1998). The LE markers can be readily detected on a genome-wide basis by using breed crosses or analysis of large half-sib families within the breed. Such genome scans require only sparse marker maps (15 to 50 cM spacing, depending on marker informativeness and genotyping costs; to detect most QTL of moderate to large effects (Darvasi et al., 1993). The LD markers can be identified using candidate genes (Rothschild and Soller, 1997) or fine-mapping approaches (Andersson, 2001). Direct markers (i.e., polymorphisms that code for the functional mutations) are the most difficult to detect because causality is difficult to prove and, consequently, a limited number of examples are available, except for single-gene traits (Andersson, 2001). Direct markers where a linkage analysis has been performed and a zero recombination rate found between the markers and the QTL, or where sequence data have verified the exact location of the genetic change in a number of individuals. Direct markers can be used across families after prediction of an allelic effect for a given genetic background. Both markers can be used in MAS programmes that incorporate other pedigree and phenotypic information for the genetic evaluation of animals (Davis and DeNise, 1998).

Direct markers can be identified by use of candidate gene approach. Candidate gene approach proposes that a significant proportion of quantitative genetic variation of a given trait is contributed by segregation of functional alleles of one or more of the candidate genes for the trait (Rothschild and Soller, 1997). Candidate genes are genes that play a role in the development or physiology of a trait of economic importance. At the DNA level, a candidate gene comprises a contiguous tract of DNA, including introns, exons, and upstream and downstream regulatory regions concerned with biosynthesis of a single protein or via alternative processing to produce related proteins. Allelic variation at a candidate gene sequence can cause a change in protein production or efficiency in a metabolic process that will influence a specific trait. The candidate gene approach can be very powerful and can detect loci even with small effect, provided that the candidate gene represents a true causative gene. However, there are often many candidate genes for the trait of interest and it may be more time-consuming to evaluate all of these than performing a genome scan. Furthermore, the candidate gene approach might fail to identify a major trait locus simply because of the gap in knowledge about a gene function. Candidate gene tests must be interpreted with caution because spurious results can occur because of linkage disequilibrium to linked or non-linked causative genes or because the significant thresholds have not been adjusted properly when testing multiple candidate genes. Once the chromosomal location of a trait locus has been determined, this information can be applied in breeding programmes by using Marker-Assisted Selection (MAS). Candidate genes can be sequenced and analyzed in animals manifesting divergent expressions of a given trait of interest. Sequence analysis provides highest resolution of DNA variation; provides the fundamental structure of the gene systems. It is a vital tool in the analysis of gene structure and expression (Drinkwater and Hazel, 1991).

Quantitative trait loci have been detected in experimental and commercial populations of cattle, swine and sheep. In dairy cattle, linked markers have been reported for milk and component yields (Georges et al., 1995) and cheese yield (Graham et al., 1984).

8. Genetic marker technologies applied in animal breeding

Recent developments in molecular biology and statistical methodologies for QTL mapping have made it possible to identify genetic factors affecting economically important traits. Such developments have the potential to significantly increase the rate of genetic improvement of livestock species, through MAS of specific loci, genome-wide selection, gene introgression and positional cloning (Andersson, 2001). Instead of conventional animal breeding programmes solely relying on phenotype and pedigree information, the incorporation of detected QTL into genetic evaluation provides a great potential to enhance selection accuracies, which expedites the genetic improvement of animal productivity (Jiang et al., 2010).

Genetic marker technologies, like MAS, parentage identification, and gene introgression can be applied to livestock selection programmes. Highly saturated genetic maps are now available for cattle, swine and sheep to provide the genetic framework for developing MAS programmes (Davis and DeNise, 1998).

8.1. Marker-assisted selections (MAS) and gene-assisted selections (GAS)

Genetic improvement involves selection of outstanding individuals from a population to produce better yields in future generations. For a long time, dairy breeders have used genetic evaluations to identify superior animals. Selective use of these animals improved phenotypic measures for milk production and milk components, especially in Holstein cattle. However, there are some limitations to selecting on predicted breeding values. This selection approach has limited ability to improve lowly heritable traits without adversely affecting production. Lowly heritable traits often include those associated with disease resistance, reproduction, duration of productive life, and some conformation traits correlated with fitness (Sonstegard et al., 2001). Most breeding schemes do not account for population effects on genetic diversity, and selection is optimized for genetic response in the next generation rather than the highest long-term response (Meuwissen, 1997). Information from genetic markers that identify desirable alleles of economically important traits could be used with breeding values to guide mating decisions, resulting in genetic gains over a broader range of traits. Additionally, MAS could be used to select the most desirable phenotypes affected by non-additive gene action or epistatic interactions between loci (Sonstegard et al., 2001). Marker-assisted selection is a selection approach in which the relative breeding value of a parent is predicted using genotypes of markers associated with the trait. However, Lande and Thompson (1990) showed that genetic information cannot entirely replace phenotypic information. They developed a model combining phenotypic and genotypic information to be used in a selection programme, in which the selection index was constructed once every three generations. Before MAS can be applied in commercial dairying, economic trait loci (ETL) must be identified, validated, and characterized for utility in improving genetic gain (Sonstegard et al., 2001).

There are three phases in the development of MAS programmes. In the detection phase, DNA polymorphisms are used as linked or direct markers to detect QTL segregating in particular populations with specific allele frequencies. One or more markers associated with QTL are identified, and the size of the QTL allele effects and the location of the QTL in the genome are estimated. In the evaluation phase, the linked markers are tested in target populations or families to determine whether the detected QTL are segregating in those populations. In the implementation phase, linked markers shown to be predictive in a population are used within families and direct markers are used across families to produce a database of genotypes. These data are combined with phenotypic and pedigree information in genetic evaluation for the prediction of genetic merit of individuals within the population (Davis and DeNise, 1998).

In livestock, there are basically four design possibilities for marker QTL linkage analysis; (i) using F2 populations crossing two similar F1 populations, or a backcross between the F1 and one of the original populations; (ii) using a half-sib sire design on which heterozygous sires for the markers are mated to a random sample of females and all the progeny is genotyped; (iii) using instead a grand-daughter design on which a sire and their sons evaluated by progeny testing are genotyped; (iv) using crosses of individuals with extreme phenotypes for one trait or trait combination. Animals from divergently selected lines or from populations with wide variation for important traits are also used. Method (i) allows detecting QTL already fixed in one breed. Methods 2 and 3 are more suitable for prediction of QTL effects for within-population selection (Montaldo and Meza-Herrera, 1998). The challenge of the design of a breeding programme is to balance selection emphasis among traits to maximize response in the overall objective. With the availability of genetic markers and tests, there is need to balance emphasis on molecular versus quantitative genetic information. This also holds for selection against genetic defects, the emphasis on which must be balanced against selection on quantitative traits. Extra genetic gains from MAS, therefore, depend on the effect of direct selection on individual loci on genetic progress at other loci (polygenes) and for other traits that affect overall genetic merit. Although tandem selection results in the most rapid fixation of the gene(s) that are targeted by the molecular score, it results in the greatest loss in response for polygenes and for traits that are not included in the molecular score and may, therefore, result in less response in the trait and the overall breeding goal. The choice between tandem and index selection (and other alternatives) also depends on other factors, like market and cost considerations. Tandem and index selection apply to the use of molecular information in a given stage of selection (Montaldo and Meza-Herrera, 1998).

Marker-assisted introgression programmes are based on tandem selection in a multigenerational backcrossing programme, in which a marker selection (MS) based on the presence of donor breed alleles at or around the target gene is used in the first selection step (foreground selection), followed by background selection on a MS based on presence or absence of recipient alleles at markers spread over the genome, on phenotype, or an index of the two (Dekkers, 2004). A major gene in another population can be introduced through the process of introgression by means of backcrosses assisted by molecular markers. In this case, it does not seem to exist advantage in using a single genetic marker information, in comparison with the use of only phenotypic information when the characteristic is continuous and the considered genetic effects are additives (Groen and Smith, 1995). Classical introgression schemes (introgressing specific QTL alleles) are most likely to be successful when combined with deliberate selection for the specific favorable alleles, known to exist in the donor line, or by selection on closely linked markers. Using genomic selection, all marker alleles in LD with favourable QTL alleles are potentially selected for; this method may, therefore, be especially relevant in situations where a number of QTL underlie the genetic variation of the trait. During the backcrossing process, donor alleles are likely to be lost or at low frequencies unless favoured by selection within the crossbred line. Crossing can be used for introgression of favourable novel alleles and may be worthwhile even when there are considerable differences in the genetic levels of the recipient and donor lines (Ødegard et al., 2009).

Whereas initial applications of MAS in livestock populations may have been on *ad hoc* bases, it is clear that successful implementation of a MAS program requires a comprehensive integrated approach that is closely aligned with business goals and markets. Implementation of MAS requires development and integration of procedures and logistics for DNA collection and storage, genotyping and storage, and for data analysis. This must be supported by a systematic approach to quality control and must support day-to-day decision making (e.g., on which animals to genotype or regenotype in case of errors, which animals to phenotype, etc.) (Dekkers, 2004). Meuwissen & Goddard (1996) showed that response to MAS is maximal at the starting generations. The decrease in response to MAS throughout the subsequent generations may result from increased frequency of recombination events that leads to linkage equilibrium and, consequently, decreases the MAS efficiency (Lahav et al., 2006). The main application and potential for use of markers to enhance genetic improvement in livestock is through within-breed selection. This requires markers that trace within-breed variability (Dekkers, 2004).

MAS/ GAS versus conventional selection methods

Conventional animals breeding programmes depend on selection programmes based on phenotypic selection where traits are measured directly and animals with superior performance in the traits are used as breeding stock where the trait is limited, like milk production, progeny test schemes have allowed the genetic merit of the sex not displaying the trait to be estimated. Several problems are associated with phenotypic selection, and include: (i) narrowing the genetic base of a population; (ii) the approach can only be applied to traits that are easily measured; and (iii) high costs. In traits that are displayed only in adults, which comprise most of the production traits, it is necessary to raise a large number of individuals for which the trait is recorded, so that a few can be chosen for breeding. In case of progeny testing for milk production, the costs are very high, as the test sires have to be raised and then the daughters themselves raised and bred before the trait can be measured and the elite sires selected (Naqvi, 2007). Marker and gene assisted selection technique can efficiency solve problems associated with the conventional selection methods.

There is considerable marketing hype associated with emerging technologies, with predictions by the patent holders that gene marker selection techniques will soon entirely replace conventional breeding methods. Nevertheless, such efficiency gains will depend on the rate of scientific advancement in gene marking. Since economically important traits in dairy cattle, like milk yield and composition, are influenced both by a combination of genes and management factors. That only a handful of the more than 30,000 genes in cattle have been marked suggests that DNA-based seed stock selection, which relies on the small number of available markers, is unlikely to produce sizeable efficiency gains in the very near future. Furthermore, scientists and industry experts are concerned that a rapid substitution of gene marker selection for conventional breeding will result in unanticipated efficiency losses in the long term bases (Akhimienmhonan and Vercammen, 2007).

Economic benefits of MAS/ GAS on improvement of livestock genetics

Molecular genetics allows studying the genetic make-up of individuals at the DNA level. The main reasons why molecular genetic information can result in greater genetic gain than phenotypic information are: (i) assuming no genotyping errors, molecular genetic information is not affected by environmental effects and, therefore, has heritability equal to 1; (ii) molecular genetic information can be available at an early age, in principle at the embryo stage, thereby allowing early selection and reduction of generation intervals; (iii) Molecular genetic information can be obtained on all selection candidates, which is especially beneficial for sex-limited traits for example milk yield, traits that are expensive or difficult to record, or traits that require slaughter of the animal (carcass traits) (Naqvi, 2007).

It is believed that MAS could be particularly profitable in dairy cattle. Because this species concentrates many conditions unfavourable to phenotypic selection and, therefore, favourable to MAS; most traits of interest are sex-limited; the generation interval is long; AI bulls should be progeny tested before extensive use, which is a long and costly step; the breeding schemes are more and more designed with bull dams selected before their first lactation on pedigree information only, in order to reduce the generation interval; last but not least, functional traits, like disease resistance or fertility, have a low heritability but are more and more important in the breeding goal (Boichard, 2002). When AI is used predominantly, the number of key animals in the breeding scheme is limited and makes MAS relatively easy to implement. Although MAS could be oriented towards increasing the genetic trend on breeding objective or modifying the breeding objective by efficiently including low heritability traits, the breeders can use it to decrease the cost of the breeding programme by reducing the number of bulls sampled (Boichard, 2002).

According to Dekkers (2004), opportunities for increases in genetic gain through MAS on a given QTL differ depending on whether the QTL is marked by LE, LD, or direct markers; whereby genetic gains from MAS are lower for LE markers than for direct markers. The difference is caused by the accuracy of estimates of the molecular score, which is lower for LE markers because of the limited information that is available to estimate effects on a within-family basis, whereas for direct markers, effects are estimated from data across families. In that study, differences were reduced but far from eliminated when marker spacing was reduced to 1 or even 0.05 cM. Greater differences between the two types of markers are expected if phenotypic and/ or genotypic data is not available on all individuals, which will limit the accuracy of molecular scores based on LE markers for individuals in families with limited data, in particular if marker-QTL distances are considerable. Furthermore, the LD markers also enable use of phenotypic and genotypic data across families to estimate marker scores but accuracies may be slightly lower than for direct markers due to incomplete marker–QTL LD and a greater number of effects that must be estimated. Accuracy of estimates of molecular scores based on data from 1,000 individuals was 0.66 and 0.79 for haplotypes of 4 and 11 markers. Increasing the number of markers from 4 to 11 increased accuracy, but to a greater degree if more progeny were evaluated. Final considerations regarding the use of LE versus LD versus direct markers involve opportunities for marketing and protection (Dekkers, 2004).

Calculating the benefit requires focus on three main aspects: where returns are realized, because this determines the value of a unit of improvement and the genetic parameters to be applied; where the technology is applied, because this determines the rate of gain and the flow of genes to the sector in which the return is gained and the direct costs of implementing the technology and; the source of returns, i.e., whether the technology affects genetic structure of the population, the estimation of genetic value, and/ or the accuracy of the estimated genetic value. This needs to be assessed in order to predict the volume of improvement that will arise from application of the technology. The impact of a genetic technology can be calculated relevant to the breeding objective for the production/ market system in which the return is realized. This is because the value of one unit change in a trait in the breeding objective is not constant across different enterprises. Associations between marker haplotypes and QTL alleles may predict performance traits, like milk quality, without the requirement for large-scale measurement of phenotypes. The benefit of application of a genetic improvement technology can be assessed by defining the net value of the improvement on an individual breeding female scale, an enterprise scale, and an industry scale. These predicted annual improvements can then be compared with the annual costs of implementing the technology and analyzed with a conventional economic analysis to determine the overall net present value and the benefit:cost or internal rate of return when the technology is applied to an industry (Davis and DeNise, 1998).

MAS divert selection emphasis away from polygenes and traits without marked QTL, and the ultimate success of MAS is determined by its impact on total genetic merit. It has also been shown that the impact of MAS on other loci and traits differs between the three selection strategies, and is greatest for tandem selection, followed by index selection, and preselection. Commercial application of MAS requires careful consideration of economic aspects and business risks. Economic analysis of MAS requires a comprehensive approach that aims to evaluate the economic feasibility and optimal implementation of MAS. Generally, implementation of MAS will have a greater impact on market share than on genetic gain. Nevertheless, it is important that economic analysis is conducted in relation to business and market realities and goals (Dekkers, 2004).

If seed stock decision makers routinely replace animals with non-conforming genes with those having conforming genes, then both the gains and level of biodiversity will diminish over time. Livestock breeding contains many public good attributes, and it is important for policy makers to properly understand these attributes before determining whether policy intervention is warranted ((Akhimienmhonan and Vercammen, 2007). One way of evaluating the success of genetic improvement is to calculate genetic trends in a population over time (Leitóna et al., 2008). Genetic selection on production traits is reducing reproductive efficiency of dairy cattle (Castillo-Juarez et al., 2000), and increasing susceptibility to some diseases which, consequently, increases the risk of culling. Functional traits, and possibly also fertility traits, should, therefore, be included as part of the breeding goal (Dal Zotto et al., 2005).

8.2. Molecular analysis of genetic diversity

The use of microsatellites in genetic distancing of breeds is gaining momentum in characterizing and better understanding of animal genetic variation. The increasing knowledge of mammalian genetic structure and the development of convenient ways of measuring that structure have opened up a range of new possibilities in the areas of animal and product identification and tracing. Parentage verification by livestock breed and registry associations has now being based on microsatellite characterization and other genetic markers rather than blood typing. The advantages of the new system are substantial.

Better precision in identification should be possible, because the number of independent loci typed can be increased at will. The value of any particular locus depends on the number and relative frequencies of the alleles present or marker identity in the population, as well as on the ease with which it can be amplified and read in the laboratory (Naqvi, 2007).

8.3. Molecular conservation

The first step in considering sustainable management or conservation of a particular population of animals is genetic characterization. How unique is it in genetic terms? How different is it from other populations? How wide or narrow and, therefore, how endangered, are its internal genetic resources? The development of efficient methods of reading the molecular structure of populations has added a totally new range of instruments that can be used for the development of rational and balanced genetic management strategies. The most widely used of these techniques is the characterization of a population at a range of microsatellite loci. The compelling need for conserving domestic species is to prevent the loss of the many differentiated populations that, because of geographic or reproductive isolation, have evolved distinct characteristics and now occupy different environmental niches. Three basic approaches can be identified for preserving genetic diversity: maintaining living herds or flocks, cryo- preserving gametes or embryos and establishing genomic libraries (Naqvi, 2007).

9. Cloning adult dairy animals

Cloning an animal is the production of a genetically identical individual, by transferring the nucleus of differentiated adult cells into an oocyte from which the nucleus has been removed. This is known as "nuclear transfer" and is how the Dolly sheep was produced. In the case of Dolly, mammary gland cells in culture from a 6-year old donor ewe where subjected to a reduction in the concentration of serum and, consequently, obliged to enter in a quiescent state of the cell cycle (G0). Nuclear transfers to enucleated oocytes, was followed by electrical pulses for fusion of the donor cell nucleus and oocyte membranes and to activate division (Wilmut et al., 1997). Use of cloning in animal genetic improvement for milk production may increase the rates of selection progress in certain cases, particularly in situations where artificial insemination is not possible, like in pastoral systems with ruminants. Cloning is another technique that raises concerns both from the ethical and practical point of view. In animals, besides the very low success rates, some abnormalities should suggest that more information is required on the consequences of such practices in humans but also in animals, before its routine use (Montaldo, 2006).

10. Transgenic dairy animals

The production of transgenic farm animals that contain exogenous DNA stably incorporated into their genome so that the 'transgene' is transmitted to the offspring in a Mendelian fashion has several applications. Besides the obvious scientific interest for the study of genes and their regulation, transgenic animal technologies have been proposed as a method to accelerate livestock improvement, by means of introducing new genes or modifying the expression of endogenous genes that regulate traits of economic importance (Wheeler, 2003) like milk production traits.

The ability to insert genes into livestock embryos, the incorporation of those genes and their stable transmission into the genome of the resultant offspring will enable major genetic advances to be realized in animal agriculture. Some of the other methods that have been used to produce transgenic animals include: (i) DNA transfer by retroviruses; (ii) microinjection of genes into pronuclei of fertilized ova; (iii) injection of embryonic stem (ES) cells and/ or embryonic germ (EG) cells, previously exposed to foreign DNA, into the cavity of blastocysts; (iv) sperm mediated exogenous DNA transfer during *in vitro* fertilization; (v) liposome-mediated DNA transfer into cells and embryos; (vi) electroporation of DNA into sperm, ova or embryos; (vii) biolistics; and (viii) nuclear transfer with somatic or embryonic cells (Wheeler, 2003). The use of the bovine α lactalbumin gene promoter and regulatory regions has great potential for studying the basic biology of milk secretion as well as for many additional applications in agriculture and biomedicine (Wheeler, 2003).

Because so many separate steps are involved in the transgenic technology, the success rates are often low usually one or two per cent. Normally about half express the transgene. In those, which do show expression, the gene may be activated in unintended tissues or at abnormal times in the animal's development. This unpredictability of gene expression is to a greater extent contributed by lack of control either of the site of integration in the host genome, or the number of copies integrated. Furthermore, transgene transmission to the next generation is sometimes abnormal. One consequence of variable expression has been to produce unacceptable side effects on the health and welfare of animals. Consumer concern from lack of convincing information on transgenics and antipathy to transgenesis is very strong in many countries, and both producers and consumers would reject a technology which had negative effects on animal welfare. Genes promoting productivity (milk yield) or reducing costs (disease resistance) are most likely to be found within the species concerned. If a gene is sufficiently well characterized to permit its use in transgenesis, then it will also be possible to genetically characterize individuals carrying the gene and to make direct selection and propagation highly efficient. In dairy animals, most consideration has been given to genes that modify fat or protein synthesis in the mammary gland (Naqvi, 2007).

Among the different applications of milk modification in transgenic animals are the following (Montaldo, 2006): (i) to modify bovine milk to make it more appropriate to the consumption of infants. Human milk lacks β-lactoglobulin, has a higher relationship of serum proteins to caseins, and has a higher content in lactoferrin and lysozyme when compared to bovine milk; (ii) to reduce the content of lactose in the milk to allow their consumption to people with intolerance to lactose; (iii) to alter the content of caseins of the milk to increase their nutritive value, cheese yield and processing properties. Research has intended to increase the number of copies of the gene of the κ- casein, to reduce the size of the micelles and modifying the κ-casein to make it more susceptible to the digestion with chymosin; and (iv) to express antibacterial substances in the milk, such as proteases to increase mastitis resistance.

10.1. Ethical issues on applications of transgenic technology

Arguments opposing animal biotechnology can be divided into two categories; (i) concerns of technological ethics that might be raised with regard to the general unintended consequences of technical change; and (ii) concerns that relate specifically to biotechnology by virtue of new techniques for moving genetic materials from one organism to another. Among arguments that relate to biotechnology in a way that does not apply generally to technical change, concerns about patenting can be treated as a special case. The ethics of biotechnology include arguments for the development of transgenic animals, as well as objections and limitations. Research on ethical issues in biotechnology can improve the evaluation and implementation of transgenics farm animals by analyzing arguments of ethical concern and by presenting logically rigorous arguments for alternative perspectives. This element of ethical concern can be interpreted as an expression of anxiety or uncertainty about the definition of the moral community and the identification of borders or limits for ethical concern. Transgenic animals reinforce a challenge to implicitly accepted borders that define the scope of the moral community in terms of the human species (Thompson, 1993).

There are two very different and unresolved conceptions of animal welfare. One conception assumes that animal welfare is optimal only when the animal is allowed to realize its "natural" potentials and live accordingly in environments that closely resemble those of the animal in a wild setting. That implies, e.g., that animals would be free roaming and competing for the feed to the extent that aggression may occur and only the strongest would receive sufficient nourishment. According to this view, if the animal is seen to be suffering for causes that are natural, then welfare is not necessarily compromised. There is debate among different nations, and also among different experts, about the need to label food that is derived from genetically modified products. The mere application of gene-technology would not on that basis alone justify the need of labelling. The foregoing points represent a summary and rough overview of the most salient ethical issues surrounding the use of transgenic dairy production. One of the important points made in this connection was that more specific assessments need to be made on a case-by-case and step-by-step basis (Kaiser, 2003).

11. Conclusions

In livestock, knowledge of effects of specific genes and gene combinations on important traits could lead to their enhanced control to create new, more useful populations. The use of specific gene information could help to increase rates of genetic improvement, and open opportunities for using additive and non-additive genetic effects of domestic species, provided wise improvement goals are used and this new technology is optimally used together with the so called 'traditional' or 'conventional' methods based on phenotypic and genealogical information (Montaldo, 2006). Success of commercial application of MAS is unclear and undocumented, and will depend on the ability to integrate marker information in selection and breeding programmes. Opportunities for the application of MAS exist, in particular for GAS and linkage disequilibrium MAS and, to a lesser degree, for linkage equilibrium MAS because of greater implementation requirements. Regardless of the strategy, successful application of MAS requires a comprehensive integrated approach with continued emphasis on phenotypic recording programmes to enable quantitative trait loci detection, estimation and confirmation of effects, and use of estimates in selection (Dekkers, 2004).

Donors and governments should fund research that examines the usefulness of gene marker technology for dairy cattle producers, find ways to educate producers about this new technology, and report to producers all third-party analysis of specific test claims. Finally, policy makers should promote the efficient commercialization of this emerging technology (Akhimienmhonan and Vercammen, 2007). A rational use of the molecular methodologies in milk production genetic improvement requires the simultaneous optimization of selection on all the genes affecting important traits in the population. The maximum benefit can be obtained when these techniques are used in conjunction with reproductive technologies like artificial insemination, and collection and production *in vitro* of embryos to accelerate genetic change (Bishop et al., 1995).

Author details

Sammy K. Kiplagat and Moses K. Limo
Department of Biochemistry and Molecular Biology, Egerton University, Egerton, Kenya

Isaac S. Kosgey
Department of Animal Sciences, Egerton University, Egerton, Kenya & Laikipia University College, Nyahururu, Kenya

12. References

Akhimienmhonan, D. and Vercammen, J. 2007. An Economic Analysis of Gene Marker Assisted Seedstock Selection. Canadian Agricultural Innovation Research Network. Number 2.

Andersson, L. 2001. Genetic dissection of phenotypic diversity in farm animals. Nat. Rev. Genet. 2: 130–138.

Ayalew, W., Rischkowsky, B., King, J.M. and Bruns, E. 2003. Crossbreed did not generate more net benefits than indigenous goats in Ethiopian smallholdings. Agricultural systems 76: 1137-1156.

Barillet, F. 2007. Genetic improvement for dairy production in sheep and goats. Small Ruminant Research 70: 60–75.

Bayer, W. and Feldmann, A. 2003. Diversity of animals adapted to smallholder system. conservation and sustainable use of agricultural biodiversity. Nat Rev Genet, 2:130-138.

Bishop, M. D., Hawkins, G. A. and Keeler, C. L. 1995. Use of DNA markers in animal selection. Theriogenology 43: 61-70.

Boichard, D., Grohs, C., Bourgeois, F., Cerqueira, F., Faugeras, R., Neau, A., Rupp, R., Amigues, Y., Boscher, M.Y. and Levéziel, H. 2003. Detection of genes influencing economic traits in three French dairy cattle breeds. Genet. Sel. Evol. 35:77-101.

Castillo-Juarez, H., Oltenacu, P.A., Blake, R.W., Mcculloch, C.E. and Cienfuegos-Rivas, E.G. 2000. Effect of herd environment on genetic and phenotypic relationships among milk yield, conception rate and somatic cell score. J.Dairy Sci. 83, 807-814.

Dal Zotto, R., Carnier, P. and Gallo, L. 2005. Giovanni Bittante2, Martino Cassandro2Genetic relationship between body condition score, fertility, type and production traits in Brown Swiss dairy cows. Italian Journal of Animal Science 4: 30-32.

Darvasi, A., and Soller, M. 1993. Optimum spacing of genetic markers for determining linkage between marker loci and quantitative trait loci. Theor. Appl. Genet. 89:351–357.

Davis, G. P. and DeNise, S. K. 1998. The impact of genetic markers on selection. Journal of Animal Science 76: 2331-2339.

Dekkers, J.C.M. 2004. Commercial application of marker- and gene-assisted selection in livestock: strategies and lessons. Journal of Animal Science 82: 313-328.

Drinkwater, R.D. and Hazel, D.J.S. 1991. Application of molecular biology to understanding genotype-environment interactions in livestock production. In: Proceedings of an International Symposium on Nuclear Techniques in Animal Production and Health. IAEA/ FAO, 15-19 April, 1991, Vienna, Australia, pp. 437-452.

Freitas, L.S., Verneque, R.S., Peixoto, M.G.C.D., Pereira, M.C., Wenceslau, R.R., Felipe, V.P.S. and Siva, M.A. 2010. Genotype-environment interactions for milk production in the Gyr (Bos indicus) dairy cattle in Brazil. Last retrieved from http://www.kongressband.de/wcgalp2010/assets/pdf/0508.pdf on 29th July, 2011.

Georges, M., Nielsen, D., Mackinnon, M., Mishra, A., Okimoto, R., Pasquino, A.T., Sargeant, L.S., Sorensen, A., Steele, M.R., Zhao, X., Womack, J.E. and Hoeschele, I. 1995. Mapping quantitative trait loci controlling milk production in dairy cattle by exploiting progeny testing. Genetics 139: 907-920.

Goddard, M.E. 2009. Fitness traits in animal breeding programs. In: van der Werf, J.H.J., Graser, H.-U., Frankham, R. and Gondoro, C. (Eds.), Adaptation and Fitness in Animal Populations. Evolutionary and Breeding Perspectives on Genetic Resource Management. Springer, The Netherlands.

Graham, E.R.B., McLean, D.M. and Zuiedrans, P. 1984. The effect of milk protein genotypes on the cheese making properties of milk and on yield of cheese. In: Proceedings of the Fourth Conference of the Australia Association of Animal Breeding and Genetics, Adelaide, Australia, pp. 36.

Groen, A.F. and Smith, C. 1995. A stochastic simulation study of the efficiency of marker-assisted introgression in livestock. Journal of Animal Breeding and Genetics 112:161-170.

Hammami, H., Rekik, B. and Gengler N. 2008. Genotype by environment interaction in dairy cattle. Biotechnol. Agron. Soc. Environ. 13: 155-164

Kaiser, M. 2003. Ethical issues surrounding the gm-animals / gm-fish production. FAO Expert Meeting 17-21 November, 200, Rome, Italy.

Karugia, T.J., Okeyo, A.M., Kaitho, R., Drucker, A.G., Wollny, C.B.A. and Rege, J.O.E. 2000. Economic analysis of crossbreeding programs in sub-Saharan Africa: a conceptual framework and Kenyan case study. Last retrieved from http://www.femi.it/web/activ/_activ.html on 28th July, 2011.

King, G. 2006. Animals and environment. Last retried from http://www.aps. uoguelph.ca/~gking/Ag_2350/animenv.htm on 28th July, 2011.

Lahav, T., Atzmon, G., Blum, S., Ben-Ari, G., Weigend, S., Cahaner, A., Lavi, U. and Hillel, J. 2006. Marker-assisted selection based on a multi-trait economic index in chicken: experimental results and simulation. Animal Genetics 37: 482–488.

Lande, R. and Thompson, R. 1990. Efficiency of marker-assisted selection in the improvement of quantitative traits. Genetics 124: 743–56.

Leitóna, B.V. and Zeledóna G.G. 2008. Genetic trends, genotype-environment interaction and inbreeding in Holstein and Jersey dairy cattle from Costa Rica. Téc Pecu Méx. 46: 371-386.

Meuwissen, T.H.E. and Goddard, M. E. 1996. The use of marker haplotypes in animal breeding schemes. Genetics, Selection, Evolution 28: 161–176.

Meuwissen, T.H. and Goddard, M.E. 1996. The use of marker haplotypes in animal breeding schemes. Genetics, Selection, Evolution 28: 161–76.

Mirkena, T., Duguma, G., Haile, A., Tibbo, M., Okeyo, A.M., Wurzinger, M. and Sölkner, J. 2010. Genetics of adaptation in domestic farm animals: A review. Livestock Science 132: 1–12.

Montaldo, H. H. 2006. Genetic engineering applications in animal breeding. Electronic Journal of Biotechnology 9 (2).

Montaldo, H.H. and Meza-Herrera, C.A. 1998. Use of molecular markers and major genes in the genetic improvement of Livestock. EJB Electronic Journal of Biotechnology 1(2).

Naqvi, A.N. 2007. Application of molecular genetic technologies in livestock production: potentials for developing countries. Advances in Biological Research 1: 72-84.

Ødegard, J., Yazdi, M. H., Sonesson, A.K. and Meuwissen. T.H.E. 2009. Incorporating desirable genetic characteristics from an inferior into a superior population using genomic selection. Genetics 181: 737–745.

Olesen, I., Gjerde, B. and Groen, A.F. 1999. Accommodation and evaluation of ethical, strategic and economic values in animal breeding goals. Book of Abstracts No. 5, EAAP, pp. 33.

Pärnal, E., Pärnal, K. and Dewi, L.A. 2003. Economic value of milk production and functional potentials for developing countries. Advances in Biological Research 1: 72-84.

Rothschild, M.F. and M. Soller. 1997. Candidate gene analysis to detect genes controlling traits of economic importance in domestic livestock. Probe 8: 13–20.

Sonstegard, T.S., van Tassel, C.P. and Ashwell, M.S. 2001. Dairy cattle genomics: tools to accelerate genetic improvement. Journal of Animal Science 79: 307-315.

Thompson, P.B. 1993. Genetically modified animals: ethical issues. Journal of Animal Science 81: 32-37.

Wheeler, M.B. 2003. Production of transgenic livestock: Promise fulfilled. Journal of Animal Science 81: 32-37.

Wilmut, I., Schnieke, A.E., Mcwhir, J., Kind, A.J. and Campbell, K.H.S. 1997. Viable offspring derived from fetal and adult mammalian cells. Nature 385: 810-813.

Chapter 2

Genetic Polymorphisms
of Some Bovine Lactogenic Hormones

Yousef Mehmannavaz and Abolfazl Ghorbani

Additional information is available at the end of the chapter

1. Introduction

1.1. The role of quantitative and molecular genetics in animal breeding

To date, most genetic progress for quantitative traits in livestock, especially for dairy cows has been made by selection on phenotype or on estimated breeding values (EBV) derived from phenotype, without knowledge of the number of genes that affect the trait or the effects of each gene. In this quantitative genetic approach to genetic improvement, the genetic architecture of traits has essentially been treated as a 'black box'. Despite this, the substantial rate of improvements that have been and continue to be achieved in commercial populations is clear evidence of the power of these approaches. The success of this approach depends on accurate information concerning data or data structure and genetic evaluation methods (Tambasco et al. 2003). The traits have high heritability and the traits can be measured on all selection candidates (males and females) are ideal situation for quantitative selection methods because accurate EBV can be obtained on all animals (Dekkers, 2004).

However, genetic progress may be further enhanced if we could gain insight into the black box of quantitative traits. Molecular genetics allows for the study the genetic make-up of individuals at the DNA level and may provide the tools to make those opportunities a reality, either by direct selection on genes that affect traits of interest major genes or quantitative trait loci (QTL) - or through selection on genetic markers linked to QTL. The main reasons why molecular genetic information can result in greater genetic gain than phenotypic information are: a) Assuming no genotyping errors, molecular genetic information is not affected by environmental effects and, therefore, has heritability equal to 1. b)-Molecular genetic information can be available at an early age, in principle at the embryo stage, thereby allowing early selection and reduction of generation intervals. C)

Molecular genetic information can be obtained on all selection candidates, which is especially beneficial for sex-limited traits, traits that are expensive or difficult to record, or traits that require slaughter of the animal (carcass traits) (Dekkers, 2004).

For the last decade, molecular genetics has lead to the discovery of individual genes or candidate genes with substantial effects on the traits of economic importance. Candidate gene strategy has been proposed by direct search for quantitative trait loci (QTL) (Tambasco et al. 2003). In other words, the genetic variation in a gene affects the physiological pathways and phenotype. Moreover, the proportion of genetic and phenotypic variation would be likely to affect the breeding strategy for improvement of important traits in the future. Genetic markers associated with traits of interest can be searched directly by applying molecular biology techniques. These techniques can identify genetic variation at specific loci and analyze the relationship between genetic variation at QTL and production traits (Arendonk et al., 1994). Application of molecular genetics for genetic improvement relies on the ability to genotype individuals for specific genetic loci. The information utility from candidate genes in breeding programs has potential to substantially enhance the accuracy of selection and increasing selection differences (Missohou et al., 2006).

1.2. Importance of genetic polymorphisms studies in dairy cattle breeding

For more than 50 years, dairy breeders have used genetic evaluations to identify superior animals. Selective use of these animals improved phenotypic measures for milk production and milk components, especially in Holstein cattle. However, there are some limitations to selecting on predicted breeding values. Most breeding schemes do not account for population effects on genetic diversity, and selection is optimized for genetic response in the next generation rather than the highest long-term response (Meuwissen, 1997). This selection approach also has limited ability to improve lowly heritable traits without adversely affecting production. Lowly heritable traits often include those associated with disease resistance, reproduction, duration of productive life, and some conformation traits correlated with fitness. Information from genetic markers that identify desirable alleles of economically important traits could be used with breeding values to guide mating decisions, resulting in genetic gains over a broader range of traits. In addition, marker-assisted selection (MAS) could be used to select the most desirable phenotypes affected by nonadditive gene action or epistatic interactions between loci. Soller and Beckmann (1983) proposed that MAS can also reduce the costs of the artificial insemination industry incurs using progeny test evaluations as the sole method for screening candidate bulls.

The most genetic improvement in dairy cattle industry through BLUP methodology has been made by selection of merited bulls. For this target, the recording of milk production traits is done for all industrial dairy cows, and breeding values are annually estimated for them. Thereafter, bulls are firstly selected based on parent EBVs, then some of them are proofed based on progeny test (at least 50 daughters) for using in wide artificial insemination system. The problems associated with phenotypic data recording such as long time and high expense and also low cooperation of some dairymen cause to less accurate estimation of breeding values and thereafter in the selection process. Thus, collection of genotypic data by molecular methods in addition to phenotypic data is necessary to improve the selection procedure.

1.3. Objectives of this chapter

Many studies have reported that the candidate genes influence milk traits in cows. In addition, some genes control more than one trait. For instance, the growth hormone (*GH*) gene influences expression of growth and milk traits. The important candidate genes in bovine somatotropic axis play a key role in productivity, metabolism, reproduction and disease resistance. Therefore, The objective of this chapter is to review and evaluate the relationships between the polymorphisms of some candidate genes related with bovine somatotropic axis consist of prolactin (PRL), growth hormone receptor (GHR) and insulin like growth factore-1 (IGF-I) which have an influence on milk production traits such as milk yield, milk fat yield, milk protein yield, milk fat percentage and milk protein percentage in cows. A study of the candidate genes for significant economic traits could be applied for a direct search of QTL in order to plan a breeding program in the future.

2. Literature review

2.1. Prolactin (PRL)

2.1.1. Prolactin introduction

Prolactin is a versatile polypeptide hormone that was first identified as a product of the anterior pituitary in 1933. It is synthesized and secreted not only in the anterior pituitary gland but also produced by numerous other cells and tissues, including the mammary gland (extrapituitary prolactin). In various classes of vertebrates more than 300 actions and activities of this multifunctional hormone have been reported (Bole-Feysot et al. 1998). Based on its genetic, structural, binding and functional properties, prolactin belongs to prolactin/growth hormone/placental lactogen family group (group I of the helix bundle protein hormones) (Boulay and Paul, 1992; Horesman , Yu-Lee, 1994).

2.1.2. Structure of prolactin hormone

The prolactin molecule is arranged in a single chain of amino acids with three intramolecular disulfide bonds between 6 cystein residues (Cooke et al., 1981). In cattle, the prolactin chain consists of 199 amino acids with a molecular mass of ~23 KDa (Wallis, 1974). Prolactin is synthesized as a prohormone. Following cleavage of the signal peptide, the length of the mature hormone is between 194 and 199 amino acids, depending on species. The signal peptide contains 30 amino acids; thus the mature bovine prolactin is composed of 199 amino acids (Freeman et al., 2000).

2.1.3. Effects of prolactin on milk production

The varied effects of prolactin have been identified on the mammary gland include growth and development of the mammary gland (mammogenesis), synthesis of milk (lactogenesis), and maintenance of milk secretion (galactopoiesis). In the 1920's it was found that extracts of the pituitary gland, when injected into virgin rabbits, induced milk production. Subsequent research demonstrated that prolactin has two major roles in milk production: a) Prolactin induces lobuloalveolar growth of the mammary gland. Alveoli are the clusters of cells in the mammary gland that actually secrete milk. b) Prolactin stimulates lactogenesis or milk production after giving birth. Prolactin along with cortisol and insulin act together to stimulate transcription of the genes that encode milk proteins. The critical role of prolactin in lactation has been confirmed in mice with targeted deletions in the prolactin gene. Female mice that are heterozygous for the deleted prolactin gene (and produce roughly half the normal amount of prolactin) show failure to lactate after their first pregnancy (Freeman et al., 2000).

2.1.4. Bovine prolactin gene structure

The Bovine Prolactin gene (bPRL) found on the chromosome 23 (23q21 position) in the bovine genome (Hallerman et al., 1998). The bPRL gene is about 10 kb in size and is composed of 5 exons and 4 introns (Camper et al., 1984). This encodes the 199 amino acids mature protein in cattle (Cooke et al., 1981).

The exons of bPRL gene (GenBank: AF426315.1) consist of exon 1: 855 to 936 nt,, exon 2: 3661-3842 nt, exon 3: 6186-6293 nt, exon 4: 8321-8500 nt and exon 5: 9129-9388 That encode the 229 residues prolactin precursor (Protein ID: "AAL28075.1) (Cao et al., 2002).

All sequences of exons 2, 3 and 4 re coding sequences (CDS), but some sequences (not all) on the exon 1 (position 909 to 936 nt) and exon 5 (position 9129 to 9320 nt) are CDS. Nucleotides 909 – 936 (on exon 1) and 3661 to 3772 (on exon 2) encoded signal peptide of prolactin hormone that is separated of hormone in maturing. Matured section of hormone was encoded by CDS of exon 2 (3723-3842 nt), exon 3 (6186-6293 nt), exon 4 (8321-8500 nt) and exon 5 (9129-9317 nt) (Cao et al., 2002).

Transcription of the prolactin gene is regulated by two independent promoter regions. The proximal 5,000-bp region directs pituitary-specific expression, while a more upstream promoter region is responsible for extrapituitary expression (Berwaer et al., 1991). The Bovine prolactin cDNA is 914 nucleotides long and contains a 687-nucleotide open reading frame encoding the prolactin prohormone of 229 amino acids (Cao et al., 2002). In the 5' flanking region of the bovine prolactin gene, a distal regulatory element was found, which enhances the basal level of expression of the gene fivefold and functions independently of position and orientation. The postulated enhancer region extends from –1175 to –996 and displays considerable sequence similarity to equivalent regions in human and rat prolactin gene promoters (Brym et al., 2007).

2.1.5. Polymorphisms of bPRL gene associated with milk production

Extensive genetic polymorphism studies were carried out, finding more than 20 SNPs within the bovine PRL structure gene sequence, although all of them were silent mutations or located within introns (Sasavage et al. 1982; Brym et al. 2005). The most important polymorphism is located on exon 4 that identified by RsaI endonuclease. This SNP (A/G) has studied by many researchers. Nevertheless, a few independent groups confirmed statistically significant associations between this SNP variants and milk production traits in dairy cattle (Chung et al. 1996; Dybus, 2002; Dybus et al. 2005; Brym et al. 2005; Mehmannavaz et al., 2010). Based on Chung et al. (1996) study, it was shown the association between RsaI- PRL polymorphism with milk fat percentage. Study of this polymorphism in Jersey cattle showed a significant association with milk fat yield and milk fat percentage (Dybus, 2002). Study of RsaI- PRL polymorphism in 125 Russian Red Pied cows (Alipanah et al., 2007) and 186 Black-and-White cows and in 138 Jersey cows (Brym et al., 2005) confirmed the relationships of this SNP with milk yield, fat yield and milk fat percentage.

In the study of Li et al. (2006) Holstein dairy cows, 5' regulation region of bovine prolactin (bPRL) gene was screened by PCR-RFLP and PCR-SSCP techniques and two mutation sites were discovered for the first time. Analysis of the association between the polymorphisms of PRL gene and milk traits showed that the XbaI-RFLP locus significantly affected milk protein and milk fat in first parity. The SSCP locus significantly affected the milk fat yield on parity 1 and milk protein yield on parity 4 (p<0.05).

The association analysis of the G/T SNP in position -485 of prolactin gene promoter in 649 Chinese Holstein cows with the milk performance traits indicated that the SNP in the promoter was significantly associated with milk yield, fat yield and protein yield and protein percentage (Feng et al., 2006).

2.3. Growth Hormone Receptor (GHR)

2.3.1. Growth Hormone Receptor introduction

Growth hormone (GH), also known as somatotropin, is a major stimulator of postnatal growth and milk production in cattle (Etherton and Bauman, 1998). At the tissue level, the GH action is mediated by a specific cell membrane receptor, the growth hormone receptor (GHR).

GHR is a member of the class I hematopoietin or cytokine/growth hormone/prolactin receptor superfamily. Members of this family include receptors for erythropoietin (EPO), granulocyte colony-stimulating factor (GCSF), granulocyte macrophage-colony stimulating factor (GM-CSF), The β-chain of interleukin (IL)-2 through (IL)-9, (IL)-11, (IL)-12, thrombopoietin, and leukemia inhibitory factor (LIF). Receptors for interferon (IFN) a/b, IFNg, and IL-10 are more distantly related and considered class II receptors in the family. The class I cytokine receptors span the membrane once and contain an extracellular region, a single hydrophobic transmembrane domain of 24 amino acids, and an intracellular region.

The extracellular and intracellular regions vary in length. The overall sequence homology of the class I hematopoietin receptors is low; however, there is 14–25% identity in approximately 200 amino acids of the extracellular domain. The family also contains a WSXWS (tryptophan, serine, any amino acid, tryptophan, serine) motif in the membrane proximal region of the extracellular domain that is present in all members except GHR. This motif in the GHR has the conserved substitutions YXXFS (where X is glycine, serine, lysine, or glutamic acid). There is also the presence of two pairs of cysteines usually found in the N-terminal region. GHR contains seven extracellular cysteins. The cytoplasmic domains of these receptors share some common motifs. There exists a membrane proximal proline rich motif referred to as Box 1. Box 1 is present in all members and consists of eight amino acids (c-X-X-X-AL-P-X-P, where c represents hydrophobic residues, X any amino acid, AL aliphatic residues, and P proline). In GHR, this sequence reads ILPPVPVP. Another motif, termed Box 2, is present in most of these receptors. It is characterized as a cluster of hydrophobic amino acids ending with one or two positively charged residues. In GHR, Box 2 is located approximately 30 amino acids toward the C-terminal from Box 1 and spans 15 amino acids (Kopchick & Andry, 2000).

Binding of GH to GHR activates the Janus kinase 2 (JAK2); activated JAK2 in turn activates signal transducer and activator of transcription 5 (STAT5) through phosphorylation; phosphorylated STAT5 translocates from the cytoplasm to the nucleus, where it binds to specific DNA regions and activates transcription (Herrington and Carter-Su, 2001). A well-known gene controlled by GH through this JAK2–STAT5 pathway is IGF-I, which is believed to mediate most of the growth-stimulating and at least part of the milk production-stimulating effect of GH (Etherton and Bauman, 1998). In addition to STAT5, GH-activated JAK2 also phosphorylates insulin receptor substrate 1, phospholipase C, and SHC protein, leading to changes in gene expression, enzymatic activity, or metabolite transport (Herrington and Carter-Su, 2001).

2.3.2. Growth Hormone Receptor structure

As predicted from its cDNA sequence (Hauser et al., 1990), the bovine GHR protein is a single-chain polypeptide of 634 amino acids, composed of an 18-AA signal peptide (not present in mature GHR protein) encoded by exon 2 of the GHR gene, a 242-AA extracellular domain encoded by exons 3 to 7, a 24-AA single transmembrane domain encoded by exon 8, and a 350-AA intracellular domain encoded by exons 9 and 10. The bovine GHR mRNA is heterogeneous in the 5'-untranslated region, due to initiation of transcription from different leader exons (or alternative exon 1) and alternative splicing (Jiang and Lucy, 2001).

2.3.3. Bovine Growth Hormone Receptor gene structure

The bovine GHR gene has been mapped to the proximal long arm of chromosome 20 in region 20q 7.1 (Viitala et al., 2006). The gene coding for bovine GHR consists of 9 exons (from 2 to 10) in the translated part and of a long 5'-noncoding region, that includes 9 untranslated exons – 1A, 1B, 1C, 1D, 1E, 1F, 1G, 1H, 1I (Chrenek et al., 1998). Exons from the untranslated region are spliced alternatively and each of them has its own transcription start site. Long of exons of 2 to 10 are 72, 66, 130, 161, 179, 166, 91, 70 and 1432, respectively. All nocleotides of exons of 3 to 9 are coding sequences, but some of sequences of 2th and 10th exons are coding. Exon 2 encodes the last 11 bp of the 59-UTR, the 18-amino-acid signal peptide, and the first 5 amino acids of the extracellular domain. Exons 3–7 encode the majority of the amino acids that make up the extracellular region. Exon 8 encodes the final 3 extracellular amino acids, a 24-amino-acid hydrophobic (transmembrane) domain, and the first 4 amino acids of the intracellular domain. Exons 9 and 10 encode the remaining 346 amino acids of the intracellular domain. Exon 10 also encodes for a 2-kb 3'-UTR.

A LINE-1 element from the family of retrotransposons, about 1.2 Kbp-long, was found upstream from exon 1A (Lucy et al., 1998). Heterogeneity in the 5' untranslated region (UTR) of the growth hormone receptor gene has been shown in different species of mammals. Nine variants of GHR mRNA have been identified in humans (V1–V9; [10]) and cattle (1A–1I; (Jiang and Lucy, 2001). In cattle, variant 1A is exclusively expressed in the liver and transcriptionally controlled by the liverenriched factor, hepatocyte nuclear factor-4 (HNF-4); (Jiang and Lucy, 2001).

2.3.4. Polymorphisms of GHR Gene in Relation with Milk Production Traits

The association of TaqI-RFLP on GHR gene with milk protein percentage in italian Holstein cows was reported by Falaki et al (1996). Study of genetic markers on GHR gene for milk traits in 128 holstein cows indicated that polymorphism GHR-AluI in 5'UTR region of GHR is associated with milk yield (Aggrey et al., 1998).

Aggrey et al. (1999) studied the association of three polymorphism of 5' UTR region of GHR (GHR-AluI, GHR-StuI and GHR-AccI) with breeding values of milk production traits in 301 Canadian Holstein bulls. They resulted that only GHR-AluI has significant effect on breeding values of milk fat yield.

The T to A substitution in exon 8 of bGHR gene results in the nonconservative replacement of a neutral phenylalanine with an uncharged but polar tyrosine residue (F279Y). The corresponding phenylalanine residue is located within the transmembrane domain 0f GHR and is conserved among all analyzed mammals. This SNP associated with a srong effect on milk yield and composition in the cows (Blott et al., 2003). The effects of this polymorphism were studied by Vittala et al., (2006) in Ayershire dairy cows and its association with milk fat percentage and milk protein percentage was showed.

A 286 bp fragment of exon 10 of bGHR in 365 Hungarian Holstein cows was amplified and genotyped by RFLP-AluI. The significant effects of this polymorphism with milk yield, milk fat percentage and milk protein percentage were shown (Kovacs, 2006).

Polymorphism analysis of A/G substitution in exon 6, detected by AluI enzyme has not any association with milk production traits (Hradecka et al., 2006) and breeding values of milk production traits (Hradecka et al., 2008).

2.4. Insulin Like Growth Factor 1 (IGF-I)

2.4.1. IGF-I introduction

Insulin-like Growth Factor I (IGF-I), also known as somatomedin C, is a member of the insulin superfamily. It was originally discovered as a mediator of growth hormone actions on somatic cell growth, but has also been shown to be an important regulator of cell metabolism, differentiation and survival (Werner et al., 1994). IGF-I is produced primarily by the liver as an endocrine hormone as well as in target tissues in a paracrine/autocrine fashion. It is found in blood and other body fluids as a complex with specific high affinity IGF binding proteins (IGFBP-1 to -6). The IGFBPs are modulators of IGF actions, which control IGF bioavailability to specific cell-surface receptors. IGF-I actions are mediated by two type I transmembrane receptor tyrosine kinases: the IGF-I receptor (IGF-I R), and the insulin receptor (INS R) that exists in two alternatively spliced isoforms (INS R-A and -B) (O'Dell and Day, 1998).

2.4.2. Protein structure of IGF-I

IGF-I is synthesized as a preprotein that is proteolytically cleaved to generate the mature protein linked by three disulfide bonds. Mature IGF-I is highly conserved among mammals, with 100% sequence identity between the human, bovine, porcine, equine and canine proteins. Mature mouse IGF-I is a non-glycosylated, 70 amino acid (aa) residue secreted polypeptide that is derived from either a 153 aa or a 159 aa preproproteins. It shares 99% and 94% aa sequence identity with rat and human IGF-I, respectively (Rinderknecht and Humbel, 1978).

Sequences of bovine and human IGF-1 is the same form and bIGF-I is a variant of hIGF-I, deleted three N terminal peptides (glycine - proline - glutamine) of it and then bIGF-I is showed as (- 3N: IGF-1) (Francis et al. 1988).

IGF-I is synthesized in the liver and multiple other tissues. It is found in blood and other body fluids as a complex with specific high affinity IGF binding proteins (IGFBP-1 to -6). The IGFBPs are expressed in specific patterns during development. They are modulators of IGF actions, which control IGF bioavailability to specific cell-surface receptors. Their functions are further regulated by IGFBP proteases, which proteolytically cleave the IGFBPs to lower the affinity with which they bind IGFs and increase IGF bioavailability. Some IGFBPs also have IGF-independent effects on cell functions. IGF-I circulates primarily as a ternary complex with IGFBP-3 or IGFBP-5 and the acid-labile subunit (ALS). Some IGF-I is also present in binary complexes with other IGFBPs. Whereas the ternary complexes are generally restricted to the vasculature, the binary complexes freely enter the tissues.

IGF-I actions are mediated by two type I transmembrane receptor tyrosine kinases: the IGF-I receptor (IGF-I R), and the insulin receptor (INS R) that exists in two alternatively spliced isoforms (INS R-A and -B). Both IGF-I R and INS R share a highly homologous structure and are ubiquitously expressed. Each receptor is derived from a precursor that is proteolytically cleaved into two disulfide-linked subunits: The extracellular and the transmembrane-subunits. Functional IGF-I receptors are tetrameric glycoproteins composed of two disulfide-linked IGF-I Rs or disulfide-linked hybrids of one IGF-I R and one INS R. Whereas IGF-I binds with high-affinity to homodimeric IGF-I R and heterdimeric IGF-I R:INS R-A or –B hybrids, high-affinity binding of insulin is observed only with dimeric INS R or IGF-I R:INS R-A hybrid but not with IGF-I R:INS R-B hybrid. The signaling responses from the various receptors are different depending whether insulin or IGF-I is used as the activating ligand. This kit demonstrates significant cross-reactivity with rat IGF-I and has been validated for the determination of relative mass values for natural rat IGF-I in cell culture supernates, rat serum and plasma. The amount of natural rat IGF-I measured is expressed as mouse IGF-I equivalent.

2.4.3. The role of IGF-I in milk production

Insulin-like growth factor I (IGF-I) is known as regulator of mammary gland development. This factor regulate the milk production through the stimulation of mitogenesis of mammary glands, prevention of apoptosis and mediatory of growth hormone function (GH) on milk synthesis (Lactogenesis) (Monaco et al, 2005).

Nutrient partitioning for lactogenesis is mediated and sustained by alterations in the growth hormone-insulin-like growth factor (GH-IGF) axis. Under physiological conditions, pituitary derived GH induces hepatic IGF-I synthesis via receptor mediated signalling (Bichell et al., 1992) and consequently systemic IGF-I negatively regulates GH production (Le Roith et al., 2001). However in situations of high nutrient demand, such as during NEB, the GH-IGF axis uncouples in the liver (Thissen et al., 1994) and this is associated with a reduction in total circulating IGF-I and elevated GH concentrations (Etherton and Bauman, 1998). The actions of GH vary considerably in different physiological states (Bell et al., 1995), however the net effect of this uncoupling during early lactation supports a facilitatory role for the indirect actions of GH on lipolysis and gluconeogenesis (Thissen et al., 1994) and attenuated growth promoting actions and support by IGF-I in peripheral tissues (Fenwick et al., 2008). In the dairy cow, the periparturient reduction in IGF-I synthesis is associated with a concomitant reduction in the liver-specific GH receptor type 1A (GHR1A) (Jiang et al., 2005).

2.4.4. Gene structure of IGF-I

The gene coding IGF-I in human is located on chromosome 12 at position 12q23 (Daughaday and Rotwein, 1989) and in the mouse and cow, it has been mapped on chromosome 10 (Shimatsu and Rotwein, 1987) and 5 (Miller et al., 1991), respectively. In humans the IGF-1 gene contains 6 exons and is about 90 kbp-long (Steenbergh et al. 1991). Due to an alternative splicing of exons 1 and 2, two different transcripts are formed: the one with exon 1 containing 1155 nucleotides (nt), while the other one, with exon 2, is shorter and contains 750 nt. Production of these transcripts is controlled by two different promoters both containing canonical regulatory sequences – TATA-box and CCAAT-box (Jansen et al. 1991). It was shown that transcripts of both classes are differentially expressed in various tissues, being, however, most abundant in liver (Wang et al. 2003). In all bovine tissues tested, the expression of IGF-1 class 1 transcript was higher than that of transcript 2. The expression of *IGF-1* was shown to be regulated both on the level of transcription and translation (Wang *et al.* 2003).

2.4.5. Polymorphisms of IGF-I Gene in relation with milk production traits

In cattle, a few polymorphisms has been identified in the nucleotide sequence of IGF-I gene that mainly associated with growth traits. But, there is only 3 studies, investigating the IGF-I polymorphisms with milk production traits. Hines et al. (1998) did not confirm any association of SnaBI-RFLP polymorphism on IGF-I gene with milk production traits. But, Siadkowska et al (2006) showed the significant association of the SNP with fat and protein of milk.

3. Methods and materials

3.1. Samples

Semen samples were collected from 282 progeny-tested Holstein bulls born from 1990 to 2006. They were obtained from Animal Breeding Center of Iran (Karaj, Iran). Genomic DNA from semen was extracted as previously described by Zadworny and Kuhenlein (1990).

3.2. Genotyping

3.2.1. Prolactin genotyping

The PCR was carried out according to Brym et al. (2005). Briefly, in 25 µL of a mix containing: 1.25 µL 20x PCR buffer; 1.3 µL dNTP (2 mM each); 70 pmol of each primer: forward 5' CCAAATCCACTGAATTATGCTT 3', reverse 5' ACAGAAATCACCTCTCTCATTCA 3'; 1.2 mM MgCl2; 0.8 unit TaqI DNA polymerase; 100-600 ng of genomic DNA; and H2O up to 25 µL. The PCR reaction was carried out in an Eppendorf thermocycler under the following conditions: initial denaturation (94ºC/3 min), 35 cycles of denaturation 94ºC/30 s), annealing (58.5ºC/30 s) and extension (72ºC/30 s), and final synthesis (72ºC/5 min). In order to genotyping of Bulls, 10 µL of PCR product was digested with 10 unit of RsaI restriction endonuclease and analysed by electrophoresis in 8% acrilamid gel with ethidium bromide.

3.2.2. Growth hormone receptor genotyping

A 836 bp fragment of 5'-flanking region of GHR (position -1866 to -1031) was amplified according to Aggrey et al. (1999). Briefly, the sequences of the forward and reverse primers were 5'-TGCGTGCACAGCAGCTCAACC-3' and 5'-AGCAACCCCACTGCTGGGCAT-3', respectively. The PCR amplification was carried out in 37 cycles at 95°C for 35 s, 66°C for 45 s, and 72°C for 60 s. The amplified DNA was digested for 12 hours at 37°C with 5 units of AluI restriction endonuclease. The digested DNA fragments were separated by electrophoresis in 8% acrylamid in 1X TAE buffer and visualized under UV light (UVIDOC).

3.2.3. Insulin like Growth Factor-I Genotyping

Detection of IGF-1 polymorphism was carried out according to Ge et al. (2001). Briefly, the 249-bp fragment of the IGF-1 gene was amplified using following primers: (Forward): 5'-ATTACAAAGCTGCCTGCCCC-3', and (Reverse): 5'-ACCTTACCCGTATGAAAGGAATATACGT-3'. The PCR amplification cycles were: 94°C for 1 min, 64°C for 45 s and 72°C for 1 min (31 cycles). The PCR amplified DNA fragment of the IGF-1 was digested at 37°C for 12 hours with 5 units of SnaBI nuclease. The digestion products were separated on 2% agarose gels in 1X TAE buffer and visualized in UVIDOC Imager.

3.3. Statistical analysis

The allele frequencies were calculated by simple allele counting according to the Hardy-Weinberg equilibrium (Falconer and Mackay 1996); the possible deviations of genotype frequencies from expectations were tested by chi-square (χ^2).

The effect of genotypes of each gene on milk production traits, namely, milk yield (kg), fat content (%), fat yield (kg), and protein yield (kg) and protein content (%) were analyzed by GLM procedure of SAS (2002). The following statistical model was used:

$$EBV_{ijk} = \mu + year_i + G_j + e_{ijk}$$

Where, EBV_{ijk} ,the estimated breeding value for milk related traits adjusted for number of daughters; μ ,the overall mean, $year_i$,the fixed effect of birth year of bulls (for genetic trends), G_j , the fixed effect of genotypes in each gene; e_{ijk} , the residual effect.

Breeding values of the bulls for milk production traits were obtained from the September 2008 Iranian Animal Breeding Center evaluations, which were based on an animal model. The model included animal effect as random effect, age of calving as covariate factor and fixed effects of herd-year-season. The reliabilities of EBVs for all the bulls were high and on average 92 %.

Average effect of allele substitution was determined by coding genotypes as 0 for low frequent homozygote genotype, 1 for hetrozigote genotype, and 2 for high frequent genotype in each gene. As described by Falconer and Mackay (1996), the regression coefficient estimates the average effect of allele substitution, or the average effect of replacing a high frequent allele with low frequent allele in each gene:

$$EBV = b_0 + b_1 (year) + b_2 (Genotypecode) + e$$

Where, EBV, the estimated breeding values as dependent variable, b_0, b_1 and b_2 representing the intercept, genetic trend and Average effect of allele substitution, respectively; year , the effect of birth years of bulls as independent variable (genetic trends), Genotypecode, assigned codes for genotypes.

For study of change trend of allelic frequencies in 18 years, the yearly Ratio of frequency of low frequent allele to high frequent allele for any gene was estimated. Then, the regression coefficient (b_1) of these yearly ratios on birth years of bulls was estimated:

$$\text{Ratio} = b_0 + b_1 (\text{year}) + e$$

4. Results and discussions

4.1. Paternal genetic trends for milk related traits

Genetic trends in the Iranian Holstein bulls were significant ($p < 0.05$) for all traits and progressive for milk, fat and protein yield but it was diminishing for fat and protein content (Table 1). The progressive and diminishing trends were resulted from two reasons, very high economic importance of milk regarding fat and protein content in the Iranian Holstein selection indices; and negative correlations between milk yield, and fat or protein content.

Traits	b	SE	P-value
Milk yield (kg)	+63.7608	5.0534	<0.0001
Fat yield (kg)	+1.5073	0.1419	<0.0001
Fat content (%)	-0.0110	0.0014	<0.0001
Protein yield (kg)	1.4181+	0.1120	<0.0001
Protein content (%)	-0.0057	0.0049	<0.0001

Table 1. Paternal Path genetic trends for milk related traits in the Iranian Holstein bulls

4.2 *Rsa*I-RFLP in Bovine PRL Gene

The transition of G into A in position 8398 creates a restriction site for RsaI endonuclease. Digestion of the 294 bp PCR product with the enzyme resulted in two restriction fragments of 162 and 132 bp for AA homozygotes, one uncut fragment of 294 bp for GG homozygotes, and all three fragments for AG heterozygotes (Fig. 1).

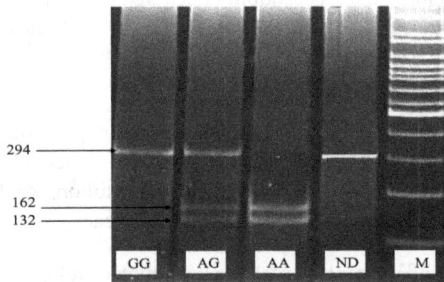

Figure 1. The 294 bp PCR products of PRL/exon4 were digested with RsaI and electrophoreses on 8% acrilamid gel. GG, AG and AA are the different genotypes of PRL/exon4. ND= Nondigested 294-bp PCR product. M= marker (100bp).

Allele frequencies were estimated in 268 bulls (0.069 and 0.931 for A and G, respectively). The frequencies of AA, AG and GG genotypes were 0.007, 0.123 and 0.870, respectively. Predicted genotype frequencies were similar to observed ones suggesting that genotype distributions were in the Hardy-Weinberg equilibrium ($\chi 2 = 0.477 < 3.82$).

Similar frequency of allele G (0.887) for Black-and-White cows and no similar frequency of allele G (0.294) for Jersey cows were reported by Brym et al. (2005). It can be explained by different history of the breeds, long-term geographical isolation, and selection towards high fat and protein contents of milk. It also indicated that PRL/exon4 SNP may be a marker of a linked SNP or locus involved in variation of milk composition.

Based on Table 2, the average allele substitution was negative and significant for milk and protein yield (p<0.05) i.e. allele G was an unfavorable allele for milk and protein yield. Brym et al. (2005) showed that PRL/exon4 SNP had a significant effect on milk yield and fat content in the first lactation. The result of present study for milk yield concurred with brym et al (2005), but results for fat content and protein yield were not similar. The mentioned authors showed that allele G is a favorable allele for fat content and unfavorable for milk yield. Similarly, the results of this study confirmed that allele G has negative effect on milk, fat and protein yield; and positive effect on fat and protein content, but it was significant only for milk and protein yield.

Traits	α	SE	P value
Milk yield (kg)	-203.4924	94.687	0.0325
Fat yield (kg)	-4.0107	2.5198	0.112
Fat content (%)	+0.027119	0.02376	0.2547
Protein yield (kg)	-4.3019	2.0935	0.0409
Protein content (%)	+.01755	0.00903	0.0530

α: Average substitution effects of allele G

Table 2. Average allele substitution effects of PRL/exon4 polymorphism for milk related traits

The coefficient of correlation between yearly EBV means of bulls with frequency Allele "A" to "G" ratio in any year was 0.104 and non significant (Table 3). This correlation showed that traditional selection programs did not affect the frequency of the PRL/exon 4 SNP. It was expected that number of allele A (favorable allele) would be increased during years, but the increasing rate was not significant. In the future, marker assisted selection based on major genes may increase the favorable allele frequency.

The results in table 3 showed that PRL gene polymorphism has no significant effect on genetic trends of milk performance traits. Genetic changes during years occurs in selected populations such as the analyzed sample of Iranian Holstein bulls, and the rate of change depend on selection strategies (selection indices, the traits accounting in indices, their economic and breeding coefficients, accuracy of estimated breeding values and etc). The effects of candidate genes or major genes on quantitative trait phenotype may be more than other genes, thus the effects of these genes on genetic trends may be more than others. Genetic trends in molecular level depend on change of frequency of genes (especially major genes) through years and the rate of independent effect of each gene. The results of the present study indicated that no significant effect of PRL polymorphism on milk related traits trends, Because of no significant trend in favorable allele frequency.

	Birth year	Milk yield	Fat yield	Fat content	Protein yield	Protein content
Ratio (A to G)	0.1069	0.0300	0.1083	0.1367	0.0254	-0.0496
p-value	0.6730	0.9058	0.6689	0.5886	0.9202	0.8449

Table 3. Coefficients of correlations between yearly EBV means of milk related traits and yearly ratio of allele "A" to "G" frequencies

4.3. *Alu*I-RFLP in Bovine GHR Gene

There were three AluI sites in the 836 bp fragment of 5' flanking region of the bovine GHR gene. The digested AluI(-/-) PCR product exhibited three fragments of 747 bp, 75 bp, and 14 bp (not detected on the gel). For the AluI(+/+) PCR product, the 747 bp fragment was cleaved into 2 fragments of 602 and 145 bp (Fig. 2). The polymorphic AluI site revealed a mutation at position -1182 (A-to-T transversion).

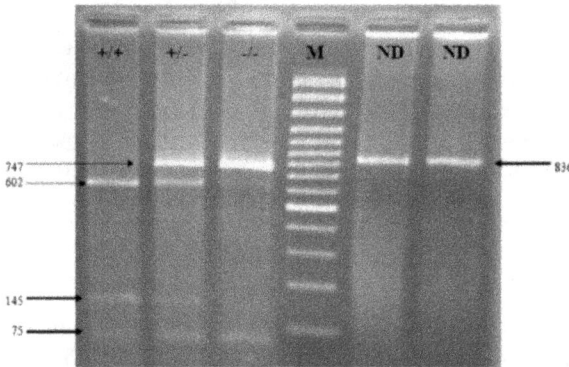

Figure 2. Digestion products of the 836 bp fragment in the 5'UTR region of growth hormone receptor gene with enzyme AluI, loaded on 8% acrilamid gel. The genotypes of AluI (-/-), AluI (+/+) and AluI (+/-) were shown in left side of gel, respectively. M: Marker (100bp), ND: undigested PCR product.

The genotype and allele frequencies at this SNP are shown in Table 4. The calculated $\chi2$ was 1.282 and it was lower than critical value of χ^2 table ($\chi^2_{\alpha=0.05,df=1}$ =3.841), then the null hypothesis did not rejected, suggesting that genotypes distributions were in the Hardy-Weinberg equilibrium.

genotypes	AluI(+/+)	AluI (+/-)	AluI (-/-)
Number of animals	84	131	67
Observed frequency	0.298	0.465	0.237
Expected frequency	0.282	0.498	0.220
Allele frequencies	AluI (+)=0.531	AluI (-)=0.469	
Calculated $\chi2$=1.282		Critical value of $\chi2$ =3.841	

Table 4. The genotype and allele frequencies of GHR gene

Least square means of genotypes are presented in table 5. Bulls with AluI(+/+) genotype had best EBV for milk yield, fat yield and protein yield, but the differences between genotypes were not statistically significant (P>0.05). on the contrary, the highest fat percent and protein percent EBVs were observed in bulls with AluI(-/-) genotypes that its difference in compared to other genotypes was significant only for fat percent (P<0.05).

Trits	Genotypes			P-value
	AluI (-/-)	AluI (+/-)	AluI (+/+)	
Milk yield	-251.85	-210.34	-85.93	0.2828
Fat yield	-1.89	-2.76	-0.545	0.7411
Protein yield	-8.87	-1.73	-1.22	0.3869
Fat percent	0.113 [a]	0.057 [ab]	0.003 [b]	0.0155
protein percent	0.0488	-0.037	-0.0253	0.283

a,ab,b: Lsmeans were signed with the different letter within any row, were differ significantly (p<0.05).

Table 5. Least square means and p-value for estimated breeding values of milk related traits in Iranian Holstein bulls based on different GHR genotypes

The average allele substitution effect AluI (+) instead of allele AluI (-), estimated by Falconer and Mckay method (1996) was 80.62, 0.588, 1.772, -0.0548 and -0.007566 for EBVs of milk yield, fat yield, protein yield, fat percent and protein percent, respectively (table 6). These results showed that AluI (+) allele may be increased the milk, fat and protein yield; on the other hand, AluI(-) allele have increment effect on fat and protein percent that only the average effect of the AluI(-) allele on fat percent was statistically significant (P<0.05).

Change in ratio of AluI (+) to AluI (-) frequencies based on birth years of bulls (18 years) was studied by fitting of a linear regression. The estimated regression coefficient and its p-value were 0.01424 and 0.017, respectively. Therefore it can be said that frequency of allele AluI (+) has been increased averagely as 1.424 % per year. Based on obtained p-value (0.017) and lack of fit test, the linear relationship between allele frequency and birth years of bulls was confirmed in the Iranian Holstein bulls.

The AluI SNP was located within the 1.2-kb LINE-1 element, a retrotransposon of viral origin, inserted in the bovine GHR gene 5' region. The frequency of AluI(-) allele in present study was 0.469, that has been reported 0.473 in Canadian Holstein cattle (aggrey et al. 1999); 0.45 and 0.64 in Lithuanian Black & White and Lithuanian Red cattle, respectively (Skinkyte et al. 2005). Hetrozigosity rate (frequency of heterozygote genotype) in Iranian Holstein bulls was 0.465 and these rates were 0.45, 0.59 and 0.43 for Canadian Holstein bulls, Lithuanian Black & White and Lithuanian Red cattle, respectively. Comparison of allele frequencies and heterozygosity rate in Canadian and Iranian Holstein bulls showed these populations are similar that may be caused by similarity in selection programs of two populations. Observed differences between Iranian and Lithuanian cattle population may be due to the studied cows in the Iranian population included bulls that were used in the selection program, while the Lithuanian cows had been selected from commercial herds.

Traits	α	SE	P-value
Milk yield (kg)	80.62	53.204	0.1336
Fat yield (kg)	0.588	1.573	0.7096
Protein yield (kg)	2.022	2.870	0.4829
Fat percent (%)	-0.0548*	0.01838	0.0038
Protein percent (%)	-0.007566	0.00499	0.1338

α: Average allele substitution effects of allele AluI (+) instead of AluI (-).

Table 6. Average allele substitution effects of GHR polymorphism on EBVs of milk related traits

As reported by Aggrey et al. (1999), the AluI (+/+) bulls had higher fat EBV than the AluI (-/-) bulls. No significant differences in this study were shown between GHR genotypes for fat EBV, but the AluI(-) allele is a favorable allele for fat percent EBV. However, it seems that this polymorphism have marked effect on milk fat (fat percent and/or fat production), and observed differences between two results may be due to fat recording accuracy in Canada and Iran. The 5' region of the GHR gene contains regulatory sequences which control the expression of GHR and interact with a large number of cis-acting and trans-acting factors (Heap et al, 1995). Modulation of the affinity of binding of any of these factors may affect GHR transcription and consequently its binding ability with GH.

The significant relation between birth years and yearly ratio of allele frequencies may be resulted of association between AluI(+) allele and paternal genetic trends of economic important traits in Iranian Holstein selection program. Although, the significant correlation of allele AluI(+) with milk traits was not confirmed in this study, but its positive effect was observed, and also its likely associations with other important traits in dairy cattle breeding, such as body conformation traits is not impossible (under studding) that these reasons may be justifying of the allele frequency change during Holstein breeding program in Iran. The decreasing trend of allele frequency AluI (-) in selected years is probably due to fat percent trait is not included in the selection program of dairy cows in Iran.

One of the advantages of breeding values as the dependent variable in the association studies of polymorphism with traits is the possibility of investigation of major gene effect on genetic trends. It is recommended the various statistical models and further research for better understanding of molecular mechanisms of genetic trend.

4.4. SnaBI-RFLP in Bovine IGF-I gene

The C/T transition at position -472 in the 5'-noncoding region of the IGF-1was first reported in Angus cattle by Ge et al. (1997) as SSCP. This mutation is at position 512 bp upstream from the ATG codon. The C→T substitution creates a SnaBI restriction site and digestion of the 249 bp PCR product with the restriction SnaBI nuclease resulted in two DNA bands (223 and 26 bp) for homozygote (TT) and three bands (249, 223 and 26 bp) for the heterozygote. The DNA amplified from homozygous (CC) animals remained undigested with SnaBI restriction endonuclease (fig. 3).

Based on Table 7, the expected genotype frequencies were not similar to observed, suggesting that genotype distribution was not in the Hardy-Weinberg equilibrium ($p<0.05$).

Based on table 8, bulls with CT genotype had higher estimated breeding values of milk and fat yield compared to CC and TT genotypes ($P<0.1$). The heterozygous bulls had higher protein yield, fat and protein content (%), but the differences between genotypes for these traits were not statistically significant ($p>0.1$).

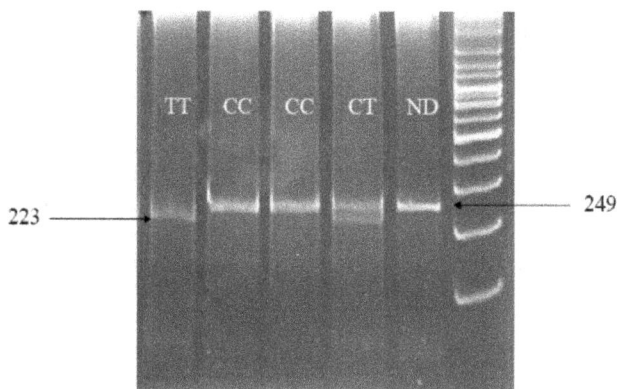

Figure 3. Acril amid gel (8%) electrophoresis showing RFLP-SnaBI in 5'-noncoding region of the bovine IGF-1 gene. TT=223 bp, CC= 249 bp, CT=249 and 223 bp, ND= undigested product and M=100 bp DNA marker. The 26 bp band was not seen in gel.

Genotypes	TT	CT	CC	Chi-square test
Number of animals	45	157	80	
Observed frequency	0.159	0.557	0.284	$\chi2=4.878$
Expected frequency	0.192	0.492	0.316	Critical Value=3.841
Alleles	T=0.438		C=0.562	

Table 7. The observed and expected genotypic and allelic frequencies of IGF-1 gene polymorphism

Traits	Genotypes			
(EBV)	CC	CT	TT	P-Values
Milk Yield (Kg)	9.8355 [b]	118.8269 [a]	-46.819 [b]	0.072
Fat Yield (Kg)	1.5159[b]	5.4918 [a]	1.1288 [b]	0.092
Protein Yield (Kg)	2.0755	3.5566	1.2167	0.302
Fat Content (%)	0.0059	0.02815	0.02599	0.4779
Protein Content (%)	0.0150	0.0099	0.0181	0.4132

Table 8. Least square means for milk production traits in Iranian Holstein bulls with different IGF-1 genotypes

The average effect of the T allele substitution was not statistically significant and that was 28.88, 0.0962, -0.468 kg for EBVs of milk, fat and protein yields, and 0.014 and 0.0019 % for fat and protein content, respectively (table 9).

The regression coefficient of yearly frequencies of heterozygous genotype on birth years of bulls were -0.0048 and this coefficient was not statistically significant (p<0.1). So, the change trend of CT genotype frequencies was not linear. No significant relations were shown between yearly means of estimated breeding values of milk related traits and yearly frequencies of CT genotype i.e. no significant relation was seen between genetic trends and IGF-1 gene (Table 10).

Trait (EBV)	b	SE	P-Value
Milk Yield (Kg)	28.88	42.806	0.500
Fat Yield (Kg)	0.09622	1.1636	0.9358
Protein Yield (Kg)	0.46796	0.9269	0.6141
Fat Content (%)	-0.01406	0.01166	0.229
Protein Content (%)	-0.00196	0.00371	0.5984

b: Linear regression coefficient estimating average substitution effects of T allele

Table 9. Average allele substitution effects of IGF-1 polymorphism in EBVs of milk related traits

Trait	b	SE	P-Value
Milk yield (kg)	-0.00007	0.000087	0.4328
Fat yield (kg)	-0.00331	0.00377	0.3925
Protein yield (kg)	-0.00285	0.00382	0.4665
Fat content (%)	0.2852	0.0337	0.5366
Protein content (%)	0.7685	0.9638	0.4369

Table 10. Effects of IGF-1 gene on genetic trend in Iranian Holstein bulls

Similar frequencies of alleles C and T in bovine IGF-1 gene were found by Hines et al. (1998), who reported an estimate of 0.55 and 0.45 for the frequency of C and T alleles in a population of Holstein cattle. Also, Li et al. (2004) reported estimates of 0.56 and 0.44 in two commercial lines of dairy cattle; respectively. However, different estimates of frequencies (0.64 for (C) and 0.36 (T) alleles) were reported by Ge et al. (2001) in Angus cattle.

The association between RFLP-SnaBI of the IGF-1 gene and milk traits was studied by Siadkowska et al. (2006), using 262 polish Holstein-Friesian cows. They did not find any differences between genotypes in daily milk yield, but CT cows yielded significantly more daily fat (+20 g) and protein (+14.5 g) than the cows with CC genotype (P<0.05). The CT genotype also appeared favorable for fat and protein content of milk. Hines et al. (1998) reported no association between IGF-1 gene RFLP-SnaBI and dairy production traits in Holstein dairy cattle. No other papers were found in the literature concerning effects of IGF-1 polymorphism on milk production traits. The effects of this SNP have been generally tested in relation to meat production traits in previous studies.

5. Conclusion

The analysis of this study confirmed that PRL, GHR and IGF-I could be a strong candidate for application in marker-assisted selection. This study did not prove a significant effect of PRL polymorphism on paternal path genetic trends for milk production traits in Iranian Holsteins. The effects of the SNP on selection indices or other traits especially conformation traits and semen related traits of bulls should be the subject of further research.

The results proved a significant effect of IGF-1 polymorphism on EBVs for milk production traits. However, the important role of IGF-1 in the meat production process is well known, thus its polymorphism effects on other traits specially conformation traits of bulls should be the subject of further research. Also, the association of gene polymorphism with genetic trend was studied in first time in this study, and understanding of molecular mechanism of genetic trend needs to additional researches.

Author details

Yousef Mehmannavaz
Department of Animal Science, Maragheh Branch, Islamic Azad University, Maragheh, Iran

Abolfazl Gorbani
Department of Animal Science, Shabestar Branch, Islamic Azad University, Shabesta, Iran

Acknowledgement

The authors thank Research unit of Islamic Azad University, Maragheh Branch for their supports in all stages of this research. Data and semen of Iranian Holstein bulls were provided through animal breeding center of Iran.

6. References

Aggrey, S.E.; Yao, J.; Sabour, M.P.; Lin, C.Y.; Zadworny, D.; Hayes, J.F. & Kuhnlein, U. (1999). Markers within the regulatory region of the growth hormone receptor gene and their association with milk-related traits in Holsteins. *Journal of Heredity*. 90, 148–151.

Aggrey, S.E.; Yao, J.; Zadworny, D.; Hayes J.F. & Kuhnlein. U. (1998). Synergism between genetic markers in the growth hormone and growth hormone receptor genes in influencing milk related traits in Holsteins. *Journal of Dairy Science*. 80: (Suppl 1) 229-232.

Alipanah, M.; Kalashnikova L. & Rodionov, G. (2007). Association of prolactin gene variants with milk production traits in Russian Red Pied cattle. *Iranian Journal of Biotechnology*. 5(3): 158-161.

Arendonk, J. V.; Tier, B. & Kinghorn, B.P. (1994). Use of multiple genetic markers in prediction on breeding values. *Genetics*, 137: 319-329.

Bell A.W. (1995). Regulation of organic nutrient metabolism during transition from late pregnancy to early lactation. *Journal of Animal Science.* 73:2804–19.

Berwaer, M.; Monget, P.; Peers, B.; Mathy-Hartert, M.; Bellefroid, E.; Davis, J.R.; Belayew, A. & Martial, J.A. (1991). Multihormonal regulation of the human prolactin gene expression from 5000 bp of its upstream sequence. *Molecular and Cellular Endocrinology.* 80: 53–64.

Bichell, D.P.; Kikuchi K. & Rotwein, P. (1992). Growth hormone rapidly activates insulin-like growth factor I gene transcription in vivo. *Molecular Endocrinology.* 6:1899–908.

Blott, S.; Kim, J.J.; Moisio, S.; Schmidt-Kuntzel, A.; Cornet, A.; Berzi, P.; Cambisano, N.; Ford, C.; Grisart, B.; Johnson, D.; Karim, L.; Simon, P.; Snell, R.; Spelman, R.; Wong, J.; Vilkki, J.; Georges, M.; Farnir, F. & Coppieters, W. (2003). Molecular dissection of a quantitative trait locus: A phenylalanine-to-tyrosine substitution in the transmembrane domain of the bovine growth hormone receptor is associated with a major effect on milk yield and composition. *Genetics.* 163:253–266.

Bole-Feysot, C.H.; Goffin, V.; Edery, M.; Binart, N. & Kelly, P.A. (1998). Prolactin (PRL) and its receptor: actions signal transduction pathways and phenotypes observed in PRL receptor knockout mice. *Endocrinology Review.* 19: 225–268.

Boulay, J.L. & Paul, W.E. (1992). The interleukin-4related lymphokines and their binding to hematopoientin receptors. *Journal of Biological Chemistry.* 267:20525-20528.

Brym P.; Kamiński, S.; Wójcik, E. (2005). Polymorphism within the bovine prolactin receptor gene (PRLR). *Animal Science Papers and Reports.* 23 : 61-66.

Brym, P.; Malewski, T.; Starzynski, R.; Flisikowski, K.; Wojcik, E.; Rusc, A.; Zwierzchowski, L. & Kaminski, S. 2007. Effect of New SNP Within Bovine Prolactin Gene Enhancer Region on Expression in the Pituitary Gland. *Biochemical Genetics.* 45:743–754.

Camper, S.; Luck, D.N.; Yao, Y; Woychik, R.P; Goodwin, R.G.; Lyons, R.H. & Rottman, F.M. (1984). Characterization of the bovine prolactin gene. *DNA.* 3: 237–249.

Cao, X.; Wang, Q.; Yan, J.B.; Yang, F.K.; Huang, S.Z. & Zeng, Y.T. (2002). Molecular cloning and analysis of bovine prolactin full-long genomic as well as cDNA sequences. *Yi. Chuan Xue. Bao.* 29(9):768-773.

Chrenek, P.; Vasicek, D.; Bauerova, M. & Bulla, J. (1998). Simultaneous analysis of bovine growth hormone and prolactin alleles by multiplex PCR and RFLP. *Czech Journal of Animal Science.* 43: 53–55.

Chung, E.R.; Rhin, T.J. & Han, S.K. (1996). Association between PCR-RFLP markers of growth hormone and prolactin genes and production traits in dairy cattle. *Korean Journal of Animal Science.* 38: 321–336.

Cooke, N.E.; Coit, D.; Shine, J.; Baxter J.D. & Martial, J.A. (1981). Human prolactin cDNA structural analysis and evolutionary comparisons. *Journal of Biological Chemistry.* 256: 4007–4016.

Daughaday, W.H.; Rotwein, P. (1989). Insulin-like growth factors I and II: Peptide messenger ribonucleic acid and gene structures, serum, and tissue concentrations. *Endocrinology Reviews.* 10: 68-91.

Dekkers, J.C.M. (2004). Commercial application of marker and gene associated selection in livestock: Strategies and lessons. *Journal of Animal Science*, 82: E313-E328.

Dybus A. 2002. Associations of growth hormone GH and prolactin PRL genes polymorphism with milk production traits in Polish Black and White cattle. *Animal Science Papers and Reports*. 20: 203–212.

Dybus, A.; Grzesiak, W.; Kamieniecki, H.; Szatkowska, I.; Sobek, Z.; Blaszczyk, P.; Czerniawska-Piatkowska E. & Zych, M.Muszynska, S. (2005). Association of genetic variants of bovine prolactin with milk production traits of Black-and-White and Jersey cattle. *Archive fur Tierzukht*. 48 : 149-156.

Etherton, T. D. & Bauman, D. E. (1998). Biology of somatotropin in growth and lactation of domestic animals. *Physiological Reviews*. 78:745–761.

Falaki, M.; Gengler, N.; Sneyers, M.; Prandi, A.; Massart, S.; Formigoni, A.; Burny, A.; Portetelle D. & Renaville, R. (1996). Relationships of polymorphisms for growth hormone and growth hormone receptor genes with milk production traits for Italian Holstein-Friesian bulls. *Journal of Dairy Science*. 79:1446-1453.

Falconer, D.S. & Mackay, T.F.C. (1996). *Introduction to Quantitative genetics*. 4th Ed, Longmans Green, Harlow. Essex, UK.

Feng, H.; Dongxiao, S.; Yu, Y.; Wang Y. & Zhang, Y. (2006). Association between SNPs within Prolactin Gene and Milk Performance Traits in Holstein Dairy Cattle. *Asian-Australian Journal of Animal Science*. 19(10) : 1384 – 1389.

Fenwick, M. A.; Fitzpatrick, R.; Kenny, D. A.; Diskin, M.G.; Patton, J.; Murphy, J. J. & Wathes, D. C. (2008). Interrelationships between negative energy balance (NEB) and IGF regulation in liver of lactating dairy cows. *Domestic Animal Endocrinology*. 34: 31–44

Francis, G.L.; Upton, M.F.; Ballard, F.J.; Mcneil K.A. & Wallace, J.C. (1988). Insulin-like growth factors 1 and 2 in bovine colostrum. *Biochemistry Journal*. 251: 95-103.

Freeman, M.E.; La Kanyicska, B.; Lerant, A. & Nagy, G. (2000). Prolactin: Structure, Function, and Regulation of Secretion. *Physiological Reviews*. 80:1523-1631.

Ge, W.; Davis, M.E.; Hines, H.C.; Irvin, K.M. & Simmen, R.C.M. (2001). Associations of a genetic marker with blood serum insulin-like growth factor-1 concentration and growth traits in Angus cattle. *Journal of Animal Science*. 79: 1757-1762.

Hallerman, E.M.; Theilmann, J.L.; Beckmann, J.S.; Soller, M. & Womack, J.E. (1998). Mapping of bovine prolactin and rhodopsin genes in hybrid somatic cells. *Animal Genetics*. 19: 123–131.

Hauser, S. D.; McGrath, M. F.; Collier, R. J. & Krivi. G. G. (1990). Cloning and in vivo expression of bovine growth hormone receptor mRNA. *Molecular and Cellular Endocrinology*. 72:187-200.

Heap, D.; Lucy, M.C.; Collier, R.J.; Boyd, C.K. & Warren, W.C. (1995). Nucleotide sequence of the promoter and first exon of the somatotropin receptor gene in cattle. *Journal of Animal Science*: 73, 1529-1534.

Herrington, J. & Carter-Su, C. (2001). Signaling pathways activated by the growth hormone receptor. *Trends in Endocrinology of Metabolism*. 12:252–257.

Hines, H.C.; Ge, W.; Zhao, Q.; Davis, M.E. (1998). Association of genetic markers in growth hormone and insulin-like growth factor I loci with lactation traits in Holsteins. *Animal Genetics*. 29(1): 69-74.

Horseman, N.D. & Yu-Lee, L.Y. (1994). Transcriptional regulation by the helix bundle peptide hormones: growth hormone, prolactin, and hematopoietic cytokines. *Endocrinology Reviews*. 15: 627–649.

Hradecka E.; Rehout V. & Citek, J. (2006). The polymorphism of growth hormone receptor gene in Holstein and Czech Pied bulls. *Acta Fyto. Zootechica*. 224-227.

Hradecka, E.; Citek, J.; Panicke, L.; RehoutV. & Hanusova, L. (2008). The relation of GH1, GHR and DGAT1 polymorphisms with estimated breeding values for milk production traits of German Holstein sires. *Czech Journal of Animal Science*. 53(6): 238–245.

Jansen, E.; Steenbergh, P.H.; LeRoitH, D.; Roberts, J.C.T. & Sussenbach, J.S. (1991). Identification of multiple transcription start sites in the human insulin-like growth factor I gene. *Molecular and Cellular Endocrinology*.78: 115-125.

Jiang, H., Lucy, M.C.; Crooker B.A. & Beal, W.E. (2005). Expression of growth hormone receptor 1A mRNA is decreased in dairy cows but not in beef cows at parturition. *Journal of Dairy Science*. 88:1370–7.

Jiang, H.L. & Lucy, M.C. (2001). Variants of the 5'-untranslated region of the bovine growth hormone receptor mRNA: isolation, expression and effects on translational efficiency. *Genetics*, 265, 45–53.

Kopchick, J.J. & Andry, J.M. (2000). Growth Hormone (GH), GH Receptor, and Signal Transduction (Minireview). *Molecular Genetics and Metabolism*. 71: 293–314.

Kovacs, K. (2006). AluI polymorphism of bovine growth hormone and growth hormone receptor genes in a Hungarian Holstein-Frisian bull dam population. *Ph.D. Thesis*. *Godollo*. Hungary.

Le Roith, D.; Bondy, C.; Yakar, S.; Liu, J.L.; Butler, A. (2001). The somatomedin hypothesis: 2001. *Endocrinology Reviews*. 22:53–74.

Li, J.T.; Wang, A. H.; Chen, P.; Li, H.B.; Zhang, C. S. & Du, L.X. (2006). Relationship between the Polymorphisms of 5' Regulation Region of Prolactin Gene and Milk Traits in Chinese Holstein Dairy Cows. *Asian-Australian Journal of Animal Science*. 19(4) : 459-462.

Lucy, M.C.; Boyd, C.K.; Koenigsfeld, A.T. & Okamura, C.S. (1998). Expression of somatotropin receptor messenger ribonucleic acid in bovine tissue. *Journal of Dairy Science*. 81: 1889–1895.

Meuwissan, T. H. E. (1997). Maximizing the response of selection with a predefined rate of inbreeding. *Journal of Animal Science*. 75:934–940.

Miller J.R.; Thomsen, P.D.; Dixon, S.C.; Tucker, E.M.; Konfortov, B.A. & Harbitz, I. (1991). Synteny mapping of the bovine IGHG2, CRC and IGF-1 genes. *Animal Genetics*. 23: 51-58.

Missohou, A.; Talaki, E. & Laminon, I.M. (2006). Diversity and genetic relationships among seven West African goat breeds. *Asian Australian Journal of Animal Science*. 19: 1245-1251.

Monaco M.H.; Gronlund, D.E.; Bleck, G.T.; Hurley, W.L.; Wheeler M.B. & Donovan, S.M. (2005). Mammary specific transgenic over-expression of insulin-like growth factor-I (IGF-I) increases pig milk IGF-I and IGF binding proteins, with no effect on milk composition or yield. *Transgenic Research.* 14(5):761-73.

O'Dell, S.D. & Day, I.N.M. (1998). Molecules in focus: Insulin–like growth factor II (IGF-II). *International Journal of Biochemistry and Cell Biology.* 30:767-771.

Rinderknecht, E. & Humbel, R.E. (1978). The amino acid sequence of human insulin like growth factor I and its structural homology with pro-insulin. *Journal of Biological Chemistry.* 253:2769-2776.

Sasavage, N.L.; Nilson, J.H.; Horowitz, S. & Rottman, F.M. 1982. Nucleotide sequence of bovine prolactin messenger RNA. *Journal of Biological Chemistry.* 257: 678–681.

Shimatsu, A.; & Rotwein, P. (1987). Mosaic evolution of the insulin like growth factors. Organization, sequence, and expression of the rat insulin-like growth factor I gene. *Journal of Biological Chemistry.* 262: 7894.7900.

Siadkowska, E.; Zwierzchowski, L.; Oprządek, J.; Strzałkowska, N.; Bagnicka, E. & Krzyżewski, J. (2006). Effect of polymorphism in IGF-1 gene on production traits in Polish Holstein-Friesian cattle. *Animal Science Papers and Reports.* 24(3): 225-237.

Skinkytė, R.; Zwierzchowski, L.; Riaubaitė, L.; Baltrėnaitė, L. & Miceikienė. I. (2005). Distribution of allele frequencies important to milk production traits in Lithuanian Black & White and Lithuanian Red cattle. *Veterinarica Zootechenica.* 31 (53):93-96.

Soller, M. & Beckmann, J. S. (1983). Genetic polymorphisms in varietal identification and genetic improvement. *Theoritical and Applied Genetics.* 67:25-29.

Steenbergh, P.H.; Koonen-Reemst, A.M.C.; Cleutjens, C.B.J.M. & Sussenbach, J.S. (1991). Complete nucleotide sequence of the high molecular weight human IGF-1 mRNA. Biochem. *Biophysics Research Communication.* 175: 507-514.

Tambasco, D.D.; Paz, C.C.P; Tambasco-Studart, M.; Pereira, A.P.; Alencar, M.M.; Freitas, A.R.; Coutinho, L.L.; Packer, I.U. & Regitano L.C.A. (2003). Candidate genes for growth traits in beef cattle cross Bos taurus × Bos indicus. *Journal of Animal Breeding and Genetics,* 120: 51-56.

Thissen, J.P.; Ketelslegers J.M. & Underwood, L.E. (1994). Nutritional regulation of the insulin-like growth factors. *Endocrinology Reviews.* 15:80–101.

Viitala, S.; Szyda, J.; Blott, S.; Schulman, N.; Lidauer, M.; Maki-Tanila, A.; Georges, M. & Vilkki, J. (2006). The Role of the Bovine Growth Hormone Receptor and Prolactin Receptor Genes in Milk, Fat and Protein Production in Finnish Ayrshire Dairy Cattle. *Genetics.* 173: 2151–2164.

Wallis M. (1974). The primary structure of bovine prolactin. *FEBS Letters.* 44: 205–208.

Wang, Y.; Price, S.E.; Jiang, H. (2003). Cloning and characterization of the bovine class 1 and class 2 insulin-like growth factor- 1 mRNAs. *Domestic Animal Endocrinology.* 25(4): 315-328.

Werner, H.; Adamo, M.; Roberts C.T. & Leroith, D. (1994). Molecular and cellular aspects of insulin-like growth factor action. *Vitamins and hormones.* 48:1-58.

Zadworny, D. & Kuhenlein, U. (1990). The identification of the Kappa-casein genotype in Holstein dairy cattle using the polymerase chain reaction. *Theoretical and Applied Genetics*. 80:631-634.

Chapter 3

Genetic Factors that Regulate Milk Protein and Lipid Composition in Goats

Marcel Amills, Jordi Jordana, Alí Zidi and Juan Manuel Serradilla

Additional information is available at the end of the chapter

1. Introduction

The mammary gland fulfills the essential role of providing all the nutrients needed to sustain the life and growth of the newborn under the form of milk, a white fluid composed primarily by water, carbohydrates, lipids, proteins and minerals. Domestication of cow, sheep and goats in the Near East 9,000 YBP and the subsequent creation of breeds specialized in milk production allowed humans to take profit of this rich source of proteins and minerals, becoming an important component of their diet either in the form of fresh milk or derived products such as cheese, yogurt, kefir, butter and many others (**Figure 1**). Because of their adaptability to a harsh climate and scarce vegetation, dairy goats occupy an important niche in the economy of tropical countries such as India, Bangladesh and Sudan, which happen to be the three main goat milk producers at a worldwide scale (FAOSTAT 2009). In Europe, France, Spain and Greece are the largest goat milk producers and, in comparison with Asian and African countries, have a much more intensified production system (FAOSTAT 2009).

Proteins and lipids are essential components of milk and they can have a very strong impact on milk nutritional and technological properties (Bauman et al. 2006). Milk casein content, for instance, is one of the main determinants of cheese yield and both traits are positively correlated (Remeuf and Hurtaud 1991). Similarly, fat content and composition are key factors determining milk and cheese attributes. In this way, milk with a low fat percentage is associated with a reduced cheese yield and firmness as well as with negative effects on flavor and color (Lamberet et al. 2001). Moreover, short-chain fatty acids (FA), such as C4:0-C12:0, have been implicated in the appearance of a rancid soapy flavor in milk, whilst the hardness and melting point of fat is largely determined by its unsaturated FA content (Fox and Sweeney 2003). Importantly, a relevant fraction (around 70%) of goat milk fat is composed by saturated FA that have detrimental effects on human health because they are associated with an increased risk of suffering cardiovascular diseases (Pfeuffer and

Schrezenmeir 2000). It is also worth to mention that hydrolysis of milk proteins releases a wide array of short bioactive peptides which might have many beneficial effects on human health, such as (i) hypotensive, antithrombotic, antioxidative, antimicrobial, and immunomodulatory activities (Fitzgerald and Meisel 2004, Korhonen and Pihlanto 2001), (ii) enhancement of mineral absorption (Meisel 1998), and (iii) antitumoral properties (Matar et al. 2003). Milk protein and fat contents vary from breed to breed due to both environmental and genetic factors (**Figure 2**)

Figure 1. Portray of a woman making butter (Paris, 1499). Source: *Compost et Kalendrier des Bergères*. Robarts Library, University of Toronto.

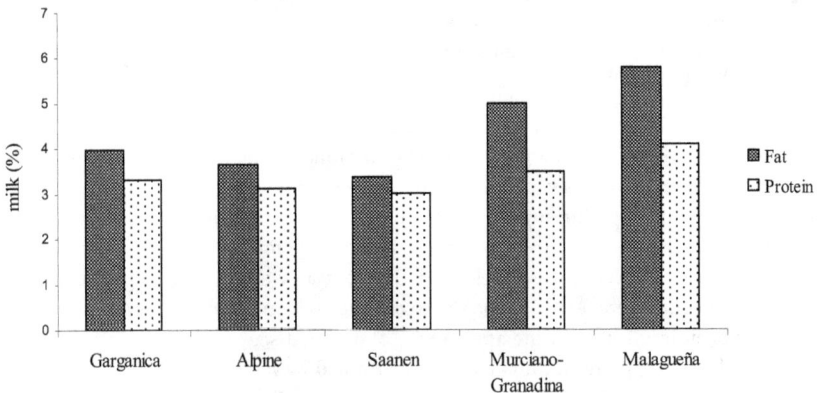

Figure 2. Milk protein and fat content in diverse goat breeds. Sources: Martini et al. (2010), Rupp et al. (2011), Fernández et al. (2005) and website of the Asociación Española de Criadores de la Cabra Malagueña (http://www.cabrama.com)

The major protein milk fraction (around 80%) is constituted by caseins, *i.e.* insoluble phosphoproteins organized in complex multi-molecular aggregates, named micelles, that at certain conditions of temperature and acidity (T ~ 20 ºC, pH = 4.6), precipitate from skim milk (Thompson et al. 1965). Besides caseins, micelles contain inorganic materials such as calcium phosphate and calcium citrate (Smiddy et al. 2006). In ruminants, four casein types have been distinguished so far (Martin et al. 2002): α_{s1}-casein (CSN1S1), α_{s2}-casein (CSN1S2), β-casein (CSN2) and κ-casein (CSN3). CSN1S1, CSN1S2 and CSN2 are considered to be calcium-sensitive caseins because they have a high phosphate content and precipitate in the presence of calcium (Grosclaude 1991). Loci encoding these three caseins have been shown to descend, through successive gene duplication events, from an ancestral locus encoding the secretory calcium-binding phosphoprotein proline-glutamine rich 1 (*SCPPPQ1*) molecule, that plays a key role in dental enamel mineralization (Kawasaki and Weiss 2003). In contrast, CSN3 is a calcium-insensitive casein responsible of micelle stabilization and whose molecular ancestry is completely different, since it evolved from the follicular dendritic cell secreted peptide (*FDCSP*) gene, which is expressed in fibroblasts producing periodontal ligament (Grosclaude 1991, Kawasaki and Weiss 2003). The enzymatic hydrolysis of CSN3 by chymosin involves the destabilization of micelles and the subsequent coagulation of milk, which is the initial step in the manufacturing of cheese (Jollès 1975, Remeuf et al. 1991).

Whey proteins represent around 20% of the total protein content of milk and their common denominator is that they remain soluble after milk coagulation (Madureira et al. 2007). Examples of whey proteins are β-lactoglobulin (BLG), α-lactalbumin (LALBA), immunoglobulins, serum albumin, lactoferrin, and lactoperoxidase. Functions of these proteins are very heterogeneous (Madureira et al. 2007), affecting processes as different as lactose synthesis (LALBA), transport of hydrophobic molecules (BLG) and microbial immunity (lactoferrin and lactoperoxidase).

Protein synthesis in the mammary gland depends on the uptake of amino acids from the circulatory system and it is controlled by lactogenic hormones (insulin, prolactin, and glucocorticoids) as well as by the blood concentrations of circulating amino acids (Weekes et al. 2006, Rhoads and Grudzien-Nogalska 2007). Milk protein gene expression is strongly affected by the reproductive cycle, being activated at mid-pregnancy, peaking during lactation and declining after weaning. Generally, the conversion efficiency of dietary nitrogen to milk proteins is relatively poor in the mammary epithelial cells of goats (25-30%) for reasons that remain to be defined (Bequette et al. 1998). The transport of amino acids, and even short peptides, through the membrane of secretory cells is facilitated by a variety of molecular systems that have been broadly reviewed by Shennan and Peaker (2000).

With regard to milk lipids, the most abundant fraction (around 95%) is constituted by triglycerides, whilst the remaining 5% encompasses free FA, phospholipids, cholesterol esters, diglycerides and monoglycerides (Harvatine et al. 2009). A distinctive feature of goat milk is that medium chain FA (*e.g.* C8, C10 and C12) and C18:1 unsaturated FA are particularly abundant (Fontecha et al. 2000). Milk FA are either synthesized by mammary epithelial cells through the lipogenic pathway or uptaken from the circulating plasma

(Bauman and Griinari, 2003). In ruminants, lipogenesis contributes short and medium-chain FA (C4:0 to C16:0) encompassing 50% of the milk FA pool (Barber et al. 1997). Ruminal fermentation provides the main precursors (*i.e.* acetate and β-hydroxybutyrate) for the *de novo* synthesis of FA (Van Soest, 1994). In contrast, long-chain FA are mostly obtained through the hydrolysis of blood lipoproteins by lipoprotein lipase (Bauman and Davis 1974). These circulating lipids, in turn, come from the mobilization of adipose tissue stores as well as from dietary lipid absorption in the digestive tract (Bauman and Griinari, 2003).

Functional genomic studies have allowed to dissect the complex network of genes that drive forward fat synthesis in the bovine mammary gland (Bionaz and Loor 2008). This network is composed by genes that participate in a wide variety of metabolic pathways related with FA uptake and transport (*e.g. LPL, VLDLR, ACSL1, CD36, FABP3*), triglyceride synthesis (*e.g. LPIN1, DGAT1, AGPAT6, GPAM*), lipid droplet formation (*e.g. XDH, BTN1A1*), lipogenesis (*e.g. ACACA, FASN*), FA desaturation (*SCD1, FADS1*) and activation (*ACSL1, ACSS2*) and membrane-associated transport of metabolites *(ABCG2)*. The multiple components of the milk fat synthesis machinery defined above are coordinated by diverse transcription factors with a well-known role on lipid metabolism such as *SREBF1, SREBF2, PPARG, INSIG1*, and *PPARGC1A* (Bionaz and Loor 2008).

2. Genetic parameters of milk protein and lipid traits in goats

Although casein content is a crucial determinant of cheese yield, its utilization as a selection criterion has been hindered by the fact that this phenotype cannot be easily measured with routine analytical methods. Advances in infrared spectroscopy techniques, however, may overcome this difficulty by providing accurate measurements of the clotting protein fraction at a reasonable cost and time expense (Díaz-Carrillo et al. 1993). Fat content, another important factor determining cheese yield, is routinely recorded and it has been included as an important selection objective in most of breeding programs (Barillet 2007).

Genetic parameters of milk traits under selection need to be accurately defined in order to implement breeding strategies (selection vs. crossbreeding), estimate breeding values and predict selection responses. Many of the heritability and genetic correlation estimates that have been reported in the scientific literature for milk protein and fat contents of dairy goats are listed in Tables 1 and 2. Large parameter ranges can be observed, a feature that might be likely explained by the fact that multiple breeds and methods have been used to estimate them. Another important difference amongst studies is the lactation time point at which phenotypes are obtained, an environmental factor that can have dramatic effects on genetic parameter estimation. With regard to the heritability of total casein and casein fraction contents, very few estimates have been reported so far. Worth to mention those obtained for casein content in the Alpine (h^2 = 0.65-0.66), Murciano-Granadina (h^2 = 0.13-0.19) and Malagueña (h^2 = 0.29) breeds (Ricordeau and Bouillon 1971, Sigwald et al. 1981, Benradi et al. 2007 and 2009). It is also necessary to highlight the study of Benradi et al. (2007), where heritabilities of CSN1S1 (h^2 = 0.25) and CSN1S2 (h^2 = 0.09) contents were estimated in the Murciano-Granadina breed.

Breed (Country)	Protein content	Fat content	Reference
Saanen (Norway)	-	0.28	Ronningen (1965)
Saanen (Norway)	-	0.40-0.59	Ronningen (1967)
Alpine (France)	0.59	0.62	Ricordeau et al. (1979)
Saanen, Alpine, Toggenburg (USA)	-	0.54	Kennedy et al. (1982)
Alpine (France)	0.52	0.50	Boichard et al. (1988)
Alpine (France)	0.49-0.53	0.46	Bouloc (1987)
Alpine (France)	0.67	0.56	Bouillon & Ricordeau (1975)
La Mancha (USA)	-	0.63	Iloeje et al. (1981)
Nubian (USA)	-	0.66	Iloeje et al. (1981)
Toggenburg (USA)	-	0.54	Iloeje et al. (1981)
Saanen (France)	0.41	0.47	Boichard et al. (1988)
Saanen (France)	0.42	0.42	Bouloc (1987)
Saanen (México)	0.38-0.63	0.32-0.64	Torres-Vázquez et al. (2009)
Saanen (South Africa)	0.44	0.21	Muller et al. (2002)
Alpine (France)	0.58	0.58	Bélichon et al. (1998)
Saanen (France)	0.50	0.60	Bélichon et al. (1998)
Local breeds (Greece)	0.51	0.38	Zygoyiannis (1994)
Murciano-Granadina (Spain)	0.25-0.47	-	Analla et al (1996)
Alpine (France)	0.66-0.85	0.58-0.77	Barbieri et al. (1995)
Verata (Spain)	0.14-0.42	0.30-0.32	Rabasco et al. (1993)
Murciano-Granadina (Spain)	0.30	-	Benradi et al. (2009)
Murciano-Granadina (Spain)	0.50	0.15	Benradi (2007)
Malagueña (Spain)	0.30	-	Benradi (2007)

Table 1. Heritability estimates of milk protein and fat content traits in goats

The main trend that emerges from these analyses is that protein, casein and fat contents have moderate heritabilities, so they can be improved at a reasonable pace by using classical selection. These traits show, in general, positive medium or high genetic correlations among themselves; being, consequently, also positive their respective correlated responses to selection. However, they display negative genetic correlations with milk yield, a circumstance that is quite unfavourable given the high economic impact of this phenotype. In spite of this, protein and fat contents are important selection criteria and they are frequently included in selection indexes (Manfredi et al. 2000; Montaldo and Manfredi 2002)

Another interesting issue that deserves to be discussed is the effect that the highly polymorphic *CSN1S1* gene has on the estimation of genetic parameters of milk traits in goats. In this way, Barbieri et al. (1995) reported that heritability of protein content was reduced from 0.66 to 0.34 and that the genetic correlation between protein yield and content dropped from 0.09 to -0.22 when the *CSN1S1* genotype was taken into account as a fixed effect. This would mean that a substantial part of the genetic variance of goat protein content is explained by the polymorphism of the *CSN1S1* gene.

An important methodological advance in the estimation of genetic parameters of dairy traits has consisted in the use of random regression models applied to test day records (Schaeffer, 2004)). This approach has allowed to obtain estimates of heritabilities at, and genetic correlations among, different timepoints throughout the lactation curve in a wide variety of populations such as Norwegian dairy goats (Andonov et al. 1998), Spanish Payoya and Murciano-Granadina goats (Menéndez-Buxadera et al. 2008, 2010) and Canadian dairy breeds of Alpine origin (Bishop et al. 1994). To illustrate this concept, evolution of variance components for milk yield, fat, protein and dry matter contents in Murciano-Granadina goats is shown in **Figure 3**. From these data it can be inferred that estimates of heritability of yields and contents of milk components do not have stable values but, on the contrary, they vary throughout the lactation curve, being more variable near parturition and drying off. Moreover, genetic correlations between adjacent records are much higher (0.70-0.99) than those between records far apart (0.00-0.40). With no doubt, this statistical methodology allows a better genetic evaluation of dairy goats anf facilitates the selection of improving genotypes for lactation persistency (the ability of a goat to maintain as high as possible milk daily yield during lactation).

Trait	Protein	Total casein	CSN1S1	CSN1S2	Fat
Protein	-	0.39-0.90	0.65	0.55	0.45-0.93
Total casein	0.88	-	0.91	0.57	0.41
CSN1S1	0.04	0.01	-	-0.39	NA
CSN1S2	0.27	0.32	0.57	-	NA
Fat	-0.015-0.54	NA	NA	NA	-

Table 2. Genetic (above diagonal) and phenotypic (below diagonal) correlations between milk fat, protein and casein traits in goats (Benradi 2007, NA = data not available)

Genetic correlations between milk components and rheological traits as well as cheese yield have been also studied by Benradi et al. (2009). These authors reported high positive genetic correlations of protein, total caseins and CSN1S1 contents with time to curdling onset, curd firmness and curdling speed and a moderate positive correlation of the aforementioned milk components with cheese yield, so confirming the important influence of protein, and especially of total caseins and CSN1S1 contents, on the efficiency of cheese manufacture and quality.

3. Genomic architecture of protein and lipid phenotypes in goats and other ruminants

The lack of appropiate molecular tools in goats, such as large microsatellite panels uniformly covering the goat genome, microarrays and high throuput genotyping SNP chips, has indeed the fine mapping of genes related with milk protein and lipid traits at a genomic scale. While dense quantitative trait loci (QTL) maps of dairy traits have been obtained in cattle and sheep, a single partial genome scan for milk yield and fat and protein contents has been performed so far (Roldán et al. 2008). However, many of the findings obtained in cattle can be probably extrapolated to goats, so they will be commented in the following paragraphs. First of all, it should be highlighted that relative chromosomal contributions to protein and fat genetic variance are quite uneven. In cattle, strong evidences of protein content QTL have been obtained on chromosomes 3, 6 and 20, whilst protein yield QTL map to bovine chromosomes (BTA) 1, 3, 6, 9, 14 and 20 (Khatkar et al. 2004). Khatkar et al. (2004) have also reported QTL hotspots for milk fat content and yield (BTA6 and BTA 14). Second, many of the reported protein and fat content QTL just segregate in specific populations or families, meaning that they have a very restricted distribution (Khatkar et al. 2004). Third, and with a few exceptions, the identity of the genes and mutations explaining these QTL remain unsolved. The BTA14 QTL is one of the few cases where the underlying causal polymorphism has been identified *i.e.* a non-conservative amino acid substitution (K232A) in the DGAT1 enzyme that explains 51% of the daughter yield deviation variance of fat percentage (Winter et al. 2002, Grisart et al. 2002). The lack of success in finding causal mutations underlying milk trait QTL might be partly explained by the low resolution of microsatellite-based QTL mapping (confidence intervals usually encompass 20 cM or more).

Low resolution QTL detection methods have been (cattle, sheep) or will be (goats) progressively replaced by high throughput single nucleotide polymorphism (SNP) genotyping platforms. In this way, the recent advent of a bovine 50K SNP BeadChip has allowed to perform genome-wide association analyses of dairy traits at an unprecedented resolution (Hayes and Goddard 2010). As an example of the power of this approach, Schopen et al. (2011) analysed the segregation of 50,228 SNP in 1,713 Dutch Holstein cows with records for milk CSN1S1, CSN1S2, CSN2, CSN3, LALBA and BLG contents. Their results showed highly significant associations between polymorphisms mapping to BTA5, 6, 11 and 14 and protein percentage. With regard to specific protein fractions, the number of associated regions varied from three (CSN2 and BLG) to 12 (CSN1S2) and the percentage of additive genetic variance explained by the joint sets of significant SNP oscillated between 25% and 35%. Strong associations were observed between SNP mapping to casein genes and the corresponding casein fraction contents, but evidences of SNP acting in *trans* were also obtained *e.g.* SNP with effects on CSN1S1 content were detected in BTA11 and BTA14 (whilst the *CSN1S1* gene resides on BTA6). As a general conclusion, valid not only for cattle but also for sheep or goats, we can state that casein content is regulated not only by polymorphisms located within the casein genes but also by mutations mapping to other loci and chromosomes.

In goats, the precise position of genes with effects on milk lipid or protein composition or their contributions to phenotypic variance are mostly unknown. One of the few exceptions is the caprine casein gene cluster, which is known to have strong effects on milk composition and that has been finely dissected at the genomic and sequence levels (Martin et al. 2002). The four caprine casein genes are tightly clustered in a 250 kb region mapping to chromosome 6, in the order *CSN1S1-CSN2-CSN1S2-CSN3* (Rijnkels et al. 2002). Genomic organization and gene structure are highly conserved in all mammals (Rijnkels et al. 2002), with the only exception of the *CSN1S2* gene that might contain from 11 to 19 exons depending on the species under consideration (Rijnkels et al. 2002). Moreover, *CSN1S2* is duplicated in human and rodents but not in ruminants.

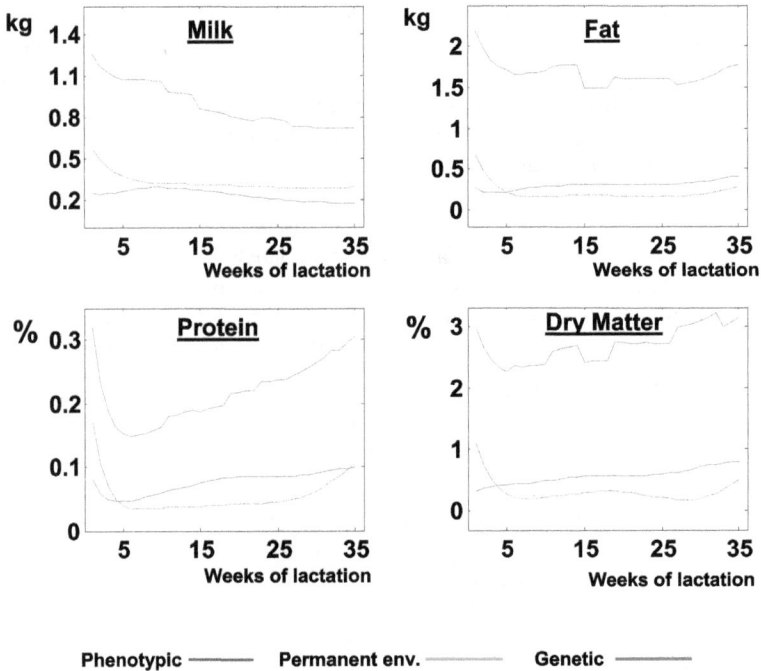

Figure 3. Variation through lactation curve of variance components for milk yield, fat, protein and dry matter contents in Murciano-Granadina goats. Source: Prof. Alberto Menéndez-Buxadera (unpublished data).

The goat *CSN1S1* gene encodes a 199 amino acid protein and it was completely sequenced by Ramunno et al. (2004). This gene is 16.7 kb long and contains 19 exons with sizes that go from 24 to 385 bp. The signal peptide and the first two amino acids of the mature protein are encoded by exon 2, whilst the stop codon is located at the boundary of exons 17 and 18. Leroux and Martin (1996) demonstrated that the goat *CSN2* gene is located 12 kb apart from the *CSN1S1* gene and that both are convergently transcribed (Leroux and Martin 1996). The transcription unit of *CSN2* contains 9 exons and encodes a 1.09 kb mRNA. The translation initiation site is located at exon 2, which encodes the signal peptide plus two amino acids of the mature protein (Roberts et al. 1992). The caprine *CSN1S2* and *CSN3* genes have been also characterized at the molecular level by Bouniol et al. (1993) and Coll et al. (1993).

The whey protein genes *LALBA* and *BLG* have been mapped to goat chromosomes 5 (Hayes et al. 1993) and 11 (Folch et al. 1994), respectively. The *LALBA* gene contains four exons and encodes a 123 amino acid protein. Structural characterization of the goat *BLG* gene has shown that it contains seven exons as previously described in other ruminant species (Folch et al. 1994). Two short interspersed nucleotide elements were found in the 3'end of the gene Moreover, a duplicated goat *BLG* pseudogene with a genomic organization very similar to *BLG* and also mapping to caprine chromosome 11 has been identified by Folch et al. (1996).

4. Candidate genes and their association with milk protein traits

From a structural and functional point of view, the casein cluster is one of the best studied genetic systems in goats. As outlined above, casein genes have been sequenced and their variability has been characterized in depth (Martin et al. 2002, Moioli et al. 2007). Even more, in several cases consistent association between this variability and milk composition traits has been firmly established (**Table 3**). In this regard, the most paradigmatic case is represented by the *CSN1S1* gene, where causal relationships between regulatory polymorphisms and CSN1S1 synthesis rate have been found. In the following sections, we will describe the variability of the casein and whey protein genes and its impact on the phenotypic variation of dairy and rheological traits.

4.1. The caprine αs1-casein gene

This locus is highly polymorphic in goats (reviewed in Martin et al. 2002 and Moioli et al. 2007), with 17 alleles (**Table 3**) that can be classified as strong (A, B1, B2, B3, B4, C, H, L and M), medium (E and I), low (F, D and G) and null (01, 02 and N). The existence of this remarkable level of variability was firstly outlined by Boulanger et al. (1984). By means of starch gel protein electrophoresis, these authors evidenced the existence of one variant A, two B variants (associated with different electrophoretic band intensities) and one variant C. These results suggested the existence of a polymorphism with quantitative effects on CSN1S1 synthesis. This interpretation was subsequently confirmed by Grosclaude et al. (1987), who identified two additional alleles (F and 0) and provided a first estimate of quantitative differences amongst *CSN1S1* genotypes based on rocket immunoelectrophoresis *i.e.* strong, medium, low and null alleles were distinguished (**Table 3**). These two studies have become classics in goats genetics because they pionereed the discovery of genetic variants with effects on milk traits in this ruminant species.

The advent of DNA-based methods allowed to characterize the specific mutations causing this variability as well as to identify new variants not detectable through electrophoresis techniques. Strong variants A, B and C were shown to differ by several amino acid substitutions *i.e.* B vs A: P16L and E77N and B vs C: H8I, R100K and T195A, but none of these polymorphisms seemed to have quantitative effects. Protein variants H and L were identified and characterized by Chianese et al. (1997) by using a variety of proteomic techniques, whilst variant M was reported by Bevilacqua et al. (2002). The main feature of the M variant is the loss of two phosphate residues in the multiple phosphorylation site consecutively to a S66L substitution. Likely, this allele emerged as a result of an interallelic recombination event between A and B2 alleles followed by a C to T transition at exon 9 (Bevilacqua et al. 2002).

Gene	Allele	Synthesis rate (g casein /L/allele)
CSN1S1	A, B1, B2, B3, B4, C, H, L, M	3.5
	E, I	1.1
	D, F, G	0.45
	01, 02, N	0
CSN1S2	A, B, C, E, F	2.5
	D	~ 1.25
	0	0
CSN2	A, A1, B, C, D, E	5
	0, 01	0

Table 3. Polymorphism of the *CSN1S1, CSN1S2* and *CSN2* genes and its relationship with the synthesis levels of the corresponding casein fractions (Martin et al. 2002, Moioli et al. 2007).

Variants E and I have been associated with an intermediate level of CSN1S1 synthesis (Boulanger 1984, Pérez et al. 1994, Chianese et al. 1997). Extensive sequencing of the E-allele revealed the existence of a 457 bp insertion at exon 19 caused by a truncated long interspersed nucleotide element (Pérez et al. 1994). This insertion might destabilize the corresponding *CSN1S1* mRNA diminishing 3-fold the synthesis of the corresponding protein (Pérez et al. 1994). Interestingly, the bovine *CSN1S1* allele G, that also contains a retrotransposon insertion at exon 19, has been associated with a lower milk CSN1S1 concentration (Rando et al. 1998). These results clearly indicate that the retrotransposon insertion is the causal mutation explaining the reduced levels of CSN1S1 in milk. However, the exact molecular mechanism by which the retrotranposon insertion represses CSN1S1 synthesis has not been elucidated yet, although a RNA interference mechanism might be suspected.

Low synthesis of CSN1S1 is explained by a defective processing of the corresponding transcript due to mutations that promote exon-skipping events and, in consequence, result in internally deleted CSN1S1 proteins (Martin et al. 2002). As much as nine different transcripts seem to be associated with allele F, the most abundant of which lacks exons 9, 10 and 11 and provokes a 37 amino acid deletion encompassing the multiple phosphorylation site (Leroux et al. 1992). A single nucleotide frameshift deletion in exon 9 (that induces the appearance of a premature stop codon at exon 12) and two 11 and 3 bp insertions at intron 9 might explain the defective processing of allele F. Similarly, the D allele is characterized by skipping of exon 9, while the G allele displays a G to A mutation in the 5′ splice site consensus sequence of intron 4 that causes the skipping of exon 4 and the synthesis of a protein lacking amino acids 14 to 26 (Martin et al. 1999).

The complete absence of CSN1S1 in milk is explained by a couple of genetic mechanisms. In the case of the 01 allele, a genomic deletion of at least 8.5 kb, that encompasses intron 12 to exon 19 of the *CSN1S1* gene, abrogates the synthesis of the corresponding protein (Cosenza et al. 2003). In close similarity with the F variant, the N allele contains a 1 bp deletion at the 23rd site of exon 9 determining a premature stop codon at exon 19. Sequencing of RT-PCR clones revealed that this variant is represented by at least 12 different transcripts lacking combinations of exons 9, 10, 11, 16 and 17 as a result of the defective processing of the mRNA. The abundances of transcripts carrying a premature stop codon are 30% and 14% for the N and F alleles, respectively (Ramunno et al. 2005). This finding might explain why the synthesis rate of the N-allele is 3-fold lower than that of the F-allele

Genotyping techniques have been developed to characterize the polymorphism of the *CSN1S1* gene in diverse goat populations. Initially, CSN1S1 variants were typed through the analysis of milk samples by SDS-polyacrylamide gel electrophoresis combined with isoelectric focusing (Grosclaude et al. 1987). The main inconvenient of this approach was that only lactating females could be typed, whilst in breeding schemes the main contributors to genetic improvement are bucks. To circumvent this problem, molecular techniques were developed. The first one was based on Southern blotting and restriction fragment length polymorphims analysis (Leroux et al. 1990). Although useful, this approach was very time consuming and not applicable to the large throughput genotyping required in breeding schemes. Later on, PCR-based techniques were published (Pérez et al. 1994, Ramunno et al. 2000) allowing the fast genotyping of the most abundant *CSN1S1* alleles.

With the aid of these molecular tools, the segregation of *CSN1S1* alleles has been studied in a wide array of breeds. Estimation of allelic frequencies in the Spanish Murciano-Granadina and Malagueña breeds showed that the E-allele was the most frequent one, followed by the B variant (Jordana et al. 1996). In contrast, in the French and Italian Saanen and Alpine breeds as well as in the French Corse breed the low content F-allele was predominant, while the E-allele would rank second (reviewed in Trujillo et al. 1998). It should be taken into account, however, that these estimates are quite outdated and that selection for *CSN1S1* variants might have changed their frequencies dramatically (at least in French breeds). In African, Canarian, Maltese and Garganica breeds, strong CSN1S1 content alleles are the most frequent ones (reviewed in Trujillo et al. 1998). A recent survey of American goat breeds highlighted the coexistence of different allelic CSN1S1 frequency patterns, with breeds in which the F (*e.g.* Alpine), E (*e.g.* Saanen and Oberhasli) and A+B alleles (*e.g.* LaMancha, Nigerian dwarf and Nubian) were predominant (Maga et al. 2009). These differences in the frequencies of *CSN1S1* variants might be explained by a combination of effects produced by genetic drift, selection and other evolutionary and demographic factors.

There is substantial evidence that the aforementioned polymorphisms not only affect the synthesis rate of CSN1S1 but also a wide array of production traits. A within-sire analysis of the progeny of five Alpine bucks revealed significant effects of *CSN1S1* genotype on milk protein and fat content, with the A-allele showing a clear superiority over E and F (Mahé et al. 1994). A similar trend was observed by Manfredi et al. (1995) when surveying 184 Alpine and 96 Saanen bucks. Moreover, Barbieri et al. (1995) demonstrated that the A-allele is associated with higher protein (AA > AE, AF > EE, EF > FF) and fat (AA, AE, AF > EE, EF) contents but also with a lower milk yield. Differences in protein content between genotypes might be in the order of 4 g/l (AA vs EE) to 6 g/l (AA vs FF). Chanat et al. (1999) offered a biological explanation to these findings by demonstrating that *CSN1S1* genotype has relevant effects on casein transport from the endoplasmic reticulum to the Golgi compartment (in goats with low or null *CSN1S1* genotypes this transport is severely impaired). An important question is if results obtained in French breeds can be safely extrapolated to breeds from other countries. In fact, milk composition is affected by many genetic and environmental factors that might differentially modulate the effects of goat *CSN1S1* genotype depending on the population under consideration. Results obtained in Spanish goat breeds are consistent with this hypothesis. Whilst significant differences were observed in the Malagueña breed when comparing milk CSN1S1 concentrations in BB (6.94 ± 0.38), BF (5.36 ± 0.22), EE (4.58 ± 0.13) and FF (3.98 ± 0.27 g/l) goats, in the case of the Murciano-Granadina breed only the BB genotype (8.50 ± 0.60 g/l) was significantly associated with increased levels of CSN1S1, whereas BF, EE and EF genotypes displayed non-significant differences when compared with each other (Caravaca et al. 2008). Even more, the *CSN1S1* genotype did not display any significant association with protein, casein or fat content (Caravaca et al. 2009). These results suggest that *CSN1S1* genotype has significant and consistent effects on the synthesis rate of the corresponding protein, but associations with other milk components might vary from breed to breed, likely due to differences in their genetic backgrounds and production systems. Interestingly, recent data suggest a certain level of dominance of strong alleles over the weak ones (Berget et al. 2010). In this way, CSN1S1 expression of goats carrying one strong and one weak allele at the *CSN1S1* locus is much more similar to those with a strong homozygous genotype than to goats with a weak homozygous genotype (Berget et al. 2010). Noteworthy, in cattle most of genetic correlations between casein fractions are negative or null (Schopen et al. 2009). As a whole, these findings suggest the existence of complex inter-loci and intra-locus interactions between casein genes that might impact their relative contributions to phenotypic variance of milk composition.

Not only fat content but milk FA composition has been reported to be affected by *CSN1S1* genotype. In this way, low-content *CSN1S1* alleles have been associated with less C8-C12 saturated FA, less stearic acid and more palmitic, linoleic and rumenic acids than their high-content counterparts (Chilliard et al. 2006). Low-content alleles have also been linked to an increased mammary desaturase activity (Chilliard et al. 2006). There is a certain controversy about the influence of *CSN1S1* polymorphism on the mRNA expression of lipid metabolism genes, with studies that support a regulatory effect (Badaoui 2008) and others that do not (Leroux et al. 2003). Since lipids and proteins are synthesized in the endoplasmic reticulum and their transport is, to a certain extent, coupled and co-regulated, it has been hypothesized that the perturbation of casein transport induced by the *CSN1S1* genotype might also alter lipid trafficking resulting in a reduced fat secretion (Ollivier-Bousquet et al. 2002).

Another trait influenced by *CSN1S1* genotype is micelle size, that happens to be lower in AA (221 nm) than in EE (265 nm) or FF (268 nm) milks (Remeuf 1993, Pirisi et al. 1994). This feature together with an augmented global protein content might explain the better coagulation properties and increased cheese yield of the AA milk (Ambrosoli et al. 1988, Vassal et al. 1994). In this way, AA milk produces a firmer curd and displays a slower coagulation time than FF milk (Ambrosoli et al. 1988). Moreover, corrected cheese yield (kg of cheese obtained from 100 kg of milk) was around 21-23 kg for AA, 20 kg for EE and 18 kg for FF goats (Vassal et al. 1994). These associations, however, may change depending on the breed under consideration. In this regard, Caravaca et al. (2011) were unable to find significant differences between the cheese yields of milks from BB, EE and FF Murciano-Granadina goats, whilst EE milk had a significantly higher curdling rate than its BB counterpart. As mentioned above, these differences amongst studies might be explained by a complex mixture of biological and technical factors.

From a sensorial point of view (Vassal et al. 1994), the AA cheese has been reported to display a higher hardness than the FF one (score of 3.23/5 vs 2.85/5), but a weaker goat flavor intensity (score of 2.10/5 vs 2.02/5). This means that the AA milk has better technological properties than the FF one in order to produce cheese but, unfortunately, the resulting product has a less intense taste and odour. It can be speculated that *CSN1S1* effects on cheese flavor might be caused by differences in the FA content and composition (goat flavor is mostly explained by the presence of volatile branched-chain 4-methyl and 4-ethyl octanoic FA) as well as in the lipolysis rate of AA vs FF milks (Chilliard et al. 2003, 2006),

4.2. The caprine α_{S2}-casein gene

Currently, five variants encoding "normal" levels of CSN1S2 have been found (**Table 3**), *i.e.* A and B (Boulanger et al. 1984), C (Bouniol et al. 1994), E (Veltri et al. 2000, Lagonigro et al. 2001) and F (Ramunno et al. 2001[a]). Two other variants D and 0 linked to reduced concentrations of CSN1S2 have also been detected (Ramunno et al. 2001[a,b]). The main feature of the D allele is a 106-nucleotide deletion, starting from the last 11 nucleotides of exon 11 and including 95 bp of intron 11, which causes the loss of three amino acid residues *i.e.* Pro122, Thr123, and Val124. Ramunno et al. (2001[a]) proposed that this variant might be associated with an intermediate level of CSN1S2 synthesis, but evidence is still preliminar and needs to be confirmed. The 0 allele contains a non-sense G>A mutation at exon 11 that changes the codon TGG (coding for Trp110) into a TAG stop codon (Ramunno et al. 2001[b]), thus hindering the synthesis of the corresponding protein. PCR-RFLP and PCR-SSCP protocols have been implemented in order to characterize *CSN1S2* variation (Ramunno et al. 2001[a,b], Chessa et al. 2008) but, to the best of our knowledge, association studies with milk and cheese traits are still lacking.

4.3. The caprine β-casein gene

Alleles at the *CSN2* locus can be classified in two main categories depending on the synthesis rate they are associated with (**Table 3**) *i.e.* alleles that are associated with "normal" concentrations of CSN2 in milk (A, A1, B, C, D, E) and the null ones (0 and 01) in which CSN2 expression is completely abrogated. Cosenza et al. (2007) have also reported a single nucleotide polymorphism at the CSN2 promoter but this variant has not been named yet. With regard to the A, B, C, D and E variants, they have been identified by means of proteomic techniques such as isoelectric focusing (Mahé and Grosclaude 1993), peptide mass fingerprinting and tandem mass spectrometry (Neveu et al. 2002), reversed phase HPLC/electrospray ionization mass spectrophotometry (Galliano et al. 2004) and immunoelectrophoretic analysis (Chianese et al. 2007). Molecular characterization of the null variants revealed that the 0 allele contains a 1 bp deletion in the 5′end of exon 7 that induces the appearance of a premature stop codon resulting in a 20-fold reduction in mRNA synthesis and the generation of a much shorter protein (72 vs 223 amino acids, Persuy et al. 1999). This variant has been found in Creole and Pyrenean goats (Persuy et al. 1999). Similarly, Ramunno et al (1995) identified a null 01 variant with a substitution at position 373 of exon 7 that converts a triplet encoding glutamine into a stop codon. This event results in a 10-fold reduction in CSN2 mRNA synthesis and yields a non-functional protein truncated at position 181 (Ramunno et al. 1995). PCR-SSCP (Chessa et al. 2005, Caroli et al. 2006), allele-specific-PCR (Ramunno et al. 1995), and PCR-RFLP (Cosenza et al. 2005) methods have been developed to identify genetic variants A, A1, C, E, and 0 at the DNA level.

4.4. The caprine κ-casein gene

The caprine CSN3 gene is extraordinarily polymorphic with 16 alleles (A, B, B′, B″, C, C′, D, E, F, G, H, I, J, K, L, and M) identified to date (Yahyaoui et al. 2003, Jann et al. 2004, Prinzenberg et al. 2005). Interestingly, most of the detected genetic variation is non-synonymous and preliminar evidence of positive selection acting on this locus has been found (Prinzenberg et al. 2005, Clop et al. 2009). This high genetic variation can be characterized with diverse molecular techniques, although multiplexed primer-extension analysis is particularly suitable in order to detect all alleles (Yahyaoui et al. 2003). The two most abundant CSN3 alleles are A and B, while the remaining ones can be considered as low-frequency variants in the majority of caprine breeds. Interestingly, Caravaca et al. (2010) have reported that the CSN3 genotype has significant effects on casein and protein contents (BB, AB > AA). Similar results have been obtained in the Orobica breed (Chiatti et al. 2007), where isolectric focusing variant B was associated with higher protein and casein contents than its A counterpart. The molecular basis of these associations is unclear because, as far as we know, the A and B variants only differ by a I119V substitution (Yahyaoui et al. 2003). In silico analyses with the Polyphen software have shown that in most mammalian species, Ile119 is a highly conserved residue, suggesting that it might have an important functional role (Caravaca et al. 2010). It is also worth to mention that associations between CSN3 genotype and rennet coagulation time (BB>AB) have been recently reported (Caravaca et al. 2011). This finding is consistent with the key role of CSN3 in the initial phase of milk rennet coagulation, where 87–90% of CSN3 is enzymatically degraded before micellar aggregation takes place. However, it should be emphasized that CSN3 genotype did not affect cheese yield, one of the main factors determining the economic income of goat farmers (at least in

Spain and other Mediterranean countries). In consequence, using *CSN3* genotype in marker-assisted selection schemes might not be advisable if cheese yield is a major breeding goal.

4.5. Casein haplotypes and their association with milk traits

The most meaningful approach to investigate the effect of casein genes on milk composition implies the genotyping of haplotypes rather than individual locus-specific alleles. However, this methodological strategy has been rarely carried out, probably because of the technical challenge of simultaneously typing such highly polymorphic loci in a fast and reliable way. Casein haplotypes have been characterized in a number of breeds from Italy (Sacchi et al. 2005, Caroli et al. 2006, Finocchiaro et al. 2008, Gigli et al. 2008), Norway (Hayes et al. 2006, Finocchiaro et al. 2008, Berget et al. 2010), Germany (Küpper et al. 2010), Czech Republic (Sztankóová et al. 2009), West Africa (Caroli et al. 2007) and India (Rout et al. 2010). However, most of these studies just report the variability of the casein cluster in selected populations rather than analysing its impact on milk quality phenotypes. An exception to this general statement is the work performed by Hayes et al. (1996). These authors genotyped 436 goats for 39 SNP distributed in the casein loci. They found higher levels of linkage disequilibrium between SNP pairs within casein loci than between casein loci, meaning that levels of intragenic recombination in casein genes are somewhat low. Moreover, they found significant associations between *CSN1S1* haplotypes and protein percentage and fat yield, as well as between *CSN3* haplotypes and fat and protein percentages (Hayes et al. 1996). In the next future, extensive characterization of the variability of the casein cluster region with next generation sequencing techniques, construction of large casein SNP panels and genotyping with high throughput platforms will be instrumental to elucidate the influence of casein haplotypes on dairy traits.

4.6. The caprine α-lactalbumin and β-lactoglobulin genes

The *LALBA* gene has been poorly characterized in goats with a few polymorphisms described to date. Cosenza et al. (2003) reported a silent SNP at exon 3 that can be analysed by PCR-RFLP (*Mva*I). Another synonymous mutation has ben found at exon 1 (Ma et al. 2010). However, to the best of our knowledge none of these polymorphisms has been associated with milk traits in goats. With regard to the *BLG* locus, most of the polymorphisms that have been found so far lie at the promoter region. In this regard, Yahyaoui et al. (2000) reported a C>T change at position -60, while Ballester et al. (2005) found 9 SNP at the promoter region and 6 silent SNP at the coding region (exons 1,2,3 and 6). Pena et al. (2000) also found two SNP at exon 7 encoding the 3'UTR. In close resemblance with results contributed by Ballester et al. (2005), Sardina et al. (2011) reported extensive variability at the *BLG* promoter, with a total of 36 SNP identified in a panel of Sicilian goats. Several of the SNP identified by Ballester et al. (2005) and Sardina et al. (2011) have been mapped to potential transcription factor binding sites, so they might be good candidates to regulate *BLG* mRNA expression. Pending tasks are to investigate the existence of quantitative differences (in terms of *BLG* mRNA) between promoter alleles as well as to find out if they have a detectable influence on milk composition.

5. Candidate genes and their association with milk fat content and composition traits

The detection of electrophoretic variants of goat milk protein genes three decades ago gave a strong impetus to the identification of the underlying mutations and the performance of association analyses with milk traits. This has resulted in the establishment of a wide catalog of polymorphisms located in the casein and whey protein genes, several of which have well demonstrated causal effects on milk composition. In comparison, the study of the genetic basis of milk fat traits is much less advanced. So far, a reduced number of candidate genes have been characterized at the molecular level and associations with milk fat content and/or composition have been reported (**Table 4**). However, causality has not been demonstrated for any of these associations, that most likely are produced by the existence of linkage disequilibrium between the analysed SNP and the true causal mutation. This contrasts with results obtained in cattle, where causal effects have been proposed (and sometimes convincingly demonstrated) for polymorphisms located at the *DGAT1* (Grisart et al. 2002, Winter et al. 2002), *ABCG2* (Cohen-Zinder et al. 2005) and *PPARGC1A* (Weikard et al. 2005) genes. An important difference between studies performed in cattle and goats is that in the latter species candidate genes were exclusively selected on the basis of physiological criteria, since positional information (*e.g.* QTL landscape of traits under study) was not available. This is an important limitation that has severely hindered progress in goat genetics research

The acetyl-CoA carboxylase α (ACACA) enzyme catalyses the carboxylation of acetyl-CoA to form malonyl-CoA, that can be used by fatty acid synthase as a substrate. This is a key rate-limiting step in the synthesis of FA (Abu-Elheiga et al. 1997). Sequence analysis of 5.5 kb of the coding region of the caprine *ACACA* gene revealed a silent SNP at exon 45 that was suggestively associated with fat yield and other traits (Badaoui et al. 2007[a]). Moreover, Federica et al. (2009) found 3 SNP at promoter III of the caprine *ACACA* gene that map to putative transcription factor binding sites but none of them displayed significant associations with lipid traits. Lipoprotein lipase is another fundamental enzyme involved in FA release and absorption through the hydrolysis of triglycerides from chylomicrons and other lipoprotein particles (Olivecrona and Olivecrona 1998). One missense polymorphism involving a S17T change has been suggestively associated with milk fat content (Badaoui et al. 2007[b]). This polymorphism is located in the signal peptide so it has been hypothesized that it might alter protein localization or expression. Variability at the goat milk fat globule epidermal growth factor and butyrophilin genes has also been associated with milk fat yield (Qu et al. 2011).

Zidi and coworkers (2010[a,b,c,d]) pionereed the study of the genetic basis of milk FA composition in goats through the analysis of several candidate genes and the performance of association analyses. As said in previous sections, milk FA composition constitutes an important set of traits with a key influence on the nutritional and technological properties of milk. Results obtained by Zidi and colleagues are summarized in **Table 4**. Worth to mention the identification of a 3-bp indel in the 3'UTR of caprine stearoyl-CoA desaturase (*SCD1*) gene, previously reported by Bernard et al. (2001), that was suggestively associated with conjugated linoleic acid and polyunsaturated FA content. This deletion is predicted to cause a dramatic change in the secondary structure of the 3'UTR so it has been hypothesized that it might exert its effect by influencing mRNA stability. In cattle, polymorphism at the *SCD1* gene has been associated with milk FA composition by several authors (Taniguchi et al. 2004, Schennink et al. 2008, Kgwatalala et al. 2009). A significant advancement in the dissection of the genetic factors that regulate milk fat content and composition in goats will necessarily involve the use of high throughput genotyping tools, such as the Illumina BeadChip that will be soon available, to type large goat populations with multiple records for these traits (throughout the lactation and/or successive lactations). This approach would allow to identify the genomic regions influencing milk lipid phenotypes, and then the daunting task of finding the causal mutations might begin with reasonable prospects of being successful.

Gene Name	Polymorphism	Association	References
Acetyl coenzyme A carboxylase α (*ACACA*)	C5493T in exon 45	fat yield, lactose content, and somatic cell count	Badaoui et al. (2007[a])
	1206 pb C/T at promoter III (locus AJ292286)	fat and protein percentages	Federica et al. (2009)
	1322 pb T/C at promoter III (locus AJ292286)	percentage, and fat and protein yields	
Growth hormone (*GH*)	SSCP patterns in exons 2, 4 and 5	milk, fat, and protein yields	Malveiro et al. (2001)
Lipoprotein lipase (*LPL*)	G50C (Ser17Thr)	milk fat content	Badaoui et al. (2007[b])
Stearoyl Co-A desaturase 1 (*SCD1*)	c.*1902_1904delTGT c.*3504G>A	trans-10, cis-12 CLA, PUFA, and total CLA	Zidi et al. (2010[a])
Malic enzyme 1 (*ME1*)	c.483C>T	C16:0 C17:0, C18:1n-9c, C20:1, SFA, MUFA and performed fatty acids.	Zidi et al. (2010[b])
	c.667G>A	C18:1n-9t, and C17:0, C18:0, C18:1n-9c, trans-10, cis-12 CLA, C20:1 and total CLA.	
	c.1200G>A	total CLA, and C17:0, C17:1, C18:0, C18:1n-9t, C18:1n-9c, cis-9, trans-11 CLA, and trans-10, cis-12 CLA.	
Hormone sensitive lipase (*LIPE*)	c.327C>A>T	C12:0 FA, C15:0 and *de novo* FA	Zidi et al. (2010[c])
	c.558C>T	fat content, trans-10, cis-12 CLA	
	c.1162G>T	C18:3n6g.	
Prolactin receptor (*PRLR*)	c. 1201G>A (R401G)	C16:1 FA and C16:1, C18:2n6c FA and PUFA	Zidi et al. (2010[d])
	c.1355C>T (T452I)	C16:1, C18:1n9t FA, SFA, MUFA, and omega 3	

Table 4. Associations between polymorphisms at candidate genes and milk fat traits

6. Conclusions

Milk protein and fat content and composition are key determinants of the nutritional and technological quality of goat dairy products as fresh milk, cheese and yogurt. Classical quantitative genetic studies have demonstrated that there is a remarkable amount of additive genetic variance for these dairy phenotypes, prompting the search of the causal mutations that explain the observed variability. These investigations have been particularly succesful when studying the genetic basis of casein concentrations in milk, since causal mutations have been identified in the goat CSN1S1 and relevant associations have been found for CSN3. These findings have allowed the implementation of marker assisted selection schemes to improve milk quality in goats. Milk fat related phenotypes have been much less studied, although we can anticipate that the development and application of high throughput genotyping and sequencing methods will revolutionize the field in the near future.

Author details

Marcel Amills and Alí Zidi
Department of Animal Genetics, Center for Research in Agricultural Genomics (CSIC-IRTA-UAB-UB), Universitat Autònoma de Barcelona, Bellaterra, Spain

Jordi Jordana
Departament de Ciència Animal i dels Aliments, Universitat Autònoma de Barcelona, Bellaterra, Spain

Juan Manuel Serradilla
Departamento de Producción Animal, Campus de Rabanales, Universidad de Córdoba, Córdoba, Spain

7. References

Abu-Elheiga L, Almarza-Ortega DB, Baldini A, Wakil SJ. 1997. Human acetyl-CoA carboxylase 2. Molecular cloning, characterization, chromosomal mapping, and evidence for two isoforms. J Biol Chem. 272:10669-10677.

Ambrosoli R, Di Stasio L, Mazzocco P. 1988. Content of α_{S1}-casein and coagulation properties in goat milk. J Dairy Sci. 71:24-28.

Analla M, Jímenez-Gamero I, Muñoz-Serrano A, Serradilla JM, Falagán A. 1996. Estimation of genetic parameters for milk yield and fat and protein contents of milk from Murciano-Granadina goats. J Dairy Sci. 79:1895-1898.

Andonov S, Kovac M, Kompan D, Dzabirski V. 1998. Estimation of covariance components for test day production in dairy goat. Proceedings of the 6th WCGALP. Armidale, Australia, pp. 145-148.

Badaoui B, Serradilla JM, Tomas A, Urrutia B, Ares JL, Carrizosa J, Sànchez A, Jordana J, and Amills M. 2007[a]. Goat acetyl-coenzyme A carboxylase α: Molecular characterization, polymorphism and its association with milk traits. J Dairy Sci. 90:1039-1043.

Badaoui B, Serradilla JM, Tomàs A, Urrutia B, Ares JL, Carrizosa J, Sànchez A, Jordana J, Amills M. 2007[b]. Identification of two polymorphisms in the goat lipoprotein lipase gene and their association with milk traits. J Dairy Sci. 90:3012-3017.

Badaoui B. 2008. Genetic factors affecting milk composition in goats. PhD Thesis, Universitat Autònoma de Barcelona, Spain.

Ballester M, Sánchez A, Folch JM. 2005. Polymorphisms in the goat β-lactoglobulin gene. J Dairy Res. 72:379-384.

Barber MC, Clegg RA, Travers MT, Vernon RG. 1997. Lipid metabolism in the lactating mammary gland. Biochim Biophys Acta 1347:101-126.

Barbieri M, Manfredi E, Elsen JM, Ricordeau G, Bouillon J, Grosclaude F, Mahé M, Bibé B. 1995. Influence du locus de la caséine α_{s1} sur les performances laitières et les paramètres génétiques des chèvres de race Alpine. Genet Sel Evol. 27:437-450.

Barillet F. 2007. Genetic improvement for dairy production in sheep and goats. Small Rum Res. 70:60-75.

Bauman DE, Davis CL. 1974. Biosynthesis of milk fat. In Lactation: A Comprehensive Treatise. Academic Press, New York, USA, pp. 31-75.

Bauman DE, Griinari JM. 2003. Nutritional regulation of milk fat synthesis. Annu Rev Nutr. 23:203-227.

Bauman DE, Mather IH, Wall RJ, Lock AL. 2006. Major advances associated with the biosynthesis of milk. J Dairy Sci. 89:1235-1243.

Bélichon S, Manfredi E, Piacère A. 1998. Genetic parameters of dairy traits in the Alpine and Saanen goat breeds. Genet Sel Evol. 30:529-534.

Benradi Z. 2007. Análisis genético de los caracteres de rendimiento, composición, reología y rendimiento quesero de la leche en las razas caprinas Malagueña y Murciano-Granadina. Master Thesis, International Center for Advanced Mediterranean Studies, Zaragoza, Spain.

Benradi Z, Ares JL, Jordana J, Urrutia B, Carrizosa J, Baena F, Serradilla JM. 2009. Genetic parameters for milk yield, composition and rheological traits in Murciano Granadina goats. Proceedings of the 60th Annual EAAP Congress, Barcelona, Spain.

Bequette BJ, Backwell FR, Crompton LA. 1998. Current concepts of amino acid and protein metabolism in the mammary gland of the lactating ruminant. J Dairy Sci. 81:2540-2559.

Berget I, Martens H, Kohler A, Sjurseth SK, Afseth NK, Narum B, Adnøy T, Lien S. 2010. Caprine CSN1S1 haplotype effect on gene expression and milk composition measured by Fourier transform infrared spectroscopy. J Dairy Sci. 93:4340-4350.

Bernard L, Leroux C, Hayes H, Gautier M, Chilliard Y, Martin P. 2001.Characterization of the caprine stearoyl-CoA desaturase gene and its mRNA showing an unusually long 3'-UTR sequence arising from a single exon. Gene 281:53-61.

Bevilacqua C, Ferranti P, Garro G, Veltri C, Lagonigro R, Leroux C, Pietrola E, Addeo F, Pilla F, Chianese L, Martin P. 2002. Interallelic recombination is likely responsible for the occurrence of a new αs1-casein variant in the goat species. Eur J Biochem. 269:1293-1303.

Bionaz M, Loor JJ. 2008. Gene networks driving bovine milk fat synthesis during the lactation cycle. BMC Genomics 9:366.

Bishop S, Sullivan BP, Shaeffer LR. 1994. Genetic evaluation of Canadian dairy goats using test day data. Proceedings of the 29th Biennial Session of the International Committee for Animal Breeding (ICAR). Ottawa, Canada, pp. 299-302.

Boichard D, Bouloc N, Ricordeau G, Piacere A, Barillet F. 1988. Genetic parameters for first lactation dairy traits in the Alpine and Saanen goat breeds. Genet Sel Evol. 21:205-215.

Bouillon J, Ricordeau G. 1975. Paramètres génétiques des performances de croissance et de production laitière chez les caprins en station de testage. Estimation des réponses directes ou in directes à la séléction. 1ères Journées de la Recherches Ovine et Caprine. INRA-ITOVIC, Paris, France, pp. 149-155.

Boulanger A, Grosclaude F, Mahé MF. 1984. Polymorphisme des caséines αs1 et αs2 de la chèvre (Capra hircus). Genet Sel Evol. 16:157-176.

Bouloc N. 1987. Adéquation du critère de selection et possibilités d'allègement du contrôle laitier chez la chèvre. Mémoire de Fin d'Etudes. INRA, Toulouse, France.

Bouniol C. 1993. Sequence of the goat αs2-casein enconding cDNA. Gene 125:235-236.

Bouniol C, Brignon G, Mahé MF, Printz C. 1994. Biochemical and genetic analysis of variant C of caprine αs2-casein (Capra hircus). Anim Genet. 25:173-177.

Caravaca F, Amills M, Jordana J, Angiolillo A, Agüera P, Aranda C, Menéndez-Buxadera A, Sánchez A, Carrizosa J, Urrutia B, Sànchez A, Serradilla JM. 2008. Effect of αs1-casein (CSN1S1) genotype on milk CSN1S1 content in Malagueña and Murciano-Granadina goats. J Dairy Res. 75:481-484.

Caravaca F, Carrizosa J, Urrutia B, Baena F, Jordana J, Amills M, Badaoui B, Sànchez A, Angiolillo A, Serradilla JM. 2009. Short communication: Effect of αs1-casein (CSN1S1) and κ-casein (CSN3) genotypes on milk composition in Murciano-Granadina goats. J Dairy Sci. 92:2960-2964.

Caravaca F, Ares JL, Carrizosa J, Urrutia B, Baena F, Jordana J, Badaoui B, Sànchez A, Angiolillo A, Amills M, Serradilla JM. 2011. Effects of αs1-casein (CSN1S1) and κ-casein (CSN3) genotypes on milk coagulation properties in Murciano-Granadina goats. J Dairy Res. 78:32-37.

Caroli A, Chiatti F, Chessa S, Rignanese D, Bolla P, Pagnacco G. 2006. Focusing on the goat casein complex. J Dairy Sci. 89:3178-3187.

Caroli A, Chiatti F, Chessa S, Rignanese D, Ibeagha-Awemu EM, Erhardt G. 2007. Characterization of the casein gene complex in West African goats and description of a new αs1-casein polymorphism. J Dairy Sci. 90:2989-2996.

Chanat E, Martin P, Ollivier-Bousquet M. 1999. αs1-casein is required for the efficient transport of β- and κ-casein from the endoplasmic reticulum to the Golgi apparatus of mammary epithelial cells. J Cell Sci. 112:3399-3412.

Chessa S, Budelli E, Chiatti F, Cito AM, Bolla P, Caroli A. 2005. Short communication: Predominance of β-casein (CSN2) C allele in goat breeds reared in Italy. J Dairy Sci. 88:1878-1881.

Chessa S, Rignanese D, Chiatti F, Radeghieri A, Gigliotti C, Caroli A. 2008. Technical note: simultaneous identification of *CSN1S2 A, B, C,* and *E* alleles in goats by polymerase chain reaction-single strand conformation polymorphism. J Dairy Sci. 91:1214-1217.

Chianese L, Ferranti P, Garro G, Mauriello R, Addeo F. 1997. Ocurrence of three novel αs1 casein variants in goat milk. Proc. Int. Dairy Fed, Fed. Int. Laiterie Semin. Milk Protein Polymorphism II. Int. Dairy Fed., Palmerston North, New Zealand, pp. 259-267.

Chianese L, Caira S, Garro G, Quarto M, Mauriello R, Addeo F. 2007. Occurrence of genetic polymorphism at goat β-CN locus. 5th Int. Symp. Challenge to Sheep and Goats Milk Sectors, Alghero, Sardinia, Italy.

Chiatti F, Chessa S, Bolla P, Cigalino G, Caroli A, Pagnacco G. 2007. Effect of κ-casein polymorphism on milk composition in the Orobica goat. J Dairy Sci. 90:1962-1966.

Chilliard Y, Ferlay A, Rouel J, Lamberet G. 2003. A review of nutritional and physiological factors affecting goat milk lipid synthesis and lipolysis. J Dairy Sci. 86:1751-1770.

Chilliard Y, Rouel J, Leroux C. 2006. Goat's alpha-s1 casein genotype influences its milk fatty acid composition and delta-9 desaturation ratios. Anim Feed Sci Technol. 131: 474-487.

Clop A, Zidi A, Amills M. 2010. Identification of positively selected sites in the goat *kappa casein (CSN3)* gene. Anim Genet. 41:332.

Cohen-Zinder M, Seroussi E, Larkin DM, Loor JJ, Everts-van der Wind A, Lee JH, Drackley JK, Band MR, Hernandez AG, Shani M, Lewin HA, Weller JI, Ron M. 2005. Identification of a missense mutation in the bovine *ABCG2* gene with a major effect on the QTL on chromosome 6 affecting milk yield and composition in Holstein cattle. Genome Res. 15:936-944.

Coll A, Folch JM, Sànchez A. 1993. Nucleotide sequence of the goat κ-casein cDNA. J Anim Sci. 71:2833.

Cosenza G, Illario R, Rando A, Di Gregorio P, Masina P, Ramunno L. 2003. Molecular characterization of the goat *CSN1S1* (01) allele. J Dairy Res. 70:237-240.

Cosenza G, Pauciullo A, Gallo D, Di Berardino D, Ramunno L. 2005. A *Ssp*I PCR-RFLP detecting a silent allele at the goat *CSN2* locus. J Dairy Res. 72:456-459.

Cosenza G, Pauciullo A, Colimoro L, Mancusi A, Rando A, Di Berardino D, Ramunno. 2007. An SNP in the goat *CSN2* promoter region is associated with the absence of β-casein in milk. Anim Genet. 38:655-658.

Díaz Carrillo E, Alonso-Moraga A, Serradilla-Manrique JM, Muñoz-Serrano A. 1993. Near Infrared calibration for goat's milk components: protein, total casein, αs-, β- and κ-caseins, fat and lactose. J Near Infrared Spectrosc. 1:141-146.

FAOSTAT. 2009 (http://faostat.fao.org/ site/ 339/ default.aspx).

Federica S, Francesco N, Giovanna de M, Carmela SM, Gennaro C, Carmela T, Bianca M. 2009. Identification of novel single nucleotide polymorphisms in promoter III of the acetyl-CoA carboxylase-{alpha} gene in goats affecting milk production traits. J Hered. 100:386-389.

Fernández C, Mocé ML, Latorre MA, Gómez E. 2005. Producción lechera de cabras Murciano-Granadinas de la región de Murcia. Ganadería 35:38-43.

Finocchiaro R, Hayes BJ, Siwek M, Spelman RJ, van Kaam JB, Adnøy T, Portolano B. 2008. Comparison of casein haplotypes between two geographically distant European dairy goat breeds. J Anim Breed Genet. 125:68-72.

Fitzgerald RJ, Meisel HM. 2003. Milk protein hydrolysis and bioactive peptides. In: Advanced Dairy Chemistry (3rd, Part B), Kluwer Academic/Plenum Publishers, New York, USA, pp. 675-698.

Folch JM, Coll A, Sànchez A. 1994. Complete sequence of the caprine β-lactoglobulin gene. J Dairy Sci. 77:3493-3497.

Folch JM, Coll A, Hayes HC, Sànchez A. 1996. Characterization of a caprine β-lactoglobulin pseudogene, identification and chromosomal localization by in situ hybridization in goat, sheep and cow. Gene 177:87-91.

Fontecha J, Ríos JL, Lozada L, Fraga MJ, Juárez M. 2000. Composition of goat's milk fat triglycerides analysed by silver ion adsorption-TLC and GC-MS. Int Dairy J. 10:119-128.

Fox PF, McSweeney PLH. 2006. Advanced Dairy Chemistry-2. Lipids, 3rd. Edition, Springer Publishers, New York, USA, pp. 801.

Galliano F, Saletti R, Cunsol V, Foti S, Marletta D, Bordonaro S, D'Urso G. 2004. Identification and characterization of a new β-casein variant in goat milk by high-performance liquid chromatography with electrospray ionization mass spectrometry and matrix-assisted laser desorption/ionization mass spectrometry. Rapid Commun Mass Spectrom. 18:1972-1982.

Gigli I, Maizon DO, Riggio V, Sardina MT, Portolano B. 2008. Short communication: casein haplotype variability in Sicilian dairy goat breeds. J Dairy Sci. 91:3687-3692.

Grisart B, Coppieters W, Farnir F, Karim L, Ford C, Berzi P, Cambisano N, Mni M, Reid S, Simon P, Spelman R, Georges M, Snell R. 2002. Positional candidate cloning of a QTL in dairy cattle: identification of a missense mutation in the bovine DGAT1 gene with major effect on milk yield and composition. Genome Res. 12:222-231.

Grosclaude F, Mahé MF, Brignon G, Di Stasio L, Jeunet R. 1987. A Mendelian polymorphism underlying quantitative variations of goat α_{s1}-casein. Genet. Sel. Evol. 19:399-412.

Grosclaude F. 1991. Structure, déterminisme génétique et polymorphisme des 6 lactoprotéines principales des bovins, des caprins et des ovins. Journeés sur la Qualité des Laits à la Production et Aptitude Fromagère. Colloque INRA-ENSA, Rennes, France.

Harvatine KJ, Perfield JW, Bauman DE. 2009. Expression of enzymes and key regulators of lipid synthesis is upregulated in adipose tissue during CLA-induced milk fat depression in dairy cows. J Nutr. 139:849-854.

Hayes HC, Popescu P, Dutrillaux B. 1993. Comparative gene mapping of lactoperoxidase, retinoblastoma, and α-lactalbumin genes in cattle, sheep, and goats. Mamm Genome 4:593-597.

Hayes B, Hagesaether N, Adnøy T, Pellerud G, Berg PR, Lien S. 2006. Effects on production traits of haplotypes among casein genes in Norwegian goats and evidence for a site of preferential recombination. Genetics 174:455-464

Hayes B, Goddard M. 2010. Genome-wide association and genomic selection in animal breeding. Genome 53:876-883.

Iloeje MU, Van Vleck LD, Wiggans GR. 1981. Components of variance for milk and fat yields in dairy goats. J Dairy Sci. 64:2290-2293.

Jann OC, Prinzenberg EM, Luikart G, Caroli A, Erhardt G. 2004. High polymorphism in the κ-casein (CSN3) gene from wild and domestic caprine species revealed by DNA sequencing. J Dairy Res. 71:188-195.

Jollès P. 1975. Structural aspects of the milk clotting process. Comparative features with the blood clotting process. Mol Cell Biochem.7:73-85.

Jordana J, Amills M, Díaz E, Angulo C, Serradilla JM, Sánchez A. 1996. Gene frequencies of caprine α$_{s1}$-casein polymorphism in Spanish goat breeds. Small Rumin Res. 20:215-221.

Kawasaki K, Weiss KM. 2003. Mineralized tissue and vertebrate evolution: The secretory calcium-binding phosphoprotein gene cluster. Proc Natl Acad Sci USA 100:4060-4065.

Kennedy BW, Finley CM, Bradford GE. 1982. Phenotypic and genetic relationships between reproduction and milk production in dairy goats. J Dairy Sci. 65:2373-2383.

Kgwatalala PM, Ibeagha-Awemu EM, Mustafa AF, Zhao X. 2009. Influence of stearoyl-coenzyme A desaturase 1 genotype and stage of lactation on fatty acid composition of Canadian Jersey cows. J Dairy Sci. 92:1220-1228.

Khatkar MS, Thomson PC, Tammen I, Raadsma HW. 2004. Quantitative trait loci mapping in dairy cattle: review and meta-analysis. Genet Sel Evol. 36:163-190.

Korhonen H, Pihlanto A. 2001. Food-derived bioactive peptides-opportunities for designing future foods. Curr Pharm Des. 9:1297-1308.

Küpper J, Chessa S, Rignanese D, Caroli A, Erhardt G. 2010. Divergence at the casein haplotypes in dairy and meat goat breeds. J Dairy Res. 77:56-62.

Lagonigro R, Pietrola E, D'Andrea M, Veltri C, Pilla F. 2001. Molecular characterisation of the goat s$_2$-casein E allele. Anim Genet. 32:391-393.

Lamberet GA, Delacroix B, Degas C. 2001. Intensité de la lipolyse initiale des laits de chèvre et perception de l'arôme "chèvre" dans les fromages. In: Proc. Technical Symp. 7th Int. Conf. Goats: Recent Advances on Goat Milk Quality, Raw Material for Cheesemaking. Poitiers, France, pp. 130-139.

Leroux C, Martin P, Mahé MF, Levéziel H, Mercier JC. 1990. Restriction fragment length polymorphism identification of goat α$_{s1}$-casein alleles: a potential tool in selection of individuals carrying alleles associated with a high level protein synthesis. Anim Genet. 21:341-351.

Leroux C, Mazure N, Martin P. 1992. Mutations away from splice site recognition sequences might cis-modulate alternative splicing of goat α$_{s1}$-casein transcripts. Structural organization of the relevant gene. J Biol Chem. 267:6147-6157.

Leroux C, Martin P. 1996. The caprine α$_{s1}$- and β-caseins genes are 12 kb apart and convergently transcribed. Anim Genet. 27 (Suppl. 2):293.

Leroux C, Le Provost F, Petit E, Bernard L, Chilliard Y, Martin P. 2003. Real-time RT-PCR and cDNA macroarray to study the impact of the genetic polymorphism at the α$_{s1}$-casein locus on the expression of genes in the goat mammary gland during lactation. Reprod Nutr Dev. 43:459-469.

Ma RN, Deng CJ, Zhang XM, Yue XP, Lan XY, Chen H, Lei CZ. 2010. S-novel SNP of α-lactalbumin gene in Chinese dairy goats. Mol Biol. (Mosk) 44:608-612.

Madureira AR, Pereira CI, Gomes AMP, Pintado ME, Malcata FX. 2007. Bovine whey proteins-Overview on the main biological properties. Food Res Int. 40:1197-1211.

Mahé MF, Grosclaude F. 1993. Polymorphism of β-casein in the Creole goat of Guadeloupe, evidence for a null allele. Genet Sel Evol. 25:403-408

Mahé MF, Manfredi E, Ricordeau G, Piacère A, Grosclaude F. 1994. Effets du polymorphisme de la caséine α_{s1} caprine sur les performances laitières: analyse intradescendance de boucs de race Alpine. Genet Sel Evol. 26:151-157.

Maga EA, Daftari P, Kültz D, Penedo MC. 2009. Prevalence of α_{s1}-casein genotypes in American dairy goats. J Anim Sci. 87:3464-3469.

Malveiro E, Marques PX, Santos IC, Belo C, Cravador A. 2001. Association between SSCPs at Algarvia goat GH gene and milk traits. Arch Zootec. 50:49–57

Manfredi E, Ricordeau G, Barbieri ME, Amigues Y, Bibé B. 1995. Genotype caséine α_{s1} et sélection des boucs sur descendance dans les races Alpine et Saanen. Genet Sel Evol. 27:451-458.

Manfredi E, Serradilla JM, Leroux C, Martin P, Sánchez A. 2000. Genetics for milk production. Proceedings 7th World Conference on Goats. Tours, France, pp. 191-196.

Martin P, Ollivier-Bousquet M, Grosclaude F. 1999. Genetic polymorphism of caseins: a tool to investigate casein micelle organization. Int Dairy J. 9:163–171

Martin P, Szymanowska M, Zwierzchowski L, Leroux C. 2002. The impact of genetic polymorphisms on the protein composition of ruminant milks. Reprod Nutr Dev. 42:433-459.

Martini M, Salari F, Altomonte I, Rignanese D, Chessa S, Gigliotti C, Caroli A. 2010. The Garfagnina goat: a zootechnical overview of a local dairy population. J Dairy Sci. 93:4659-4667.

Matar C, LeBlanc JG, Martin L, Perdigon G. 2003. Biologically active peptides released in fermented milk: Role and functions. In: Handbook of Fermented Functional Foods. Functional Foods and Nutraceuticals Series. CRC press, Florida, USA, pp.177-201.

Meisel H. 1998. Overview on milk protein-derived peptides. Int Dairy J. 8:363-373.

Menéndez-Buxadera A, Romero F, González O, Arrebola F, Molina A. 2008. Estimación de parámetros genéticos para la producción de leche y sus componentes en la raza caprina Payoya mediante técnicas de regresión Aleatoria. Inf Tec Agr. 104:127-132.

Menéndez-Buxadera A, Molina A, Arrebola F, Gil MJ, Serradilla JM. 2010. Random regression analysis of milk yield and milk composition in the first and second lactations of Murciano-Granadina goats. J Dairy Sci. 93:2718-2726.

Moioli B, D'Andrea M, Pilla F. 2007. Candidate genes affecting sheep and goat milk quality. Small Rumin Res. 68:179-192.

Montaldo HH, Manfredi E. 2002. Organization of selection programs for dairy goats. Proc. 7th World Congress on Genetics Applied to Livestock Production. Montpellier, France, pp. 187-194.

Muller CJC, Cloet SWP, Schoeman SJ. 2002. Estimation of genetic parameters for milk yield and milk composition of Saanen goats. Proc. 7th World Congress on Genetics Applied to Livestock Production. Montpellier, France, pp. 259-262

Neveu C, Mollé D, Moreno J, Martin P, Léonid J. 2002. Heterogeneity of caprine β-casein elucidated by RP-HPLC/ MS: Genetic variants and phosphorylation. J Protein Chem. 21:557-567.

Olivecrona G, Olivecrona T. 1998. Clearance of artificial triacylglycerol particles. Curr Opin Clin Nutr Metab Care 1:143-151.

Ollivier-Bousquet M. 2002. Milk lipid and protein traffic in mammary epithelial cells: joint and independent pathways. Reprod Nutr Dev. 42:149–162.

Pena RN, Sánchez A, Folch JM. 2000. Characterization of genetic polymorphism in the goat β-lactoglobulin gene. J Dairy Res. 67:217-224.

Pérez MJ, Leroux C, Bonastre AS, Martin P. 1994. Occurrence of a LINE sequence in the 3' UTR of the goat α_{s1}-casein E-encoding allele associated with reduced protein synthesis level. Gene 147:179-87.

Persuy MA, Printz C, Medrano JF, Mercier JC. 1999. A single nucleotide deletion resulting in a premature stop codon is associated with marked reduction of transcripts from a goat β-casein null allele. Anim Genet. 30:444-451.

Pfeuffer M, Schrezenmeir J. 2000. Bioactive substances in milk with properties decreasing risk of cardiovascular diseases. Br J Nutr. 84 (Suppl 1):S155-159.

Pirisi A, Colin O, Laurent F, Scher J, Parmentier M. 1994. Comparison of milk composition, cheesemaking properties and textural characteristics of the cheese from two groups of goats with a high or low rate α_{s1}-casein synthesis. Int Dairy J. 4:329-345.

Prinzenberg EM, Gutscher K, Chessa S, Caroli A, Erhardt G. 2005. Caprine κ-casein (CSN3) polymorphism: new developments in molecular knowledge. J Dairy Sci. 88:1490-1498.

Qu Y, Liu Y, Ma L, Sweeney S, Lan X, Chen Z, Li Z, Lei C, Chen H. 2011. Novel SNPs of butyrophilin (BTN1A1) and milk fat globule epidermal growth factor (EGF) 8 (MFG-E8) are associated with milk traits in dairy goat. Mol Biol Rep. 38:371-377.

Rabasco A, Serradilla JM, Padilla JA, Serrano A. 1993. Genetic and non-genetic sources of variation in yield and composition of milk in Verata goats. Small Rum Res. 11:151-161.

Ramunno L, Mariani P, Pappalardo M, Rando A, Capuano M, Di Gregorio P, Cosenza G. 1995. Un gene ad effetto maggiore sul contenuto di caseina β nel latte di capra, XI Convegno ASPA, Grado, Italy, pp. 185-186.

Ramunno L, Cosenza G, Pappalardo M, Pastore N, Gallo D, Di Gregorio P, Masina P. 2000. Identification of the goat CSN1S1F allele by means of PCR-RFLP method. Anim Genet. 31:342–343.

Ramunno L, Cosenza G, Pappalardo M, Longobardi E, Gallo D, Pastore N, Di Gregorio P, Rando A. 2001[a]. Characterization of two new alleles at the goat CSN1S2 locus. Anim Genet. 32:264-268.

Ramunno L, Longobardi E, Pappalardo M, Rando A, Di Gregorio P, Cosenza G, Mariani P, Pastore N, Masina P. 2001[b]. An allele associated with a non-detectable amount of α_{s2}-casein in goat milk. Anim Genet. 32:19-26.

Ramunno L, Cosenza G, Rando A, Illario R, Gallo D, Di Berardino D, Masina P. 2004. The goat α_{s1}-casein gene: gene structure and promoter analysis. Gene 334:105-111.

Ramunno L, Cosenza G, Rando A, Pauciullo A, Illario R, Gallo D, Di Berardino D, Masina P. 2005. Comparative analysis of gene sequence of goat *CSN1S1* F and N alleles and characterization of *CSN1S1* transcript variants in mammary gland. Gene 345:289-99.

Rando A, Di Gregorio P, Ramunno L, Mariani P, Fiorella A, Senese C, Marletta D, Masina P. 1998. Characterization of the *CSN1A G* allele of the bovine αs1-casein locus by the insertion of a relict of a long interspersed element. J Dairy Sci. 81:1735-42.

Remeuf F, Cossin V, Dervin C, Lenoir J, Tomassone R. 1991. Relations entre les caractères physico-chimiques des laits et leur aptitude fromagère. Le Lait 71:397-421.

Remeuf F, Hurtaud C. 1991. Relationship between physicochemical traits of milk and renneting property. In: Qualité des Laits à la Production et Aptitude Fromagère, INRA-ENSAR, Rennes, France, pp. 1-7.

Remeuf F. 1993. Influence du polymorphisme génétique de la caséine αs1 caprine sur les caractéristiques physico-chimiques et technologiques du lait. Le Lait 73:549-557.

Rhoads RE, Grudzien-Nogalska E. 2007. Translational regulation of milk protein synthesis at secretory activation. J Mammary Gland Biol Neoplasia 12:283-292.

Ricordeau G, Bouillon J, Sánchez F, Mocquot JC, Lajous A. 1979. Amélioration génétique de caprins. Facteur favorisant ou limitant le progrès génétique. 5ème Journées de la Recherche Ovine et Caprine. INRA-ITOVIC, Paris. pp. 403-426.

Rijnkels M. 2002. Multispecies comparison of the casein gene loci and evolution of casein gene family. J Mammary Gland Biol Neoplasia 7:327-345.

Roberts B, DiTullio P, Vitale J, Hehir K, Gordon K. 1992. Cloning of the goat beta-casein-encoding gene and expression in transgenic mice. Gene. 121:255-262.

Roldán DL, Rabasa AE, Saldaño S, Holgado F, Poli MA, Cantet RJ. 2008. QTL detection for milk production traits in goats using a longitudinal model. J Anim Breed Genet. 125:187-193.

Ronningen K. 1965. Cases of variation in the flavor intensity of goat milk. Acta Agric Scand. 15: 301-342

Ronningen K. 1967. A study of genetic parameters for milk characteristics in goats. Meld Norg Landbr Hogsk. 46:1-17.

Rout PK, Kumar A, Mandal A, Laloe D, Singh SK, Roy R. 2010. Characterization of casein gene complex and genetic diversity analysis in Indian goats. Anim Biotechnol. 21:122-134.

Rupp R, Clément V, Piacere A, Robert-Granié C, Manfredi E. 2011. Genetic parameters for milk somatic cell score and relationship with production and udder type traits in dairy Alpine and Saanen primiparous goats. J Dairy Sci. 94:3629-3634.

Sacchi P, Chessa S, Budelli E, Bolla P, Ceriotti G, Soglia D, Rasero R, Cauvin E, Caroli A. 2005. Casein haplotype structure in five Italian goat breeds. J Dairy Sci. 88:1561-1568.

Sardina MT, Rosa AJ, Davoli R, Braglia S, Portolano B. 2011. Polymorphisms of β-lactoglobulin promoter region in three Sicilian goat breeds. Mol Biol Rep. (in press).

Schaeffer LR. 2004. Applications of random regression models in animal breeding. Livest Prod Sci. 86:35-45.

Schennink A, Heck JML, Bovenhuis H, Visker MHPW, van Valenberg HJF, van Arendonk JAM. 2008. Milk fatty acid unsaturation: genetic parameters and effects of stearoyl-CoA desaturase (*SCD1*) and acyl CoA: diacylglycerol acyltransferase 1 (*DGAT1*). J Dairy Sci. 91:2135-2143.

Schopen GC, Heck JM, Bovenhuis H, Visker MH, van Valenberg HJ, van Arendonk JA. 2009. Genetic parameters for major milk proteins in Dutch Holstein-Friesians. J Dairy Sci. 92:1182-1191.

Schopen GC, Visker MH, Koks PD, Mullaart E, van Arendonk JA, Bovenhuis H. 2011. Whole-genome association study for milk protein composition in dairy cattle. J Dairy Sci. 94:3148-3158.

Shennan DB, Peaker M. 2000. Transport of milk constituents by the mammary gland. Physiol Rev. 80:925-951.

Sigwald JP, Ricordeau G, Bouillon J, Grappin R, du Sartel C. 1981. Paramétres génétiques et phenotypiques de la valeur fromagère du lait de chévre: richesse en matiéres azotées totales et coagulables. 6éme Journées de la Recherche Ovine et Caprine, ITOVIC-SPEOC, Paris, pp. 415-426.

Smiddy MA, Martin JE, Kelly AL, De Kruif CG, Huppertz T, 2006. Stability of casein micelles cross-linked by transglutaminase. J Dairy Sci. 89:1906-1914.

Sztankóová Z, Mátlová V, Kysel'ová J, Jandurová OM, Ríha J, Senese C. 2009. Short communication: Polymorphism of casein cluster genes in Czech local goat breeds. J Dairy Sci. 92:6197-6201.

Taniguchi M, Utsugi T, Oyama K, Mannen H, Kobayashi M, Tanabe Y, Ogino A, Tsuji S. 2004. Genotype of stearoyl-CoA desaturase is associated with fatty acid composition in Japanese Black cattle. Mamm Genome 15:142-148.

Thompson MP, Tarassuk NP, Jenness R, Lillevik HA, Ashworth US, Rose D. 1965. Nomenclature of the proteins of cow's milk-second revision: report of the committee on milk protein nomenclature, classification, and methodology of the manufacturing section of ADSA for 1963-64. J Dairy Sci. 48:159-169.

Torres-Vázquez JA, Valencia-Posadas M, Castillo-Juarez H, Montaldo HH. 2009. Genetic and phenotypic parameters of milk yield, milk composition and age at first kidding in Saanen goats from Mexico. Livest Sci. 126:147-153.

Trujillo AJ, Jordana J, Guamis B, Serradilla J, Amills M. 1998. Review: Polymorphism of the caprine α_{s1}-casein gene and its effect on the production, composition and technological properties of milk and on cheese making and ripening. Food Sci Technol Int. 4: 231-237.

Van Soest PJ. 1994. Nutritional Ecology of the Ruminant. 2nd. Ed. Cornell University Press, USA, pp. 476.

Vassal L, Delacroix-Buchet A, Bouillon J. 1994. Influence des variants AA, EE et FF de la caséine α_{s1} sur le rendement fromager et les caractéristiques sensorielles des fromages traditionnels : prèmieres observations. Le Lait 74:89-103.

Veltri C, Lagonigro R, Pietrollà E, D'Andrea M, Pilla F, Chianese L. 2000. Molecular characterisation of the goat α_{s2}-casein E allele and its detection in goat breeds of Italy. 7th International Conference on Goats, Tours, France, pp. 727.

Weekes TL, Luimes PH, Cant JP. 2006. Responses to amino acid imbalances and deficiencies in lactating dairy cows. J Dairy Sci. 89:2177-2187.

Weikard R, Kühn C, Goldammer T, Freyer G, Schwerin M. 2005. The bovine PPARGC1A gene: molecular characterization and association of an SNP with variation of milk fat synthesis. Physiol Genomics 21:1-13.

Winter A, Kramer W, Werner FAO, Kollers S, Kata S, Durstewitz G, Buitkamp J, Womack JE, Thaller G, Fries R. 2002. Association of a lysine-232/alanine olymorphism in a bovine gene encoding acyl-CoA:diacylglycerol acyltransferase (*DGAT1*) with variation at a quantitative trait locus for milk fat content. Proc Natl Acad Sci USA 99:9300-9305.

Yahyaoui MH, Pena RN, Sánchez A, Folch JM. 2000. Rapid communication: polymorphism in the goat β-lactoglobulin proximal promoter region. J Anim Sci. 78:1100-1101.

Yahyaoui MH, Angiolillo A, Pilla F, Sànchez A, Folch JM. 2003. Characterization and genotyping of the caprine κ-casein variants. J Dairy Sci. 86:2715-2720.

Zidi A, Fernández-Cabanás VM, Urrutia B, Carrizosa J, Polvillo O, González-Redondo P, Jordana J, Gallardo D, Amills M, Serradilla JM. 2010[a]. Association between the polymorphism of the goat stearoyl-coA desaturase 1 (*SCD1*) gene and milk fatty acid composition in Murciano-Granadina goats. J Dairy Sci. 93:4332-4339.

Zidi A, Serradilla JM, Jordana J, Carrizosa J, Urrutia B, Polvillo O, González-Redondo P, Gallardo D, Amills M, Fernández-Cabanás VM. 2010[b]. Association analysis between goat malic enzyme 1 (*ME1*) genotype and dairy traits. Animal 4: 1953-1957.

Zidi A, Fernández-Cabanás VM, Carrizosa J, Jordana J, Urrutia B, Polvillo O, González-Redondo P, Gallardo D, Amills M, Serradilla JM. 2010[c]. Genetic variation at the goat hormone-sensitive lipase (*LIPE*) gene and its association with milk yield and composition. J Dairy Res. 77:190-198.

Zidi A, Serradilla JM, Jordana J, Carrizosa J, Urrutia B, Polvillo O, González-Redondo P, Gallardo D, Amills M, Fernández-Cabanás VM. 2010[d]. Pleiotropic effects of the goat prolactin receptor genotype on milk fatty acid composition. Domest Anim Endocrinol.39: 85-89.

Zygoyiannis D. 1994. A study of genetic and phenotypic parameters for milk yield and milk characteristics in indigenous and crossbred goats in Greece. Wld Rev Anim Prod. 29:29-37.

Chapter 4

Application of Milk Proteins Genetic Polymorphism for Selection and Breeding of Dairy Cows in Bulgaria

Peter Hristov, Denitsa Teofanova, Ivan Mehandzhiyski,
Lyuben Zagorchev and Georgi Radoslavov

Additional information is available at the end of the chapter

1. Introduction

Main goals of the dairy cattle breeding are the search of an economically efficient way of the improvement of milk production and the qualitative milk traits. Selection and breeding of animals with desirable genotypes is of crucial importance for the genetic improvement of dairy cows.

Milk protein genetic polymorphisms of the genus *Bos* provoke a significant scientific interest, mainly associated with their evolution, population structure, breeding and hybridization. Over the last decades, studies have been concentrated on the influence of the genetic variants of the major milk proteins on the quantitative and qualitative milk traits and their technological properties (Di Stasio & Mariani, 2000; Martin et al., 2002).

Cow milk contains two classes of specific proteins, i.e. the group of caseins and the group of whey proteins. This division has been based on the milk proteins behaviour at pH 4.6. The former class, which is about 80% of the protein content in bovine milk, is precipitated at pH 4.6 (isoelectric pH). It contains four caseins, i.e. αs1- (CSN1S1), αs2-, β- and κ-casein (CSN3). The latter class, which constitutes about 20% of milk proteins, is soluble under those conditions. It contains two main whey proteins, i.e. α-lactalbumin and β-lactoglobulin (LGB) as well as some other proteins presented by insignificant concentrations (Fox & McSweeney, 1998). Each of the above mentioned proteins is presented by at least two genetic variants. Genetic variants of milk proteins differ from each other by one or more aminoacid residues in the polypeptide chains, which is due to various types of mutations in the genes encoding them. There are several methods for genotyping milk protein polymorphisms; however, the most frequently applied one is the PCR-RFLP assay. The alleles of a particular gene can be identified through their restriction profile.

Most of the studies have been focused on CSN1S1 and CSN3 of the group of caseins and LGB of the whey proteins. These proteins have a great effect on milk production and milk constituents (Erhardt, 1996). The genetic variants of the CSN3 gene have a dominant role due to the protein influence of this protein on the formation, structure and stabilization of the casein micelles with respect to milk technological properties and cheese production (Farrell et al., 1996). In bovine milk, the CSN3 protein is calcium-insensitive because it contains only one phospho-seryl residue and prevents the precipitation of the other three caseins. Some milk proteins are potential allergens (especially CSN1S1) because they are missing in human milk (EFSA, 2004; Crittenden & Bennett, 2005). LGB is the major whey protein of ruminant species. Its biological functions are not well studied but there are data that it may have a role for the phosphate metabolism in the mammary gland (Hill et al., 1997). LGB also have a role for the transport of retinol and fatty acids in the gut because it seems to be resistant to gastric digestion *in vivo* and remains intact after it passes through the stomach (Yvon et al., 1984). There are numerous reports on the polymorphism of the milk protein genes from various areas of the world. However, data for Bulgarian cattle populations are scarce. Such studies have been initiated by our team (Zlatarev et al., 2008). Subsequently, several cow breeds were examined (Hristov et al., 2011a, b). Our recent investigations have been focused on the Bulgarian Rhodopean Cattle Breed (BRC), which is of high significance due to its long-term exploitation and the relative independence from environmental conditions. It is incomparable with any other breeds in Bulgaria with respect to the milk yield, fat and protein contents, viability, use continuance and fertility. Because of all these features, the BRC breed is a valuable gene fund for the country and the region and its preservation and improvement are of significant interest.

The selection and breeding of the BRC breed (and all the other breeds in Bulgaria) with respect to the milk production and milk quality have been accomplished mainly on the basis of phenotypic features. The goal of the analysis of the polymorphisms of the milk protein genes is to obtain data applicable as additional criteria in the marker-assisted selection and breeding. This is a reliable approach, which may facilitate and accelerate the selection process, increasing the qualitative and quantitative characters of the farm production and the competitiveness of the milk industry. This strategy will also increase the economic efficiency of the dairy cattle farming and genetic improvement of dairy cattle populations.

2. Genotyping of Bulgarian Rhodopean Cattle Breed and Shorthorn Rhodopean Cattle Breed

The Shorthorn Rhodopean Cattle Breed (SRC) is one of the two native cow breeds in Bulgaria. Its population is threatened by extinction. It is considered as one of the last forms of the prehistoric European cattle breeds together with Albanian, Illyric Dwarf and Montenegro cattle. Due to the fact that the gene fund of highly productive dairy cows is getting narrower in the course of years of selection, this breed is a genetic resource for the enrichment of other breeds used in the region. On the other hand, it is of great importance to preserve its own population structure as a native cattle breed for Bulgaria. The SRC breed is also a basic breed for creation of the BRC, officially declared as a new race in 1989 (Nickolov,

1999). Therefore, the application of highly efficient molecule markers for genotyping, genetic identification and marker-assisted selection is very important for the evaluation and preservation of the biodiversity of these native Bulgarian cow breeds. Consequently, studies of milk protein genes polymorphism and their application for the identification of individuals with desirable characteristics as well as for the selection and breeding programs is the main focus of the recent work of our team.

2.1. Genotypes and allelic forms identification

Each of the studied genes is presented by at least two variants, which are genetically determined by autosomal and codominant alleles. The absence of genetic dominance is useful because the homozygous individuals give only one variant for each protein in the electrophoregram, while heterozygous ones give both variants. Thus, the estimation of the gene frequencies for a population is easy. The variants can be detected by both protein electrophoresis, isoelectric focusing (IEF) and analysis of DNA. At DNA level, polymorphisms are due either to single nucleotide substitutions or to DNA re-arrangement phenomena. The differences between genotypes could be identified using specific restriction enzymes. Particular alleles could be evidenced on the electrophoresis gel with different band length (PCR-RFLP assay).

2.1.1. CSN1S1 gene

The four casein genes are localized in a 250 kb cluster (Ferretti et al., 1990; Threadgill & Womack, 1990) situated in the Chromosome 6 (Hayes et al., 1993a; Popescu et al., 1996). Their order is CSN1S1, CSN2, CSN1S2 and CSN3. This gene cluster is also referred to as the casein locus (Martin et al., 2002) or as the super locus (Freyer et al., 1999). The genomic DNA encoding the CSN1S1 milk protein is about 17.5 kb. A recent review of the milk protein nomenclature (Caroli et al., 2009) indicates nine genetic variants of the CSN1S1 gene (A, B, C, D, E, F, G, H and I) in the genus *Bos* (Table 1). For this gene, the most common allele is B followed by C. These allele forms can be found in all cattle breeds.

The genotype identification in our recent study was performed using PCR-RFLP assay. Total DNA for PCR reaction was extracted from blood samples by the application of GeneJet™ Genomic DNA Purification Kit (Fermentas). For the amplification of the polymorphic region of the CSN1S1 gene, primers described by Koczan et al. (1993) were used. They covered parts of the 5'-flanking region and exon 1 (in total 310 bp fragment). The restriction profiles after digestion with *Tsp45I* specific endonuclease showed particular genotyping differences.

Totally, 87 animals of the BRC breed were examined for variants of the CSN1S1 gene. Three genotypes were obtained, two homozygous (BB and CC) and one heterozygous (BC). About 71% of the animals (62 cows) were heterozygous and their RFLP profiles showed three electrophoretic bands (310bp, 214 bp and 96 bp). Only 26% (23 cows) were homozygous BB animals and two electrophoretic bands were characteristic for them (214 bp and 96 bp). The homozygous CC genotype was presented by the lowest frequency (2%), which could be pointed out as an insignificant presence. There were only two cows found with that genotype expressed, with one unrestricted fragment on the electrophoregram (310 bp). The RFLP profile of the CSN1S1 gene is shown on Figure 1.

Gene	CSN1S1 variants								
Protein	B	A	C	D	E	F	G	H	I
14891–14929									
14–26		Del.							
17383	GCC			ACC					
53	Ala			ThrP					
17377–17400									
51–58								Del.	
18901	CAA				AAA				
59	Gln				Lys				
18923	TCG					TTG			
66	SerP					Leu			
19836	GAA								GAT
84	Glu								Asp
26181	GAA		GGA		GGA				
192	Glu		Gly		Gly				

Table 1. Position within the gene and mature protein of the CSN1S1 genetic variants in the genus *Bos* (Caroli et al., 2009). Del. – deletion. In bold – non-synonymous mutations.

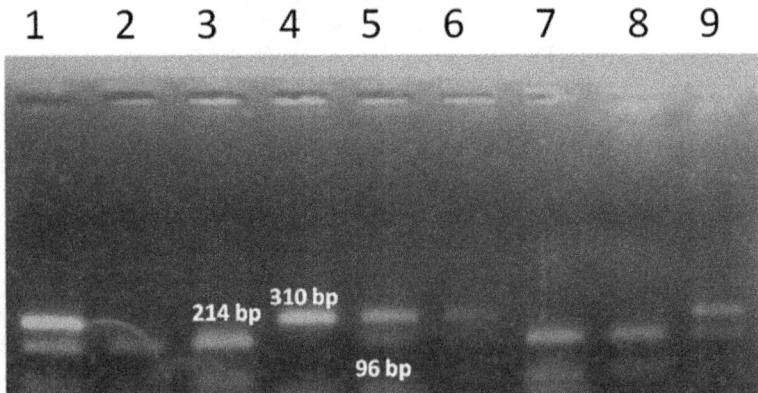

Figure 1. PCR-RFLP assay (2% agarose gel electrophoresis) of CSN1S1 gene of BRC breed after restriction of the polymorphic region with *Tsp45I* restrictase. 1, 5, 6, 9 – BC genotype; 2, 3, 7, 8 – BB genotype; 4 - CC genotype. The size of the restriction fragments is shown (white letters).

Genotype frequencies were estimated after a direct count. On the other hand, allelic frequencies were calculated from the observed genotype frequencies.

The genotypic and allelic frequencies for the CSN1S1 gene are shown on Table 2. For the CSN1S1 gene in BRC breed, it is obvious that the B allele frequency is predominant in comparison with the C allele. This finding is in agreement with previous studies, which have defined the B allele as being the most frequent in many cattle breeds (Beja-Pereira et al., 2003).

The validity of Hardy-Weinberg equilibrium for the population was evaluated using χ^2 test (Preacher, 2001). The observed and the expected genotype frequencies were of similar values, thus confirming the validity of Hardy-Weinberg equilibrium for the BRC population. The prevailing frequency of the B allele and the heterozygous BC genotype for the CSN1S1 gene allowed the assumption that animals possessing the BB and/or the BC genotypes have been used during the selection and reproduction of the BRC breed. The extremely low frequency of the homozygous CC individuals corroborated with the above-mentioned assumption.

Gene	Genotype	Genotype frequencies		Allele frequencies	χ^2	p-value
		Observed	Expected			
CSN1S1	BB	0.264	0.385	B – 0.621 C - 0.379	0.26 NS	0.88
	CC	0.023	0.144			
	BC	0.713	0.471			
CSN3	AA	0.318	0.345	A – 0.587 B – 0.413	0.01 NS	0.99
	BB	0.143	0.170			
	AB	0.540	0.485			
LGB	AA	0.395	0.439	A – 0.686 B – 0.314	0.13 NS	0.94
	BB	0.023	0.099			
	AB	0.581	0.416			

Table 2. Genotype and allele frequencies for the CSN1S1, CSN3 and LGB genes in the Bulgarian Rhodopean Cattle. NS - non-significant differences.

The SRC was genotyped as well (Table 3). A total of 38 animals were studied for the polymorphisms of the CSN1S1 gene and almost half of them (20 cows) were homozygous by the B allele (about 53%). This contrasted with the BRC breed where the dominant genotype was heterozygous. For the examined native breed, low and almost insignificant frequency (13%) of the CC genotype was estimated, similarly to the BRC breed. The allelic frequencies of the BRC and the SRC were similar but the B allele was slightly prevailing in the BRC. This was mostly due to the fact that Jersey cow, for which the B allele is highly frequent (Miciński et al., 2007), has been used as the basic cattle breed for selection and improvement of the BRC breed during the past 50 years.

Gene	Genotype	Genotype frequencies		Allele frequencies	χ^2	p-value
		Observed	Expected			
CSN1S1	BB	0.526	0.486	B - 0.697 C – 0.303	0.04 NS	0.98
	CC	0.132	0.091			
	BC	0.342	0.422			
CSN3	AA	0.237	0.213	A – 0.461 B – 0.539	0.01 NS	0.99
	BB	0.316	0.291			
	AB	0.447	0.497			
LGB	AA	0.063	0.250	A – 0.500 B – 0.500	0.56 NS	0.75
	BB	0.063	0.250			
	AB	0.875	0.500			

Table 3. Genotype and allele frequencies for the CSN1S1, CSN3 and LGB genes of Shorthorn Rhodopean Cattle. NS - non-significant differences.

The observed the expected genotype frequencies were with similar values, thus confirming the validity of Hardy-Weinberg equilibrium for the SRC population.

2.1.2. CSN3 gene

The genomic DNA encoding the CSN3 milk protein is about 13 kb. Recently, 14 genetic variants have been identified (A, A^I, B, B^2, C, D, E, F^1, F^2, G^1, G^2, H, I and J) (Caroli et al., 2009) for the gene encoding the CSN3 protein. Their characteristic nucleotide and aminoacid substitutions are shown on Table 4. The A and the B alleles are the most frequent among all the species of the genus *Bos* (Neelin, 1964; Woychik, 1964).

The total DNA extraction and the PCR amplification of the CSN3 gene were performed in the same way and under the same conditions as described in section 2.1.1. For amplification of the polymorphic region of the CSN3 gene (located between exon 4 and intron 4), primers described by Medrano & Cordova (1990a) were used (in total 350 bp fragment). For the RFLP assay, *Hinfl* specific restrictase was used. For the BRC breed, 63 animals were genotyped by the CSN3 gene. There were 34 heterozygous individuals (AB), which represented more than half of the sample (c. 54%). Four electrophoretic bands characterized that genotype (266 bp, 134 bp, 132 bp and 84 bp). Twenty animals (c. 32%) were identified as homozygous on the A allele (AA), which was visualized with three electrophoretic bands (134 bp, 132 bp and 84 bp). The homozygous BB animals were few (9 cows, c. 14%); they were identified by the presence of two electrophoretic bands, i.e. 266 bp and 84 bp (Figure 2).

From the frequency data for the CSN3 gene (Table 2), it is obvious that the A allele frequency is higher in comparison with that of B allele. This finding is in agreement with previous results, which have recognised the B allele as exhibiting lower frequency in the majority of cattle breeds (Tsiaras et al. 2005; Heck et al., 2009).

Gene	CSN3 variants													
Protein	A	A¹	B	B²	C	D	E	F¹	F²	G¹	G²	H	I	J
12690	CGC								CAC					
10	Arg								His					
12940	ACT		ACC											
93	Thr		Thr											
12950	CGT									TGT				
97	Arg									Cys				
12951	CGT				CAT	CAT								
97	Arg				His	His								
12971	TCA												GCA	
104	Ser												Ala	
13065	ACC									ATC	ATC			
135	Thr									Ile	Ile			
13068	ACC		ATC	ATC	ATC									ATC
136	Thr		Ile	Ile	Ile									Ile
13096	ACT						ACG							
145	Thr													
13104	GAT		GCT	GCT	GCT			GTT			GCT			GCT
148	Asp		Ala	Ala	Ala			Val			Ala			Ala
13111	CCA	CCG												
150	Pro													
13119	ATT		ACT											
153	Ile		Thr											
13124	AGC					GGC								?
155	Ser					Gly								Arg
13162	ACT		ACC								ACC			
167	Thr													
13165	GCA		GCG	GCG	GCG	?					GCG			
168	Ala													

Table 4. Position of the CSN3 genetic variants within the gene and the mature protein in the genus *Bos* (Caroli et al., 2009). Mutations are presented in bold. F¹ = F of Sulimova et al. (1992); F² = F of Prinzenberg et al. (1996), GenBank no. AF123250; G¹ = G of Erhardt (1996), Prinzenberg et al. (1996),

GenBank no. AF123251; G^2 = G of Sulimova et al. (1996); D = GenBank No. AJ619772. ? = Information not available.

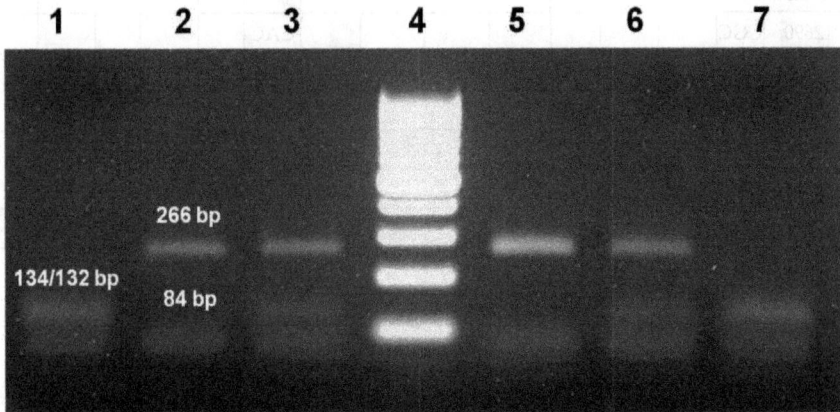

Figure 2. PCR-RFLP assay (2% agarose gel electrophoresis) of the CSN3 gene after restriction of the polymorphic region with *Hinfl* restrictase. 1, 7 – AA genotype; 2, 5 – BB genotype; 3, 6 - AB genotype; 4 – GeneRuler™ 100 bp Ladder Plus (Fermentas). The size of the restriction fragments is shown (white letters).

The validity of Hardy-Weinberg equilibrium for BRC population was confirmed by the almost equal values of the observed and the expected genotype frequencies. The dominant frequency of the A allele and the heterozygous AB genotype allowed the assumption that animals with the AA and/or the AB genotypes have been used during the selection and reproduction of the breed. The extremely low frequency of the homozygous BB individuals also agreed with the above-mentioned assumption.

The CSN3 gene polymorphism was also studied for the SRC population (Table 3). A total of 36 animals were genotyped. The presence of the homozygous AA (9 cows) and BB (10 cows) representatives was found to be with almost equal frequencies (25% and 28%, respectively). The heterozygous AB genotype was presented by the highest frequency (47%). Regardless of the comparatively small sample, the present study allowed to conclude that the homozygous genotypes are almost uniformly distributed among the SRC population. This is in contrast with the results for BRC breed, where the frequency of the AA genotype is double than that of the BB genotype (Table 2). For the CSN3 gene, the observed and the expected genotype frequencies were also similar, thus confirming the validity of Hardy-Weinberg equilibrium for the SRC population.

2.1.3. LGB gene

The LGB gene is mapped on Chromosome 11 (Hayes and Petit, 1993b). The genomic DNA encoding LGB milk whey protein is much smaller (c. 4 kb) than that encoding the two caseins described. The LGB is the major whey protein of ruminants and is also present in the milks of many other species. Its physiological function has been reported to be implicated in hydrophobic ligand transport and uptake, enzyme regulation and the neonatal acquisition of passive immunity (Kontopidis et al., 2004). A recent review of the milk protein nomenclature (Caroli et al., 2009) indicated eleven LGB genetic variants (A, B, C, D, E, F, G, H, I, J and W) in the genus *Bos* (Table 5). For this gene, similarly to CSN3, the most common alleles are A and B. These allele forms can be found in all cattle breeds. The E allele is also one of the most common alleles of the LGB gene in the genus *Bos*; however, it is not specific for *Bos taurus* but occurs in other *Bos* spp. (*Bos grunniens* and *Bos javanicus*).

The total DNA extraction and the PCR amplification of the LGB gene were performed following the same protocol and under the same conditions as described in section 2.1.1. For amplification of the polymorphic region of the LGB gene (located between exon 4 and intron 4), primers described by Medrano & Cordova (1990b) were used (in total 252 bp fragment). For the RFLP assay, *HaeIII* specific restrictase was used.

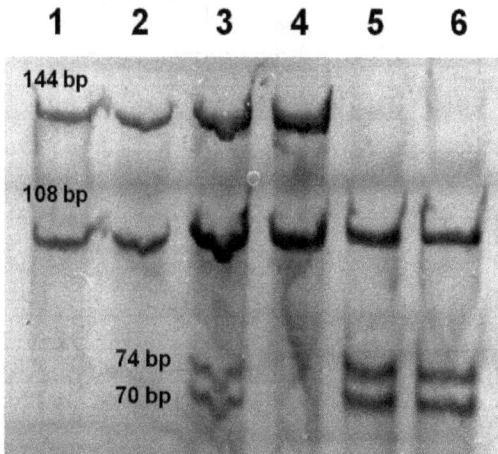

Figure 3. PCR-RFLP assay (10% acrylamide gel electrophoresis) of the LGB gene after restriction of the polymorphic region with *HaeIII* restrictase. 1, 2, 4 – AA genotype; 5, 6 – BB genotype; 3 - AB genotype. The numbers above the bands show the size of restriction fragments.

For the BRC breed, 86 animals were genotyped by the LGB gene. Fifty animals were recognised as heterozygous (AB) and they prevailed in the population (c. 58%). The AB genotype was characterized by four electrophoretic bands, i.e. 144 bp, 108 bp, 74 bp and 70 bp (Figure 3). The homozygous on the A allele (AA) animals (34 cows) were recorded with frequency of c. 40%; electrophoretically, that genotype was visualized with two bands (144 bp and 108 bp). The lowest frequency (c. 2%) was exhibited by the homozygous BB animals (2 cows), which were identified by three electrophoretic bands (108 bp, 74 bp and 70 bp).

From data shown on Table 2 about the genotype and the allele frequencies of both CSN3 and LGB genes, it is obvious that the A allele is prevailing in comparison with the B allele for both genes. These data are in agreement with previous observations for the majority of the cattle breeds (Tsiaras et al. 2005; Heck et al., 2009).

Gene	LGB variants										
Protein	B	A	C	D	E	F	G	H	I	J	W
3065	GAG			CAG							
45	Glu			Gly							
3080	CCT					TCT					
50	Pro					Ser					
3098	ATC										CTC
56	Ile										Leu
3109	CAG		CAT								
59	Gln		His								
3982	AAC	AAT			?	?	?	?	?	?	?
63	Asn										
3984	GGT	GAT						GAT			
64	Gly	Asp						Asp			
4003	AAG							AA?			
70	Lys							Asn			
4027	ATC						ATG				
78	Ile						Met				
5174	AAT	AAC	?	?	?	?	?	?	?	?	?
88	Asn										
5233	GAG								GGG		
108	Glu								Gly		
5263	GCC	GTC						GTC			
118	Ala	Val						Val			
5962	CCG									CTG	
126	Pro									Leu	
5970	GAC					TAC					
129	Asp					Tyr					
6280	GAG				GGG	GGG	GGG				
158	Glu				Gly	Gly	Gly				

Table 5. Position within the gene and the mature protein of the LGB genetic variants in the genus *Bos* (Caroli et al., 2009). In bold – mutations; ? – information not available.

The observed and the expected genotype frequencies were with similar values, which confirmed the validity of Hardy-Weinberg equilibrium for the BRC population with respect to the LGB gene. The genotype profiles of both CSN3 and LGB milk protein genes confirm the assumption that the assisted selection used animals possessing the AA and/or the AB genotypes during the reproduction of the BRC breed.

The SRC population was also genotyped with regards to the LGB gene polymorphism (Table 3). A total of 32 animals were studied. The homozygous AA (2 cows) and BB (2 cows) representatives were with the same low frequency (c. 6%). The representation of the heterozygous AB genotype was found to be high (c. 88%). From the present study, it is obvious that the homozygous genotypes are uniformly distributed among the SRC breed.

That is in contrast with the results obtained for the BRC breed with respect to the LGB gene, where the frequency of the AA genotype is double than that of the BB genotype (Table 2). It is also in contrast with the published data (Tsiaras et al. 2005; Heck et al., 2009). These results coincided with those for the CSN3 gene. For the LGB gene, the observed and the expected genotype frequencies were with similar values, thus confirming the validity of Hardy-Weinberg equilibrium for the SRC population.

Genetic variations can be detected at the phenotypic level by various protein identification techniques as well. These techniques could be, e.g., acrylamide electrophoresis in denaturing (SDS PAGE) or native conditions, IEF, chromatography etc. The electrophoretic and IEF methods, mainly used for routine typing at the protein level, only allow the detection of variations resulting in aminoacid substitutions altering the electric charge, the molecular weight or the isoelectric point of proteins. For the screening of breeds and populations at the phenotypic level, if milk is available, profiling at the protein level by IEF is recommended because the method is cheap and fast as well as gives a simultaneous picture of the phenotype expression of the main milk protein genes (Caroli et al., 2009).

Profiling of the BRC population genotypes with respect to the LGB gene was performed with IEF and 2D PAGE as well for confirmation of the obtained results.

The total bovine milk protein was separated by SDS PAGE (Figure 4), showing several differences, both in quantity and quality, of the available proteins. The predominant LGB and casein fractions were identified as compared to standards (Sigma). According to the standard, the LGB fraction was determined to be around 15 kDa (Figure 4, line 2), a little lower than expected. Previous reports showed an expected molecular mass of c. 18 kDa (Farrell et al., 2004). The casein fraction was determined to be between 25 and 35 kDa as compared to the alpha-S casein standard (Figure 4, lane 1). The differences between LGB protein isoforms (A, B and AB) could be observed on 14%-SDS PAGE (Figure 4, lanes 3-5) as a difference in molecular mass. As the A and B alleles lead to a difference of around 100 Da in the molecular masses of the resulting proteins and two bands could be observed for the LBG standards, we speculated that the milk proteins shown were from homozygous A (Figure 4, lane 3), homozygous B (Figure 4, lane 4) and heterozygous AB (Figure 4, lane 5) animals, respectively.

Further experiment showed that a one-dimensional IEF in a wide (3-10) pH gradient (Figure 5) is not sufficient for the successful profiling of the LGB and casein fractions. The reason is that both LGB and caseins, along with other milk proteins, are characterized by isoelectric point (pI) around 4.0 to 5.5 (Farrell et al., 2004).

Therefore, we conducted 2D PAGE analyses combining a vertical IEF and 14%-SDS PAGE for the second dimension, allowing a successful resolving of the predominantly low-molecular weight LGB and caseins by pI and molecular weight (Mw) (Figure 6). The results from 2D PAGE confirmed what was observed on SDS PAGE as two spots, differing by pI (4.9 – 5.1) and Mw (15 and 14.8 kDa) could be distinguished in the presumable heterozygous animals (Figure 6c) and one spot was observed in homozygous animals (Figure 6a and 6b). The casein fraction in all samples was shown to be highly heterogeneous and difficult to interpret. Further experiments, involving IEF in a narrow pH gradient should be performed in order to characterize the casein fraction.

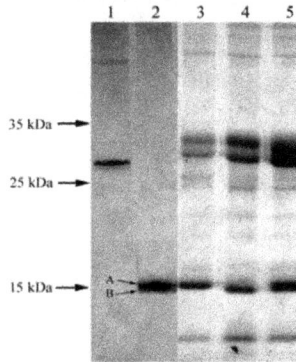

Figure 4. 14%-SDS PAGE (Coomassie staining) of bovine milk proteins. 1 - alfa-S casein standard; 2 - LGB standard; 3-5 - bovine milk protein samples. A and B indicate the A and the B alleles of the LGB gene.

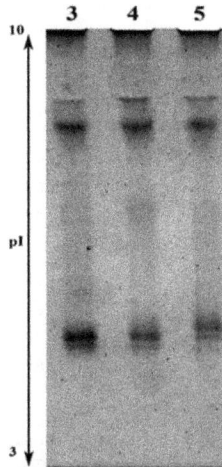

Figure 5. Vertical IEF (Coomassie staining) in a wide (3-10) pH range. Lanes 3-5 corresponds to the lanes in Figure 4.

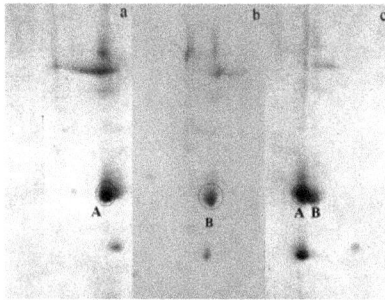

Figure 6. 2D PAGE (Coomassie staining) of milk protein samples. 6a - homozygous AA; 6b - homozygous BB; 6c - heterozygous AB animals.

3. Milk protein genes polymorphism and cattle selection and breeding

One of the most important effects of the milk protein polymorphisms on milk traits of economic importance is their relation to the technological properties of milk. The research was mainly concentrated on genetic variants of the CSN1S1, CSN3 and LGB genes. Some variants of these genes have a significant influence on the production of cheese, human nutrition etc. The constant monitoring of the milk protein variations in various breeds of cattle is an essential practice aiming to increase the frequency of genetic variants with favourable effects and to adapt their utilization for marker-assisted selection and breeding of cows. The application of genetic markers in dairy cattle breeding is a new stage in the selection practice in Bulgaria. If the cattle screening for desirable allelic forms is successful, it will have a direct scientific and practical value for stock-breeding farms in both public and private sectors. The marker-assisted selection is reliable and economically efficient way for increasing the farm production. The advantages of that approach are directly related to the genetic improvement of cattle populations, the increase of quality and quantity of production, the decrease of production loss and the improvement of the competitiveness of Bulgarian milk industry. Regardless multiple studies on milk protein genes polymorphism application for marker-assisted selection of diary cattle, studies for Bulgarian cattle populations are scarce. The first focused research for Bulgaria has been initiated by our team (Zlatarev et al., 2008). Subsequently, results were published concerning genetic variants of milk protein genes, their relationships with milk traits and importance for selection and breeding for a Bulgarian breed, i.e. the Black Pied Cattle (Hristov et al., 2011a). Recent investigations concern the Bulgarian Rhodopean Cattle Breed (BRC).

Although the SRC breed was genotyped by all the three important milk protein genes (CSN1S1, CSN3 and LGB), the relationships between recorded genetic variants and the milk quantitative and qualitative features were not examined. This is due to the specific way of the maintenance of this breed, which is targeting only the SRC population gene fund preservation, without its participation in milk industry.

With respect to the milk traits for the BRC breed, the milk productivity, the butter milk, the fat content and the protein content were examined monthly for 300 days of lactation. Data were analyzed by Statistical tool Descriptive statistics (Microsoft Excel, 2007) for each studied milk protein gene. The calculated mean values (shown as mean value ± SE) for milk productivity and qualitative traits were compared within each genotype.

3.1. CSN1S1 gene polymorphism

With respect to the importance of the CSN1S1 gene polymorphism for the milk production, it was found that the heterozygous BC animals had the highest values (3877.32 ± 114.67 kg). This exceeded with c. 12% the milk yield of the CC homozygous animals (3412.00 kg ± 103.09 kg) and with 7% that of the BB homozygous cows (3600.81 ± 153.79 kg). Similar results were obtained for butter milk data, where the BC animals had better values and the lowest values were those of the CC cows (BC - 179.93 ± 5.12 kg; BB - 170.06 ± 8.00 kg; CC - 167.01 ± 7.35 kg). These observations allowed the assumption for the superiority of the B allele of the CSN1S1 gene relative to both above-mentioned milk features. The milk fat and protein contents were affected mainly by the CC genotype. The values of the protein contents (CC - 3.72 ± 0.04%; BB - 3.68 ± 0.06%; BC - 3.63 ± 0.03%) were similar and only a slight superiority of the CC genotype was detected. The differences were more obvious with respect to the fat contents (CC - 4.88 ± 0.05%; B -, 4.72 ± 0.08%; BC - 4.66 ± 0.04%). With respect to the fat and protein contents, there was predominance of the C allele of the CSN1S1 gene. The results about qualitative and quantitative milk traits were summarized as follow: milk production and milk butter, BC>BB>CC; fat and protein contents, CC>BB>BC (Figure 7).

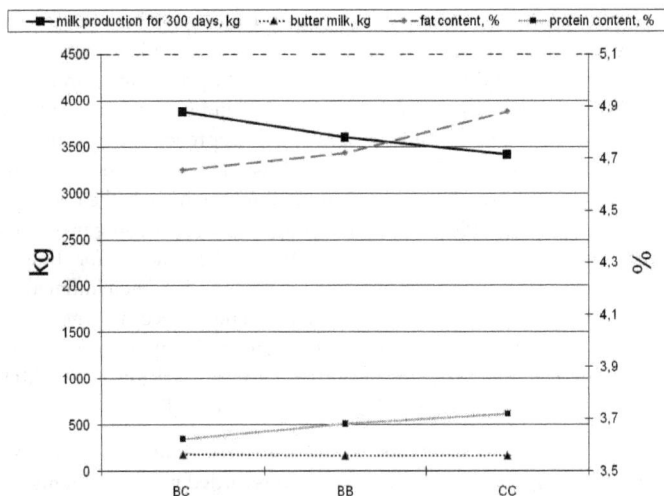

Figure 7. Influence of the CSN1S1 gene polymorphism on the milk production and the milk quality traits in cows of Bulgarian Rhodopean Cattle Breed. BB, CC, BC – genotypes.

The correlations between the CSN1S1 gene polymorphism and the milk traits obtained by other researchers have not been straightforward, partly due to the differences in parameters used and/or depending on cattle breeds. E.g., the CSN1S1 BB genotype correlated with higher milk production in some cases (Ng-Kwai-Hang et al., 1984; Aleandri et al., 1990; Sang et al., 1994) but there was also an evidence for the superiority of the hetrozygous BC genotype (Micinski et al., 2007). Our results support the positive effect of the BC genotype on the milk yield being about 9.5% higher than the homozygous genotypes. In general, the results presented and the published data confirm the dominance of the B allele over the C allele relative to the milk production. No publications were found about the influence of the genotypes of the CSN1S1 gene on the butter milk values; however, the present study revealed a positive effect of the B allele of this gene.

The data concerning the protein content are controversial. According to some reports, the BB genotype is linked to high protein content (Ng-Kwai-Hang et al., 1984; Aleandri et al., 1990; Sang et al., 1994) but the same genotype was associated with low protein values in other studies (Ng-Kwai-Hang et al., 1986, 1992). Micinski et al. (2007) reported that the CSN1S1 BC genotype affected the increase of the protein content of milk. Our observations coincided with data presented by Pečiulaitienė et al. (2007) that had demonstrated the superiority of the CC genotype relative to the protein content.

With regard to the fat content, all publications claim that the homozygous BB genotype is associated with higher values (Micinski et al., 2007; Kamiński, 1996; Pečiulaitienė et al., 2007). This contrasted with the present results exhibiting c. 4% of higher fat content in the milk of the CC animals compared to the BB cows from BRC breed.

3.2. CSN3 gene polymorphism

The present results for the milk productivity show that the heterozygous AB cows have c. 600 kg higher milk yield (4112.00 ± 149.40 kg) than the homozygous BB animals (3495.00 ± 290.40 kg) and c. 300 kg more than the AA representatives (3838.20 ± 160.22 kg). These data indicate the superiority of the A allele with respect to the milk productivity. Similar trends were also observed for the butter milk (AB - 191.45 ± 6.97 kg; AA - 180.27 ± 7.28 kg; BB - 161.17 ± 10.55 kg). Fat (AA - 4.70 ± 0.06%; BB - 4.66 ± 0.18%; AB - 4.66 ± 0.05%) and protein contents (AA - 3.66 ± 0.06%; AB - 3.63 ± 0.04%; BB - 3.56 ± 0.10%) were with similar mean values among the three genotypes. Nevertheless, there was a slight predominance of the AA genotype. The summarised results for the qualitative and quantitative milk traits were as follows: milk production and milk butter, AB>AA>BB; fat and protein contents, AA>AB≥BB (Figure 8).

Available data for the relationships between variants of the milk protein genes and the milk traits are contradictory. Usually, these relationships depend on cattle breeds and the country of origin. Some studies claim that the BB genotype was associated with higher (Van Eenennaam & Medrano, 1991) or lower (Bovenhuis et al., 1992) milk yield whereas other studies indicated no effect (Ng-Kwai-Hang et al., 1990; Lundén et al., 1997). The present results support data by Bovenhuis et al. (1992) suggesting a 15% decrease of the milk production of the BB homozygous animals compared to the AB heterozygous cows. According to our observations, there is a slight difference between butter milk mean values of the AB genotype and the BB genotype of the CSN3, with the latter being about 16% higher. Most of the previous authors (van Eenennaam & Medrano, 1991; Lundén et al., 1997) reported insignificant differences between the variants of the CSN3 gene or a slight prevalence of the AA genotype (Miciński et al., 2007).

With respect to the fat and the protein contents, the differences between the three genotypes were insignificant. This allows concluding that the CSN3 gene polymorphisms have no effect on these two milk traits. There is only a slight prevalence of the AA genotype as it concerns these milk traits, i.e. less than 1% for the fat content and about 3% for the protein content. Data about the fat content are in contrast with previous studies where an advantage of the AB genotype compared to the AA and BB genotypes has been shown but are in agreement relative to the protein content (Miciński et al., 2007).

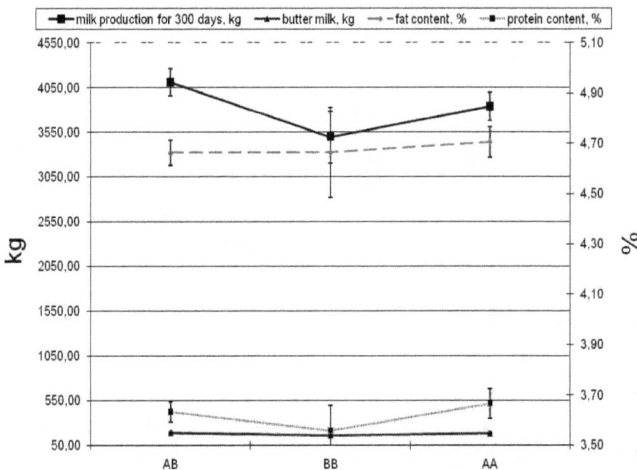

Figure 8. Influence of the CSN3 gene polymorphism on the milk production and the milk quality traits in cows of the Bulgarian Rhodopean Cattle Breed. AA, BB, AB – genotypes.

3.3. LGB gene polymorphism

The milk productivity of the homozygous BB animals was the highest one (4240.50 ± 33.50 kg). This is c. 660 kg more than that of the homozygous AA cows (3581.48 ± 154.14 kg). The heterozygous AB genotype defined an intermediate level of the milk production (3955.24 ± 125.45 kg). These results indicated the superiority of the B allele with respect to the quantitative milk traits. The fat content exhibited a similar trend (BB - 4.79 ± 0.02%; AA - 4.70 ± 0.05%; AB - 4.69 ± 0.05%). The differences were more obvious for the butter milk mean values (BB - 203.00 ± 1.00 kg; AB - 184.30 ± 5.57 kg; AA - 168.40 ± 7.26 kg). The protein content (AA - 3.72 ± 0.04%; BB - 3.60 ± 0.23%; AB - 3.58 ± 0.03%) was with almost equal mean values among the three genotypes. Nevertheless, there was a slight predominance of the AA genotype. The summarised results about qualitative and quantitative milk traits (Figure 8) were as follows: milk production and milk butter, BB>AB>AA; fat content, BB>AA>AB; protein content, AA>BB>AB.

Our results for the influence of the LGB gene polymorphism on the milk traits are more unambiguous than those for the effect of the CSN3 gene. In studies of the LGB genotype effects on the milk production, several authors reported no significant associations (Lundén et al., 1997; Ojala et al., 1997). Nevertheless, there are also reports for the positive influence on the milk quantity of all the genotypes, i.e. AA (Aleandri et al., 1990; Bovenhuis et al., 1992), AB (Pupkova, 1980) or BB (Jairam & Nair, 1983). The present results show that the BB genotype determines higher milk production.

Previous studies suggested the advantage of the AA genotype of the LGB gene on the protein content (Aleandri et al., 1990; Bovenhuis et al., 1992), are similar to our results. That is the only milk feature that is affected by the AA genotype of LGB gene. This result coincides with that for the CSN3 gene.

Positive effects of the BB genotype on the fat content (Aleandri et al., 1990; Bovenhuiset al., 1992; Hill, 1993) and of the AA genotype on the butter milk (Miciński et al., 2007) have been reported. Our observations for BRC breed are similar and have revealed the favourable influence of the BB homozygous genotype on these milk traits. The differences were much more significant relative to the butter milk than to the fat content. The distinction between fat content values associated with the BB and AB genotype is c. 2% whilst between the butter milk values are 17%.

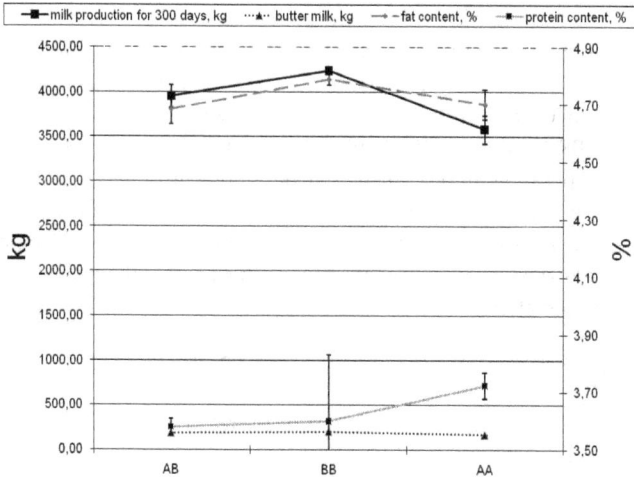

Figure 9. Influence of the LGB gene polymorphism on the milk production and the milk quality traits in cows of the Bulgarian Rhodopean Cattle Breed. AA, BB, AB – genotypes.

The analysis of the genetic polymorphisms of the CSN1S1, CSN3 and LGB genes of the Bulgarian Rhodopean Cattle Breed revealed their influence on the quantitative and qualitative milk traits. This allows selection of proper animals (cows and bulls) with desirable genotypes and increasing the frequency of favourable alleles within the population, thus improving milk composition and increasing economic efficiency of dairy cattle farms.

4. Future research

The research plans of our team include various aspects of the milk protein genotyping. Primary experiments will be related to milk proteomics. The casein fraction in all the samples revealed to be highly heterogeneous and difficult to interpret with the IEF technique are to be studied in a narrow pH gradient and with 2D PAGE method in order to achieve their better characterization. The CSN1S1, CSN3 and LGB proteins and their variants will be separated and quantified by reversed-phase high-performance liquid chromatography for higher precision and resolution. At the genomics level, these protein genes will be examined in the light of the single nucleotide polymorphism analysis and the population sequence datasets will be completed for the BRC and the SRC breeds aiming population profiling of these native Bulgarian cattle breeds. The genetic diversity of milk proteins could serve as a criterion of selection and as an informative marker in studies of the phylogenetic relationships and evolution of breeds. Finally, similar investigations will be carried out at the protein and DNA levels for other Bulgarian indigenous cattle breeds as well as for small ruminants, which have not been studied in Bulgaria up to date.

5. Conclusion

The present studies bring forward a novel approach for genotyping of two Bulgarian native cattle breeds specific for the Rhodope mountain area, i.e. Bulgarian Rhodopean Cattle (BRC) and Shorthorn Rhodopean Cattle (SRC). They also show the relationships between the genotypes of the BRC breed and qualitative and quantitative milk traits. This gives the opportunity for application, improvement and development of genetically-based marker-assisted selection and breeding of cows with desirable genotypes and increasing of dairy farms economic and financial gains. The BRC and SRC breeds are of high significance for the Bulgarian cattle biodiversity and their genotyping with respect to the studied genes could be utilized also for preservation of that native breed's gene fund.

Author details

Peter Hristov, Denitsa Teofanova and Georgi Radoslavov
Institute of Biodiversity and Ecosystem Research, Bulgarian Academy of Sciences, Bulgaria

Ivan Mehandzhiyski
Agricultural and Stockbreeding Experimental Station, Agricultural Academy, Bulgaria

Lyuben Zagorchev
Sofia University "St. Kliment Ohridski", Faculty of Biology, Bulgaria

Acknowledgement

This study was supported by grant YRG No 02/23 28.12.2009 from the National Science Fund of the Bulgarian Ministry of Education, Youth and Science, Sofia, Bulgaria.

6. References

Aleandri, R., Buttazzoni, L.G., Schneider, J.C., Caroli, A., & Davoli, R. (1990). The effects of milk protein polymorphisms on milk components and cheese-producing ability *Journal of dairy science*. Vol.73, No.2, pp. 241–255, ISSN 0022-0302

Beja-Pereira, A., Luikart, G., England, P.R., Bradley, D.G., Jann, O.C., Bertorelle, G., Chamberlain, A.T., Nunes, T.P., Metodiev, S., Ferrand, N., & Erhardt, G. (2003). Gene-culture coevolution between cattle milk protein genes and human lactase genes. *Nature genetics*. Vol.35, No.4, pp. 311-313, ISSN 1061-4036

Bovenhuis, H., Van Arendonk, J.A.M., & Korver, S. (1992). Associations between milk protein polymorphisms and milk production traits. *Journal of dairy science*. Vol.75, No.9, pp. 2549–2559, ISSN 0022-0302

Caroli, M., Chessa, S., & Erhardt, G.J. (2009). Milk protein polymorphisms in cattle: effect on animal breeding and human nutrition. *Journal of dairy science*. Vol.92, No.11, pp. 5335-5352, ISSN 0022-0302

Crittenden, R.G. & Bennett, L.E. (2005). Cow's milk allergy: A complex disorder. *Journal of the American College of Nutrition*. Vol.24, No.6 Suppl, pp. 582S-591S, ISSN 0731-5724

Di Stasio, L., & Mariani, P. (2000). The role of protein polymorphism in the genetic improvement of milk production. *Zootecnica e Nutrizione Animale*. Vol.26, No.3, pp. 69-90, ISSN 0390-0487

EFSA. (2004). Opinion of the Scientific Panel on Dietetic Products, Nutrition and Allergies on a request from the Commission relating to the evaluation of goats' milk protein as a protein source for infant formulae and follow-on formulae. *The EFSA Journal*. Vol.30, pp. 1–15, ISSN 1831-2357

Erhardt, G. (1996). Detection of a new κ -casein variant in the milk of Pinzgauer cattle. *Animal Genetics*. Vol.27, No.2, pp. 105–107, ISSN: 1365-2052

Farrell, H.M., Kumosinski, J.T.F., Cooke, P.H., King, G., Hoagland, P.D., Wickham, E.D., Dower, H.J., & Groves, M.L. (1996). Particle sizes of purified κ-casein: Metal effect and correspondence with predicted three-dimensional structures. *Journal of protein chemistry*. Vol.5, No.5, pp. 435-445, ISSN 0277-8033

Farrell, H.M.Jr., Jimenez-Flores, R., Bleck, G.T., Brown, E.M., Butler, J.E., Creamer, L.K., Hicks, C.L., Hollar, C.M., Ng-Kwai-Hang, K.F., & Swaisgood, H.E. (2004). Nomenclature of the proteins of cows' milk—Sixth revision. *Journal of dairy science*. Vol.87, No.6, pp. 1641-1674, ISSN 0022-0302

Ferretti, L., Leone, P., & Sgaramella, V. (1990). Long range restriction analysis of the bovine casein genes. *Nucleic acids research*. Vol.18, No.23, pp. 6829-6833, ISSN 0305-1048

Fox, P.F., & McSweeney, P.L.H. (1998). *Dairy Chemistry and Biochemistry*, Blackie Academic & Professional, ISBN 0412720000, London, UK

Freyer, G., Liu, Z., Erhardt, G., & Panicke, L. (1999). Casein polymorphism and relation between milk production traits. *Journal of Animal Breeding and Genetics.*Vol.116, No.2, pp. 87–97, ISSN 0931-2668

Hayes, H., Petit, E., Bouniol, C., & Popescu, P. (1993a). Localisation of the alpha-S2-casein gene (CASAS2) to the homologous cattle, sheep and goat chromosome 4 by in situ hybridisation. *Cytogenetics and cell genetics.* Vol.64, No.3-4, pp. 281-285, ISSN 0301-0171

Hayes, H.C., & Petit, E.J. (1993b). Mapping of the β-lactoglobulin gene and of an immunoglobulin M heavy chain-like sequence to homoeologous cattle, sheep, and goat chromosomes. *Mammalian Genome.* Vol.4, No.4, pp. 207–210, ISSN 0938-8990

Heck, J.M.L., Schennink, A., van Valenberg, H.J.F., Bovenhuis, H., Visker, M.H.P.W., van Arendonk, J.A.M., & van Hooijdonk, A.C.M. (2009). Effects of milk protein variants on the protein composition of bovine milk. *Journal of dairy science.* Vol.92, No.3, pp. 1192-1202, ISSN 0022-0302

Hill, J.P. (1993). The relationship between β-lactoglobulin phenotypes and milk composition in New Zealand dairy cattle. *Journal of dairy science.* Vol.76, No.1, pp. 281–286, ISSN 0022-0302

Hill, J.P., Thresher, W.C., Boland, M.J., Creamer, L.K., Anema, S.G., Manderson, G., Otter, D.E., Paterson, G.R., Howe, R., Burr, R.G., Motion, R.L., Windelman, A., & Wickham, B. (1997). The polymorphism of the milk protein β-lactoglobulin, In: *Milk composition, production and biotechnology.* Welch, R.A.S et al., (Ed.), pp. 173-213, ISBN: 0-85199-161-0, CAB International, Wallingford, UK

Hristov, P., Teofanova, D., & Radoslavov, G. (2011). Effects of genetic variants of milk protein genes on milk composition and milk yield in cows of the Bulgarian black pied cattle. *Comptes Rendus de L'Academie Bulgare des Sciences.* Vol.64, No.1, pp. 75-80, ISSN 0366-8681

Hristov, P., Teofanova, D., Mehandzhiyski, I., Yoveva, A., & Radoslavov, G. (2011). Selection of dairy cows with respect to kappa-casein gene polymorphism. *Journal of Mountain Agriculture of the Balcans.* Vol.14, No.4, pp. 667-677, ISSN 1311-0489

Jairam, B.T., & Nair, P.G. (1983). Genetic polymorphisms of milk proteins and economic characters in dairy animals. *Indian Journal of dairy science.* Vol.53, No.1, pp. 1–8, ISSN 0019-5146

Kamiński, S. (1996). Bovine kappa-casein (CASK) gene – molecular nature and application in dairy cattle breeding. *Journal of Applied Genetics.* Vol.37, No.2, pp. 179-196, ISSN 1234-1983

Koczan, D., Hobom, G., & Seyfert, H.M. (1993). Characterization of the bovine αS1-casein gene C allele based on a MaeIII polymorphism. *Animal Genetics.* Vol.24, No.1, p. 74, ISSN 0268-9146

Kontopidis, G., Holt, C., & Sawyer, L. (2004). Invited review: beta-lactoglobulin: binding properties, structure, and function. *Journal of dairy science.* Vol.87, No.4, pp. 785-796, ISSN 0022-0302

Lunde´n, A., Nilsson, M., & Janson, L. (1997). Marked effect of blactoglobulin polymorphism on the ratio of casein to total protein in milk. *Journal of dairy science.* Vol.80, No.11, pp. 2996–3005, ISSN 0022-0302

Martin, P., Szymanowska, M., Zwierzchowski, L., & Leroux, C. (2002). The impact of genetic polymorphisms on the protein composition of ruminants milks. *Reproduction Nutrition Development.* Vol.42, No.5, pp. 433-459, ISSN 0926-5287

Medrano, J.F., & Cordova, E.A. (1990a). Polymerase chain reaction amplification of bovine β-lactoglobulin genomic sequences and identification of genetic variants by RFLP analysis. *Animal biotechnology.* Vol.1, No.1, pp. 73-77, ISSN 1049-5398

Medrano, J.F., & Cordova, E.A. (1990b). Genotyping of bovine kappa-casein loci following DNA sequence amplification. *Biotechnology.* Vol.8, No.2, pp. 144-146, ISSN 0733-222X

Miciński, J., Klupczyński, J., Mordas, W., & Zablotna, R. (2007). Yield and composition of milk from Jersey cows as dependent on the genetic variants of milk proteins. *Polish Journal of Food and Nutrition Sciences.* Vol.57, No.3(A), pp. 95-99, ISSN 1230-0322

Neelin, J.M. (1964). Variants of κ-casein revealed by improved starch gel electrophoresis. *Journal of dairy science.* Vol.47, No.5, pp. 506–509, ISSN 0022-0302

Ng-Kwai-Hang, K.F., Hayes, J.F., Moxley, J.E., & Monardes, H.G. (1984). Association of genetic variants of casein and milk serum proteins milk, fat, and protein production in dairy cattle. *Journal of dairy science.* Vol.67, No.4, pp. 835-840, ISSN 0022-0302

Ng-Kwai-Hang, K.F., Hayes, J.F., Moxley, J.E., & Monardes, H.G. (1986). Relationships between milk protein polymorphisms and major milk constituents in Hoistein-Friesian cows. *Journal of dairy science.* Vol.69, No.1, pp. 22–26, ISSN 0022-0302

Ng-Kwai-Hang, K.F., Monardes, H.G., & Hayes, J.F. (1990). Association between genetic polymorphism of milk proteins and production traits during three lactations. *Journal of dairy science.* Vol.73, No.12, pp. 3414–3420, ISSN 0022-0302

Ng-Kwai-Hang, K.F., & Grosclaude, F. (1992). Genetic polymorphism of milk proteins. In: *Advanced Dairy Chemistry.* Fox, P.F. (Ed.), Proteins Vol.I, pp. 405–455, Elsevier Applied Science, ISBN 0-306-47271-6, London, UK

Nickolov, V. (1999). Bulgarian Rhodopean Cattle. *World Jersey Bulletin.* April, 24.

Ojala, M., Famula, T.R., & Medrano, J.F. (1997). Effects of milk production genotypes on the variation of milk production traits of Holstein and Jersey cows in California. *Journal of dairy science.* Vol.80, No.8, pp. 1776–1785, ISSN 0022-0302

Pečiulaitienė, N., Miceikienė, I., Mišeikienė, R., Krasnopiorova, N., & Kriauzienė, J. (2007). Genetic factors influencing milk production traits in Lithuanian dairy cattle breeds. *ŽEMĖS ŪKIO MOKSLAI.* Vol.14, No.1, pp. 32–38, ISSN 1392-0200

Popescu, C.P., Long, S., Riggs, P., Womack, J., Schmutz, S., Fries, R., & Gallagher D.S. (1996). Standardization of cattle karyotype nomenclature: Report of the committee for the standardization of the cattle karyotype. *Cytogenetics and cell genetics.* Vol.74, No.4, pp. 259-261, ISSN 0301-0171

Preacher, K.J. (2001). Calculation for the chi-square test: An interactive calculation tool for chi-square tests of goodness of fit and independence [Computer software]. Available from http://quantpsy.org.

Prinzenberg, E.-M., Hiendleder, S., Ikonen, T., & Erhardt, G. (1996). Molecular genetic characterization of new bovine κ-casein alleles CSN3-F and CSN3-G and genotyping by PCR-RFLP. *Animal Genetics.* Vol.27, No.5, pp. 347–349, ISSN 0268-9146

Pupkova, G.V. (1980). Milk protein polymorphism and milk production of Estonian Black Pied cows. *Journal of dairy science*. Vol.45, pp. 6620, ISSN 0022-0302

Sang, B.C., Ahn, B.S., Sang, B.D., Cho, Y.Y., & Djajanegara, A. (1994). Association of genetic variants of milk proteins with lactation traits in Holstein cows. *Proceeding of 7th AAAP Animal Science Congress*, Vol.2, pp. 211–217, ISBN 955-1136-00-4, Bali, Indonesia

Sulimova, G.E., Sokolova, S.S., Semikozova, O.P., Nguet, L.M., & Berberov, E.M. (1992). Analysis of DNA polymorphisms of clustered genes in cattle: Casein genes and genes of the major histocompatibility complex (BOLA). *Cytology and genetics*. Vol.26, No.5, pp. 18-26, ISSN 0564-3783

Sulimova, G.E., Badagueva, I.N., & Udina, I.G. (1996). Polymorphism of the κ-casein gene in subfamilies of the Bovidae. *Genetika (Moscow)*. Vol.32, No.11, pp. 1576–1582, ISSN 0016-6758

Threadgill, D.W., & Womack, J.E. (1990). Genomic analysis of the major bovine milk proteins genes. *Nucleic acids research*. Vol.18, No.23, pp. 6935–6942, ISSN 0305-1048

Tsiaras, A.M., Bargouli, G.G., Banos, G., & Boscos, C.M. (2005). Effect of kappa-casein and beta-lactoglobulin loci on milk production traits and reproductive performance of Holstein cows. *Journal of dairy science*. Vol.88, No.1, pp. 327-334, ISSN 0022-0302

Van Eenennaam, A., & Medrano, J.F. (1991). Milk protein polymorphisms in California dairy cattle. *Journal of dairy science*. Vol.74, No.5, pp. 1730–1742, ISSN 0022-0302

Woychik, J.H. (1964). Polymorphism in κ-casein of cow's milk. *Biochemical and biophysical research communications*. Vol.16, No.3, pp. 267-271, ISSN 0006-291X

Yvon, M., Van Hille, I., & Pellisier, J.P. (1984). In vivo milk digestion in the calf abomasum. II. Milk and whey proteolysis. *Reproduction Nutrition Development*. Vol. 24, No.6, pp. 835-843, ISSN 0926-5287

Zlatarev, S., Hristov, P., Teofanova, D., & Radoslavov, G. (2008). Impact of genetic polymorphism of kappa-casein and beta-lactoglobulin loci on milk production traits in cows of the Bulgarian Rhodopean cattle. *Comptes Rendus de L'Academie Bulgare des Sciences*. Vol.61, No.12, pp. 1577-1582, ISSN 0366-8681

Chapter 5

Detection of QTL Underlying Milk Traits in Sheep: An Update

Juan José Arranz and Beatriz Gutiérrez-Gil

Additional information is available at the end of the chapter

1. Introduction

The worldwide production of sheep milk was 9,246,922 tonnes in 2009, whereas dairy cows produced 583,401,740 tonnes (FAOSTAT, at http://faostat.fao.org; accessed July 2011). The dairy sheep production is based on the utilization of local breeds of small or moderate effective and well adapted to local conditions in the different Mediterranean countries. This peculiarity contrasts with the "Holsteinization" which is observed in dairy cattle, where a single breed with small effective number and breed in very similar conditions throughout the world is responsible for most of the production of milk at the global level. Sheep milk is mainly addressed to the manufacturing of high quality artisanal cheese in most of the cases commercialised under Protected Designation of Origin (PDO) or other quality labels. Traditionally, classical dairy sheep breeding schemes have been mainly focused on increasing production traits. As in dairy cattle, the successful implementation of classical selection models on dairy sheep breeding schemes has led to an increased milk yield and milk fat and protein contents. In the case of dairy sheep the time and degree of selection progress are greatly variable depending on the population that the selection scheme is addressed to, especially when compared with the global/homogenous genetic improvement of the world-wide spread Holstein cattle breed. Nowadays, however, dairy sheep industry needs to face new challenges such as offering healthy and attractive products to consumers at the same time that providing local farmers the ability to keep their competitiveness. With this purposes, new selection objectives taking into account animal's overall health, mammary and body conformation traits need to be defined with the aim of ensuring the progress of the dairy sheep industry. Also traits such as milk fatty acid composition may be of great interests from the consumer point of view.

At the same time that the initial idea of addressing new selection objectives in dairy sheep selection schemes has become a requirement for the dairy sheep industry, great advances have taken place in the field of sheep genomics, following those reached on other species, especially human, mouse and cattle. From the initial studies on milk protein polymorphisms (Reviewed by Barillet et al., 2005), and the first description of a QTL in a dairy sheep breed (Diez-Tascón et al., 2001), some few reports of genome scans to identify QTL influencing milk production traits in this species can be found (Gutiérrez-Gil et al., 2009a; Raadsma et al., 2009). The lower number of projects searching milk-related QTL in sheep when compared with the number of studies carried out in dairy cattle can be partially explained by the funding limitations that exist, for example, to establish a sheep experimental population as has been done in cattle in several occasions (Gutiérrez-Gil et al., 2009c; Eberlein et al., 2009). Other reasons contributing to the limited number of genome scans carried out in sheep are the great diversity of productive breeds and the different management systems that can be found in this species, which suggest that the implementation of molecular information in different sheep populations will not be as straightforward as in the case of dairy cattle. On the other hand, the size of the experimental designs based on the analysis of commercial sheep populations has important inherent limitations related to the power of the experiment due to reduced family sizes, as artificial insemination is not an overspread practice in sheep as it is in dairy cattle.

This chapter is intended to review the population designs (experimental crosses and commercial populations) used to map QTLs in dairy sheep. Also a review on the genetic markers and maps used in both whole genome studies and candidate gene approaches will be achieved. Finally the main results obtained by different projects searching for genes underlying milk production traits will be covered.

2. Genetic determination of milk production traits in sheep

2.1. Current selection objectives

In the dairy sheep industry, milk yield is the first criterion to establish milk payment systems. Hence, increasing milk yield is still the first selection objective for dairy sheep breeds. Cheese yield is directly influenced by milk protein and fat contents, which are traits showing negative genetic correlations with milk yield. This has led some dairy sheep breeding schemes (Lacaune, French Pyrenean and Churra) to take into account, at some extend, milk composition traits when estimating selection breeding values (Carta et al., 2009). On the other hand, quality and sanitary safety of products are nowadays major challenges for dairy sheep breeders. In dairy sheep subclinical mastitis occurrence has been estimated at between 16 and 35%, representing one of the most important reasons for culling prematurely. In addition to the economical losses due to medical treatment and decreased milk production, subclinical mastitis in dairy sheep is linked to the presence of contaminants in milk (pathogens or antibiotics). Somatic cell score (SCS), which has been shown as an accurate indirect measure to predict udder infection, may be used for selection in favour of mastitis resistance. Up to day, SCS is included as a selection objective for increased mastitis resistance only in the Lacaune breeding schemes (Barillet, 2007).

In addition, we should take into account that based on the European Decision Nº 100/2003 (European Commission, 2003) the most important dairy sheep breeds in Europe have been subjected during the last years to national breeding programs aiming at increasing resistance against scrapie (Carta et al., 2009). This is a transmissible spongiform encephalopathy (TSE), akin to bovine spongiphorm encephalopathy (BSE) in cattle and Creutzfeldt–Jakob disease (CJD) in humans. Due to the potential public health risk and based on known association between some allelic variants identified in the ovine *PRNP* gene and the susceptibility/resistance status of this disease in sheep, the goal of these programmes is to increase the frequency of the most resistant allele (ARR).

Genetic parameters for milk production traits in sheep have been widely studied and show a similar pattern than in cattle. On a lactation basis, moderate heritabilities for milk yield have been reported in Lacaune (0.30; Barillet, 1997) and Sarda (0.30; Sanna et al., 1997), whereas the estimates are slightly lower in Spanish breeds such as Manchega (0.18; Serrano et al., 1996), Latxa (0.21; Legarra & Ugarte, 2001) and Churra (0.26; Othmane et al., 2002). Protein and fat yield heritabilities range from 0.16-0.18 (Latxa; Legarra & Ugarte, 2001) to 0.32-0.35 (Manech). The estimates for milk protein and fat percentages show higher heritability estimates than the yields (Serrano et al., 1996; Barillet & Boichard, 1987; Sanna et al., 1997), especially in the Lacaune and Sarda breeds for which these estimates are close or even higher than 0.5-0.55 (Barillet & Boichard, 1987; Sanna et al., 1997). In general milk fat percentage shows a lower heritability than protein percentage in all the breeds, with one of the lowest estimates being found in Churra sheep (0.10; Othmane et al., 2002). The same set of studies show that the genetic correlations between milk yield and milk fat and protein yields are close to one, whereas there are negative genetic correlations between milk yield and milk protein percentage (range -0.32/-0.53) and between milk yield and milk fat percentage (-0.27/-0.63). For SCC in dairy sheep heritability estimates reported range between 0.06 and 0.18 (Othmane et al., 2002; Rupp et al., 2003; Legarra & Ugarte, 2005).

2.2. Opportunities derived from the use of molecular genetics

The genetic parameter estimates summarized above support that substantial genetic gains can be achieved on milk production traits following classical genetic selection based on standardized national recording systems. However, selection based on the phenotype is hampered for these traits due to the repetitive nature of the phenotype, which requires multiple measurements for an accurate description. The inclusion of composition traits as selection goals that some breeding schemes have already performed seems to be recommendable for other breeds based on the negative genetic correlation between these traits and milk yield. Increasing mastitis resistance through classical selection seems, however, more difficult to reach due to the low heritability estimates described for this trait. On this regard, it must be taken into account that marker- or gene- assisted selection (Dekkers, 2004) could be of great assistance to speed up selection response for milk production traits and to reach appreciable genetic gains regarding ewes' mastitis resistance. For this to be a reality we need to identify the genetic basis of the phenotypic variation observed in dairy sheep flocks. Although this may involve complex mechanisms regarding the control of gene expression and epigenetics (Jaenisch & Bird, 2003), up to now the main efforts of the sheep research community have been addressed to identify the genes influencing these quantitative traits, which are called *Quantitative trait loci* (QTL). Like most of the traits of interest in livestock species, milk traits in sheep are classical quantitative traits often described by the infinitesimal model, which assumes that the number of loci showing effect on them is infinitely large (Lynch & Walsh, 1998). Although this model produces good predictions of short-term selection response, the QTL mapping experiments have shown that the number of genes controlling the phenotypic variance of these traits is limited. A study in cattle has shown that about 30 QTL were likely to be segregating for milk production traits in a half-sib Holstein population (Chamberlain et al., 2007). In dairy sheep populations, where selection programmes started more recently and with a lower selection pressure than in dairy cattle, we would expect a slightly higher number of segregating loci to explain the phenotypic variation observed in the flocks, although always limited. With the aim of identifying these loci or QTL, low-medium genome scans based on the analysis of microsatellite markers have been undertaken in dairy sheep since 2001. These studies will be referred from now on as classical QTL mapping experiments and the description of their methodology, experimental designs and main results will be the main objective of the chapter. In addition, we must consider the great advances that are taking place nowadays in the field of sheep genomics such as the ongoing progress of the sheep genome sequencing project (International Sheep Genomics Consortium et al., 2010) and the availability of the Ovine SNP50BeadChip (Illumina). The search of QTL based on high-throughput genotyping assays will be commented in the last section of the chapter.

3. Analytical tools for classical detection of QTL for milk traits in sheep

QTL mapping experiments require the analysis of genetic markers across appropriate mapping populations to follow the segregation of markers and QTL within family structures. The development of the sheep linkage maps including an increasing number of markers has been an irreplaceable tool for the studies aiming the detection of QTL for milk traits in sheep.

In this section a brief description of the elements required for classical QTL mapping is presented (genetic markers, phenotypes, experimental design and statistical methods) with reference to the different approaches used in the different QTL studies reported so far for sheep milk traits. A detailed review of classical QTL mapping experiments for milk traits carried out in sheep and their results is provided in the next section.

3.1. Genetic markers and linkage maps

Most classical QTL mapping experiments carried out in livestock species, including sheep populations, have been based on the analysis of microsatellite markers across the resource mapping population. Microsatellites are a subclass of eukaryote tandemly repeated DNA that contains very short simple sequence repeats such as (dCdA)n, (dG-dT)n (Weber & May, 1989). Microsatellite blocks are polymorphic in length among individuals of the same species and therefore represent a vast pool of potential genetic markers. These genetic markers show a high level of polymorphism, and are uniformly spaced throughout the genome at every 30-60 kb (Weber & May, 1989). Function for microsatellites is unknown, but it has been proposed that they serve as hot spots for recombination or participate in gene regulation (Hamada et al., 1984). However, they are mainly considered as "neutral phenotypic markers". The availability of the Polymerase Chain Reaction (Mullis et al., 1986) allowed the development of linkage maps of acceptable density in livestock species (Crawford et al., 1994).

Over the 1990s decade, great efforts were made by the international sheep genetics research community to develop a useful linkage map (Crawford et al., 1994; De Gortari et al., 1998). The establishment of several mapping flocks comprising three-generation full-sibling families allowed the construction of low-density autosomal (Crawford et al., 1994) and X chromosome (Galloway et al., 1996) linkage maps. The second generation map comprised 512 loci with an average spacing of 6 cM (de Gortari et al., 1998). In 2001, an enhanced linkage map of the sheep genome comprising 1093 loci was published (Maddox et al., 2001). This medium-density linkage map included 550 new loci merged with the previous sheep linkage map. The average spacing between markers for this map was 3.4 cM with an average of 8.3 cM between highly polymorphic autosomal loci. The third generation sheep linkage map shows strong links to the cattle linkage map, with 572 of the loci common to both maps. Also 209 of the loci mapped by Maddox et al. (2001) could be mapped in the goat map (Maddox et al., 2005). This third-generation linkage map has been expanded periodically with the different versions being available in the Australian Sheep Gene Mapping Web Site at http://rubens.its.unimelb.edu.au/~jillm/jill.htm. The latest available version, v5, includes 2515 loci, 344 of which are linked to a described gene, whereas most of the reported QTL mapping experiments in sheep have taken versions 4.0 to 4.7 as reference.

Also the development of comparative maps between sheep and other species such as cattle and human at the different stages of the progress on the sheep linkage map (http://rubens.its.unimelb.edu.au/~jillm/jill.htm) has provided great assistance to identify candidate genes for the QTL mapped in the different experiments.

3.2. Experimental designs

Detection of QTL is based on the study of segregation of a heterozygous QTL within a family through the analysis of informative markers. As in other livestock populations, the experiments designs used to map the genes underlying milk traits in sheep are based on the use of experimental crosses or, alternatively, in the analysis of commercial populations.

3.2.1. Cross-bred experimental populations

In this case, the experimental design is based on the establishment of experimental populations by crossing two breeds that show clear phenotypic differences for the traits of interest. By crossing founders of these divergent founder "lines" the resulting F1 individuals can be used to generate large segregating F2 or backcross (BC) populations. The complete resource population is genotyped for the genetic markers that will be used to identify QTL, whereas phenotypic measures of the traits to study are recorded from the second generation animals (F2 or BC). Once the experimental population is established, many phenotypes for a wide-range of economical interesting traits are usually recorded in the cross-bred animals. The fact that the animals are reared in an experimental farm make easy obtaining phenotypes that could not be routinely recorded in commercial flocks. The power of these designs is maximized when the founder "lines" show alternative and fixed or nearly fixed alleles for the mutation that underlie the genetic control of the traits under study. Also the management of these populations under properly controlled environmental conditions may drastically reduce the environmental noise with the consequent increase of QTL mapping accuracy. The main disadvantages of these experimental designs are the time-consuming and expensive efforts required for the establishment of the experimental populations, and the fact that the results may not be directly implemented in commercial populations (Georges, 1998). Experimental cross populations have been widely used in pig and chicken populations. Also in cattle, there are some examples exploiting this kind of design by crossing the highly specialized dairy Holstein breed with a beef producer breed although published results are relative to other traits different to dairy traits (Eberlein et al., 2009; Gutiérrez-Gil et al., 2009c). In sheep, genome scans for mapping QTL influencing milk traits based on experimental crosses have been performed using a Sarda x Lacaune backcross population (Carta et al., 2002; Barillet et al., 2006) and a Awassi × Merino backcross family population (Raadsma et al., 2009). Also a genome scan based on a backcross pedigree using dairy East Friesian rams and non dairy Dorset ewes has been reported (Mateescu & Thonney, 2010).

The backcrosss Sarda x Lacaune resource population mentioned above was established in the framework of a European funded project (*GeneSheepSafety*, QL K5 CT2000 0656). Fourteen AI elite Lacaune rams were crossed with Sarda ewes to produce F1 rams. Ten of these F1 rams and 10 different Lacaune sires were mated to Sarda ewes to obtain 980 backcross females. Family size ranged from 76 to 121. Phenotypes were recorded for many traits from the second-generation backcross ewes. In addition to classical milk production traits, other phenotypes of interest in dairy sheep production such as milk fatty acid composition, kinetics of milk emission, udder morphology and resistance to mastistis and nematode infections were measured in this cross-population (Casu, 2004; Barillet et al., 2005; 2006).

Another remarkable research programme for QTL detection based on an experimental ovine population has been performed in Australia (CRC funded project). This experiment was based on an extreme breed back-cross and inter-cross design between Awassi fat-tail sheep and Merino superfine and medium wool sheep (Raadsma et al., 1999). Due to the extreme difference between these two types of sheep in a range of production characteristics, this experiment aimed at identifying quantitative trait loci for wool, meat and milk production. Both super-fine and medium-wool Merinos were used in the present resource (Raadsma et al., 2009). This resource population was developed in three phases, coinciding with different stages of research. In Phase I, four sires from an imported strain of improved dairy Awassi were crossed with 30 super-fine and medium-wool Merino ewes. Four resulting F1 sires (AM) were backcrossed to 1650 fine and medium-wool Merino ewes, resulting in approximately 1000 second-generation backcrosses (AMM). In Phases II and III additional crosses both within and across families were performed resulting in third- and fourth-generation animals. From the whole project a total of 2,700 progeny were produced over 10 years, representing four generations. In the genome scan for milk production traits reported by Raadsma et al. (2009) the QTL analysis was based on the information from the 172 ewe second-generation AMM progeny of one of the F1 sires.

Based on the increasing interest in the United States on sheep milk production, an experimental population of animals has been established by crossing East Friesian rams and Cornell Dorset ewes. This population has been established specifically to map QTL for milk production by crossing four East Friesian rams and 37 Dorset ewes to generate 44 F1 ewes, which were subsequently mated to 11 East Friesian rams to create 92 backcross ewes (Mateescu &Thonney, 2010).

An additional project exploiting line divergence between Awassi and Merino breeds has been performed in Hungary (FVM 46040/2003 project). In this case, the experimental population was initiated by crossing Awassi rams and Hungarian Merino ewes and then matting females in each subsequent generation back to new groups of purebred Awassi rams (Árnyasi et al., 2009). A total of 258 ewes with different proportions of Merino-Awassi genetic component were used to perform an association analysis with 13 microsatellite markers distributed on OAR6 (Árnyasi et al., 2009). This chromosome was selected as candidate for milk traits due to the large number of QTL detected in the orthologous bovine chromosome, BTA6, and because previous studies in sheep had identified significant effects on this chromosome (Díez-Tascón et al., 2001; Schibler et al., 2002). It should be noted that the casein cluster is located in this chromosome.

3.2.2. Outbred pedigrees

These experimental designs take advantage of the particular structure of some livestock commercial populations, such as those of dairy cattle and sheep where the use of artificial insemination (IA) results in large families of paternal half-sib families where the segregation of markers and QTL can be studied. In the classical "Daughter design", which was initially proposed by Neimann-Sörensen & Robertson (1961), the milk production records are obtained from the daughters, whereas all the population, including the sires, is genotyped for the genetic markers. In dairy cattle, where the IA has been extensively used for a large number of years, the "Granddaughter design" was suggested as a method to increase the statistical power of the "Daughter design" (Weller et al., 1990). In this case the availability of half-sib families of evaluated bulls allows the identification of QTL by genotyping only the founder sires and the bulls of each family whereas from the third-generation cows only the milk production phenotypes that are used to evaluate the second-generation bulls are obtained. This design allows a substantial reduction of animals to be genotyped, which is the most expensive part of the project, to reach an equivalent statistical power than the "Daughter design". Also getting biological samples for DNA extraction from the bulls localized in few insemination centres may be easier than collecting daughter samples from a wide range of different flocks. However, the selection of the design to be applied depends on the characteristic of the population to be studied. Hence, despite the increased power that the "Granddaughter design" may offer, the daughter design may be more adequate for populations with a recent implementation of IA due to the lack of large half-sib families of evaluated sires.

Compared with the experimental cross designs QTL mapping experiments carried out in commercial populations may seem to show important limitations. For example marker and QTL heterozygosity are likely to be reduced when compared with a cross-population and, moreover, they may very between the different families. Also different QTL influencing the same traits may segregate in different families, which increases complexity of results (Georges, 1998). In addition, the phenotypic records obtained in field conditions may show the influence of environmental factors that should be taken into account in the QTL analysis model (year, season, flock, etc). As an advantage, the phenotype collection, at least, for milk traits, is usually performed on the basis of a national recording system, and therefore not additional funding efforts are required to obtain phenotypic measurements. However, the main advantage of these designs is that the QTL detected in a livestock commercial population could be subjected to a more straight forward "Marker Assisted Selection" as these QTL represent the allelic variants actually segregating in the studied population. Additional confirmation studies on independent sampling populations, however, are needed before trying to implement those results in different commercial sheep populations.

Both, the granddaughter and the daughter designs have been used to map QTL for milk traits in sheep through the 26 sheep autosomes (Barillet et al., 2006). French populations of Lacaune and Manech breeds have been studied following a granddaughter design, based on the large scale AI and progeny test performed as part of the breeding programmes of these breeds (Barillet et al., 2005). A total of 700 and 83 AI rams distributed in 22 families (18 in Lacaune and 4 in Manech breeds) were included in this QTL mapping population. Family size averaged 36 sons per sire and ranged from 24 to 56. The sons had 89 daughters on average (Barillet et al., 2005).

Churra sheep is one of the most important dairy sheep breeds in Spain. The selection programme of this population was started in 1986 (de la Fuente et al., 1995). Since them, flocks included in the Churra Selection Nucleus have made use of AI. Based on the daughter design a preliminary analysis focused on sheep chromosome 6 analysed 726 ewes distributed in 14 flocks and belonging to 8 half-sib families of the Selection Nucleus (Díez-Tascón et al., 2001). Based on the same design a genome scan included in the framework of the European funded *GeneSheepSafety* project was planned to scan the 26 ovine autosomes. A total of 1421 ewes sired by 11 IA rams and distributed among 17 different flocks were used in this project for detection of QTL underlying classical milk traits (Gutiérrez-Gil et al., 2009a). Other traits of interest in dairy sheep production such as morphology traits (Gutiérrez-Gil et al., 2008; 2011) and resistance to mastitis and nematode gastrointestinal infections (Gutiérrez-Gil et al., 2007; 2009b) were also recorded in the commercial population genotyped. The average family size in this daughter design was around 110, ranging from 47 to 223 daughters per sire.

Other QTL study for milk traits based on the daughter design has been performed in the Latxa breed, which is also an important dairy sheep located in the North of Spain. With the aim of reducing the genotyping effort, a "selective DNA pooling" approach was followed in this breed to search for QTL on sheep chromosome 6 (OAR6) (Rendo et al., 2003).

3.3. Phenotypes and dependent variables

The phenotypes considered in most of the QTL mapping studies mentioned above are the classical production traits, milk yield (MY), milk protein yield (PY), milk fat yield (FY), milk protein percentage (PP), milk fat percentage (FP) and somatic cell score (SCS).

The dependent variables analysed for QTL detection are, in general, the measurements of the traits of interest adjusted for the specific environmental effects that show influence on the traits. This adjustment can be performed previously to the QTL analysis, or alternatively the fixed factors to be taken into account can be included in the QTL model when running the analyses. In commercial populations, where the trait is routinely recorded for other purposes and not only for the QTL mapping experiment, the phenotypic measures used in the QTL analysis can be breeding value estimates or deviations from the mean population. Using estimated breeding values (EBV) all the relationships among the animals are taken into account (Israel & Weller, 1998). Hence, the variability of the daughter productions is reduced and the estimated effect of a given QTL is reduced, which may make difficult QTL identification. Because of that several QTL studies use non biased measurements of the animals' productions such as the Yield Deviation (YD) or the Daughter Yield Deviation (DYD) (Israel & Weller, 1998). YDs are weighted averages of ewe's lactation yields minus solutions for management group, herd-sire, and permanent environmental effects (VanRaden & Wiggans, 1991). These are the quantitative measures that have been used in the QTL mapping experiments performed in Churra sheep (Díez-Tascón et al., 2001; Gutiérrez-Gil et al., 2007; 2009a). Because rams do not have yield deviations, DYD, which are adjusted for mates' merit, can provide a usefull, unregressed measure of daughter performance to use in grand-daughter designs (VanRaden & Wiggans, 1991). These are the dependent variables used in the analyses performed in the Lacaune-Manech Granddaughter design (Barillet et al., 2006). In a commercial population of Latxa sheep, Rendo et al. (2003) used EBVs for milk production as quantitative measurements of the screening performed on OAR6. (Table 1).

In the experimental populations established by crossing divergent breeds the quantitative measures involve, in general, the phenotypic records adjusted for the corresponding fixed factors (Carta et al., 2002; Raadsma et al., 2009), whereas some other of these experiments use EBVs as the milk yield EBV analysed by Mateescu & Thonney (2010).

3.4. Statistical methods for QTL mapping

Most of the QTL mapping experiments for milk traits in sheep have followed an interval mapping approach for detection of QTL. Only the analysis performed on OAR6 in a backcross Awassy X Hungarian merino population (Árnyasi et al., 2009) has analysed one locus at a time on each performance trait using a likelihood ratio test (Shaw, 1987) or a regression procedure (Ostergard et al., 1989). By adapting the method of LOD scores used in human genetic linkage analysis studies, Lander & Botstein (1989) proposed interval mapping to solve the problems shown by the initial QTL mapping experiments which studied single genetic markers one-at-a-time (Sax, 1923; Soller & Brody, 1976). Compared with these methods, interval mapping has been shown to provide some additional power and much more accurate estimates of QTL effect and position and to be relatively robust to failure of normality assumptions (Lander & Botstein, 1989; Knott & Haley, 1992). In the method of interval mapping the intervals between pairs of flanking markers in a linkage map are explored in turn for evidence of the presence of a QTL at various positions between the markers. Hence, the construction of a linkage map with the markers to analyse is required before performing the QTL mapping. The method described by Lander & Botstein (1989) is based in the maximum-likelihood analysis of the data. The likelihood ratio test is performed at regular intervals along the chromosome (e.g. 1 -cM), with the peak value representing the most likely position of a QTL. The significance threshold suggested is that applied in human genetics, LOD score ≥ 3 (Lander & Botstein, 1989). The disadvantage of maximum likelihood based methods for interval mapping is their computational complexity. Haley & Knott (1992) proposed a regression method applied to interval mapping. This methodology allows analysing more complex models for example to test the data for the presence of two or more linked or interacting QTL (Haley & Knott, 1992). In this case, the position which gives the best fitting model (i.e. produces the smallest residual mean square) gives the most likely position of a QTL and the best estimates of its effect.

However, the application of interval-mapping approaches to data from crosses between outbred lines would lead to the same situation observed in analyses within outbred populations that is that the power to detect a QTL varies from interval to interval depending upon the markers flanking that interval. This can lead to biases in the estimated position and effect of a QTL (Knott & Haley, 1992). Based on this, regression methods taking into account information from all of the informative markers in a linkage group (multimarker regression) were proposed for crosses between outbred lines (Haley et al., 1994) and half-sib outbred populations (Knott et al., 1996). Also Georges et al. (1995) presented a maximum-likelihood approach to QTL detection for use in half-sib populations by using information from all markers in a linkage group simultaneously, but analysed families separately. The method described by Knott et al. (1996) first calculate transmission probabilities in two-generation half-sib families and, second allow the linkage phase to differ from family to family. This analysis method is the one followed for milk traits QTL mapping in the commercial sheep populations previously described. The analyses performed in Spanish Churra sheep were performed using HSQM (Coppieters et al., 1998) which implements the multimarker regression method described by Knott et al. (1996) (Díez-Tascón et al., 2001; Gutiérrez-Gil et al., 2007, 2009a). In the Lacaune and Manech French populations, the QTL detection was carried out according to the methodology proposed by Knott et al. (1996) and Elsen et al. (1999) by within-sire linear regression (Barillet et al., 2006).

The analyses carried out in the Sarda X Lacaune cross population were performed with the INRA QTLMap software, which implements the methodology proposed by Elsen et al. (1999) by within-sire linear regression. The multimarker regression method for cross populations described by Haley et al. (1994) and implemented in the web-accessible programs QTL Express or GridQTL (Seaton et al., 2002; 2006; https://greidqtl.cap.ed.ac.uk/gridsphere) has been used to analyse the East Friesian X Dorset and Awasi x Merino backcross populations (Mateescu & Thonney, 2010; Raadsma et al., 2009). This later population was also analysed using a QTL maximum likelihood procedure suitable for the backcross design named QTL-MLE (Raadsma et al., 2009).

To determine chromosome-wise significance thresholds, most of the studies here referred have followed the permutation approach (Churchill & Doerge, 1994) for each trait and each chromosome using 10,000 permutations. Different methods have been used, however, to take into account the testing for 26 chromosomes in the genome scans reported. Hence, Bonferroni corrections have been used in some of the cases such as Churra sheep analysis of SCS (Gutiérrez-Gil et al., 2007), and the Sarda X Lacaune population (Barillet et al., 2006). Following the method described by Harmegnies et al. (2006) genome-wide permutations were implemented in the Churra sheep analysis for milk production traits (Gutiérrez-Gil et al., 2009a), In addition to the genome-wise significant QTL, these authors also considered the genome-wise suggestive linkage level, for which one false positive is expected in a genome scan (Lander & Kruglyak, 1995). On the other hand, Raadsma et al. (2009) adopted a false discovery rate (FDR) method (Benjamini & Hochberg, 1995) to adjust P-values for all traits to control for genome-wise error rates for the results obtained with the QTL-MLE analysis method. In most of the QTL experiments referred herein, the estimation of the confidence interval for the detected QTL was obtained by the bootstrapping method described by Visscher et al. (1996) (Gutiérrez-Gil et al., 2007; 2008; Barrillet et al., 2006; Mateescue & Thonney, 2010).

A summary of the populations studied for mapping of QTL for milk traits in sheep is provided Table 1. The experimental design, the number of markers analysed, the phenotypic traits analysed, and the statistical methods used for each of the QTL mapping experiments are detailed.

4. QTL mapping results for milk traits in sheep

Based on the available scientific literature and the information stored in *Sheep*QTLdb (http://www.animalgenome.org/cgi-bin/QTLdb/OA/index), we present in this section a brief description of the QTL results reported so far for classical milk production traits in the different sheep populations previously mentioned (MY, PY, FY, PP, FP and SCS), considering the analysis of lactational or test-day records for the same phenotype as the same trait. Figure 1 shows a graphical representation of the milk QTL identified for these traits through the linkage analyses performed on the experimental populations previously described. Just to note that only across-family significant QTL have been considered, and therefore some effects identified in the within-family analyses are not here commented. Apart of these QTL detected on the basis of multimarker regression linkage analyses,

Árnyasi et al. (2009) and Calvo et al. (2004a, 2004b, 2006) performed association analyses for milk production traits on selected candidate chromosomes, OAR6 and OAR1 respectively. These results are mentioned in the candidate gene approach section. The map positions of the different QTL represented in Figure 1 are based on the position suggested by the Australian linkage map version 5.0 for http://rubens.its.unimelb.edu.au/~jillm/jill.htm) for the closest marker reported for each QTL.

	Breed – Design Population	Target chromosomes (Nb of markers)	Analysis method	Reference
Experimental cross-bred populations	Sarda X Lacaune (980 BC ewes)	1 to 26 (127 microsatellites)	Multimarker regression for half sibs[1]	Carta et al. (2003)
	East Friesian X Dorset (92 BC ewes)	1 to 26 (99 microsatellites)	Multimarker regression for cross populations [2]	Mateescu & Thonney (2010)
	Awasi x Merino (172 ewes)	1 to 26 (200 microsatellites)	Maximum Likelihood[3] Multimarker regression for cross populations [2]	Raadsma et al. (2009)
Commercial populations	Awassi x Hungarian Merino (258 ewes)	OAR6 (13 microsatellites)	Association analysis	Amyasi et al. (2009)
	Churra DD (1421 ewes)	1 to 26 (181 microsatellites)	Multimarker regression [1]	Gutiérrez-Gil et al., (2007; 2009a)
	Churra DD (726)	OAR6 (11 microsatellites)	Multimarker regression [1]	Díez-Tascón et al. (2001)
	Lacaune-Manech GDD (783 AI rams – 89 daughter on average)	1 to 26 (163 microsatellites)	Multimarker regression [1,4]	Barillet et al. (2006)

[1]Knott et al. (1996); [2]Haley et al. (1994); [3]Raadsma et al. (2009); [4]Elsen et al. (1999)

Table 1. Characterization of QTL mapping experiments performed in sheep for milk traits.

As a general observation, the QTL reported in the different experiments are not coincident among each other. This contrasts with the results published in dairy cattle (reviewed by Khatkar et al., 2004 and Smaragdov et al., 2006; see also CattleQTLdb at http://www.animal genome.org/cgi-bin/QTLdb/BT/index), and suggests a high diversity of causal mutations or QTN underlying milk production traits segregating in the different sheep populations. Attention, however, should be driven to chromosomes OAR3 and OAR20, where several studies have found significant linkage associations with milk production traits (Figure 1).

4.1. Milk yield

QTL affecting milk yield (MY) have been identified in different sheep populations, with no substantial coincidences regarding QTL location among these reports. Raadsma et al. (2009) described QTL for total milk in OAR2, OAR3, OAR20 and OAR24 segregating in the Awassi x Merino cross population. All these QTL were identified with the regression and maximum-likelihood analysis methods used by these authors. Focusing on the results of the QTL-MLE analysis, the most interesting QTL are those on OAR3 and OAR20 which reached genome-wide significance and mapped close to other QTL influencing other milk yield traits (Raadsma et al., 2009). In the Sarda X Lacaune population significant linkage associations with MY were found on OAR3, OAR4, and OAR20 (Barillet et al., 2006). These QTL were coincident with other genetic effects influencing PY and FY. Another QTL identified in this population for MY on OAR16 was closely linked to a QTL for FP (Barillet et al., 2006). In the commercial population of Churra sheep, a genome-wise suggestive QTL was detected for MY in the proximal end of OAR23. The effect detected seems the result of a pleiotropic QTL influencing also PY and FY (Gutiérrez-Gil et al., 2009a). The genome scan performed in the Friesian X Dorset backcross population studied by Mateescu & Thonney (2010) identified two chromosome-wise significant QTL for MY on OAR2 and OAR18, the later of these being also associated with effects on PY.

4.2. Protein percentage and protein yield

The initial analysis performed of chromosome 6 in Churra sheep allowed the detection of a putative QTL for PP showing chromosome-wise significance, which mapped close to the casein gene cluster region (Díez-Tascón et al., 2001). This QTL was not identified, however, in the genome scan performed in an extended design of this population some years later (Gutiérrez-Gil et al., 2009a). In this genome scan, the most significant QTL, which reached genome-wise significance, was that identified for PP on OAR3 in the second half of the chromosome, close to marker KD103. One other suggestive QTL for PP was identified in this analysis in the proximal region of OAR2. The genome scan performed in the Awassi x Merino Backcross population only detected a significant QTL for PP in the first half of OAR7 (Raadsma et al., 2009). The Sarda x Lacaune genome scan identified a genome-wise significant QTL for PP in the first third of OAR1 and another genome-wise suggestive QTL on the second half of OAR7. Three genome-wise suggestive QTL for PP were found in the Lacaune-Manech population on OAR2, OAR5 and OAR9. For PY, apart of the QTL linked to MY QTL previously mentioned, a genome-wise suggestive QTL was detected in Churra sheep, in the second half of OAR1 (Gutiérrez-Gil et al., 2009a).

4.3. Fat percentage and fat yield

In the Sarda x Lacaune experimental population, a genome-wise significant QTL for FP was found on OAR20, whereas genome-wise suggestive linkage associations were reported on OAR3, OAR7 and OAR16. Close to the OAR20 QTL detected in that population, Gutiérrez-Gil et al. (2009a) reported a genome-wise suggestive QTL for FP in Churra sheep. In this population another suggestive QTL for FP was detected at the distal region of OAR2. In the Lacaune-Manech French genome scan genome-wise suggestive QTL for FP were detected on OAR1, OAR9 and OAR10. Raadsma et al. (2009) detected significant QTL for FP on OAR3, and OAR25, whereas a suggestive QTL was found on OAR8 for this trait. The first of these QTL, detected on OAR3, seems to be the result of a pleiotropic QTL linked to marker *DIK4796* that affects several milk production traits. As mentioned earlier, QTL for FY have been detected jointly with MY and PY QTL on OAR3, OAR4, and OAR20, in the Sarda X Lacaune population (Barillet et al., 2006), on OAR3 and OAR20 in the Awassi x Merino cross (Raadsma et al., 2009) and on OAR23 in Churra sheep (Gutiérrez-Gil et al., 2009a). Apart of these, the FY trait has shown genome-wise suggestive linkage associations on OAR14 in the Sarda x Lacaune population (Barillet et al., 2006), on OAR16 in the Lacaune-Manech commercial French population and on OAR25 in Churra sheep (Gutiérrez-Gil et al., 2009a).

4.4. Somatic cell score

Although this trait is not directly related with milk production its relation with subclinical mastitis incidence makes it of interest when trying to enhance productivity of sheep flocks. Hence, most of the genome scans performed for milk traits in this species have also studied this phenotype. The whole genome screening performed in Spanish Churra sheep identified a single genome-wise suggestive QTL for SCS on OAR20 (Gutiérrez-Gil et al., 2007). The best position suggested for this QTL is close to marker *OLADRBPS*, which is located in the major histocompatibility complex (MHC). In the Sarda x Lacaune population, several traits related to mastitis resistance were recorded (SCS in parity 1 to 4, SCS considered as a repeated trait within lactation and SCS considered as a repeated traits across four lactations). For these traits, together with 11 genome-wise suggestive QTL, two genome-wise significant QTL were found on OAR6 and OAR13 (Barillet et al., 2006). In the Lacaune-Manech population a suggestive genome-wise significant QTL was found on OAR14 for SCS (Barillet et al., 2006). In the case of the Awassi X Merino cross population a significant QTL for SCS was found on OAR14, whereas two suggestive QTL for this trait were reported on OAR17 and OAR22 (Raadsma et al., 2009).

4.5. Other traits of interest in dairy sheep

Apart of the classical milk yield and composition traits and SCS, QTL have been reported for other traits directly related to milk production. Raadsma et al. (2009) analysed the useful yield content. Some of the QTL detected for this trait were detected in the same regions that other milk production traits on OAR3 and OAR25. These authors also identified suggestive QTL for useful yield content on OAR6 and OAR9. In the Awassi x Merino cross population analysed by these authors QTL have been also been identified for milk lactose yield on OAR2, OAR3, OAR15, OAR20 and OAR24. These QTL were coincident with QTL influencing other milk traits (Raadsma et al., 2009).

Total lactation performance and length of lactation also have a significant economical impact on dairy sheep industry. Traits related with lactation persistency and extended lactation in sheep have been analysed for QTL detection in the Awassi x Merino backcross population (Jonas et al., 2011). These authors identified five genome-wise significant QTL for these traits on OAR3 (fat persistency), OAR10 (extended lactation somatic cells) and OAR11 (extended lactation milk, milk persistency and extended lactation protein) together with five other suggestive QTL. Interestingly, on OAR11 where the most significant QTL were identified in this study, no other QTL for milk production traits had been described before. The lack of coincidence between the QTL identified for lactation persistency and extended lactation suggests that lactation persistency and extended lactation do not have a common genetic background.

In the last years the evident relationship between human health and animal fat content in the diet may have a negative impact on the consumption of sheep cheese because of its high fat content. Sheep milk has a high level of saturated fatty acids (SFA) when compared to polyunsaturated (PUFA) and monounsaturated FA (MUFA). However, sheep milk has some other beneficial components for human health such as ω_3-fatty acids and the conjugated linoleic acid (CLA). Considering genetics to improve the fatty acid (FA) composition of milk sheep and due to the difficulties to measure these traits in commercial populations, the detection of genetic markers associated with these traits would be of great interest. Analysing the experimental Sarda x Lacaune backcross resource population, Carta et al. (2008) identified several chromosome-wise QTL for milk fatty acid composition across the sheep autosomes. The most significant QTL were found on OAR11 and OAR6. Because of its interest in relation to human health benefices, it is worth mentioning the QTL identified for CLA content on AOR4, OAR14 and OAR19, whereas QTL for the ratio CLA/Vaccenic acid, which is an indicator of the proportion of CLA that is synthesised in the mammary gland from its precursor, were found on the same region of OAR4 and on OAR22. For the latter of this QTL the *SCD* (stearoyl-CoA desaturase) gene was suggested as functional and positional candidate (Carta et al., 2006). Based on the Sarda x Lacaune genome scan results, and the location of candidate genes related to milk fatty acid composition metabolic pathways, linkage analyses for these traits were performed on OAR11 and OAR22 in a commercial population of Spanish Churra sheep including 15 half-sib families (García-Fernández et al., 2010a; 2010b). These studies assessed the role of candidate genes in relation to fatty acid composition and will be described later.

Other traits of interest in dairy sheep are those related to udder and type morphology. Udder related traits may be considered the most important functional traits in dairy sheep, as they determine the machine milking efficiency of the animal (Labussière, 1998) and have a substantial effect on its functional lifetime (Casu et al., 2003). QTL for udder traits have been reported in the Sarda x Lacaune population, the Lacaune-Manech families (Barillet et al., 2006) and in Churra dairy sheep (Gutiérrez-Gil et al., 2008). In the Sarda x Lacaune population a detailed study of udder morphology traits was performed, with digital picture measures being recorded in the cross-bred animals and a large list of udder morphology related traits being analysed for QTL detection. Genome-wise significant QTL were detected on OAR3, OAR4, OAR9, OAR14, OAR16, OAR20, OAR22 and OAR26. In the Lacaune-Manech population, genome-wise suggestive QTL were identified on OAR6 and OAR17 for udder cleft (Barillet et al., 2006). In this population, QTL were also detected for traits related to the kinetics of milk emission which are also directly related with the machine milking ability of the ewes. The difficulties to measure these traits in commercial populations make these analyses of great value for the sheep research community. Genome-wise suggestive QTL for these traits were detected on OAR9, OAR11, OAR15, OAR17 And OAR20, with the most significant QTL influencing maximum milk emission flow on OAR11.

In the commercial population of Spanish Churra sheep, a genome scan for udder morphology traits assessed according to the 9-point linear scale described by de la Fuente et al. (1996) identified chromosome-wise significant QTL on OAR7, OAR14, OAR15, OAR20 and OAR26. The most significant of these QTL was that identified in the proximal end of OAR7 for teat placement (Gutiérrez-Gil et al., 2008). In the same population QTL have also been identified for body conformation, which are functional traits of great interest in dairy sheep because its correlation with functionally productive life and overall productivity of the flock (Vukasinovic et al., 1995). Genome-wise suggestive QTL for these traits were identified on OAR2, OAR5, OAR16, OAR23 and OAR26 (Gutiérrez-Gil et al., 2011). The QTL reported on OAR16 was the most significant of these linkage associations and influenced the rear legs-rear view trait, which is related to leg conformation.

For dairy sheep reared on grazing systems, gastrointestinal nematode (GIN) parasite infections are diseases with a great impact on animal health and productivity. The selection of resistant animals to these infections might be a possible strategy for a sustainable control of this problem. Because of the low to moderate heritability estimated for indicator traits in adult dairy sheep (Gutiérrez-Gil et al., 2010) and the difficulties of routine collection of phenotypic indicators detection of genetic markers associated with these traits could be use to improve the efficiency of classical breeding. In Churra Spanish sheep, the genome scan reported by Gutiérrez-Gil et al. (2009b) describes a genome-wise significant QTL on OAR6 influencing faecal egg count. Four other chromosome-wise significant QTL were found for faecal egg count and anti-Teladorsagia circumcincta Larvae IV IgA levels. Many QTL for resistance to GIN infection have also been identified in the Sarda x Lacaune backcross population (Moreno et al., 2006; Barillet et al., 2006). Apart of several regions showing suggestive significance, genome-wise significant QTL were detected on OAR2, OAR3, OAR6, OAR12 and OAR13 (Barillet et al., 2006). QTL for host resistance against GIN parasite infections have been mapped in other non-dairy sheep populations (Davies et al., 2006; Coltman et al., 2001; Beh et al., 2002: Dominik, 2003).

Selection programs towards scrapie resistance are being implemented in many European countries based on the genotyping of the PRNP gene alleles classically associated with the disease (Hunter, 1997). However, selection in favour of the ARR/ARR genotype could induce a simultaneous change in production traits due to either the pleiotropy of the PRPN gene or genetic linkage with dairy QTL. This possibility has been studied in East Friesian (Vries et al., 2005) and Churra (Álvarez et al., 2006) sheep breeds. The association and linkage analyses performed in Churra sheep suggested that increasing the ARR frequency in the Churra population will not have an adverse effect on selection for milk traits included in the breeding objectives of this breed. Similar conclusion was drawn by De Vries et al. (2005), who did not identify any significant association between the prion protein genotypes and milk performance, type or reproduction traits. QTL other than *PRNP* for scrapie resistance have been reported on OAR6 and OAR18 in Romanov sheep (Moreno et al., 2008; 2010).

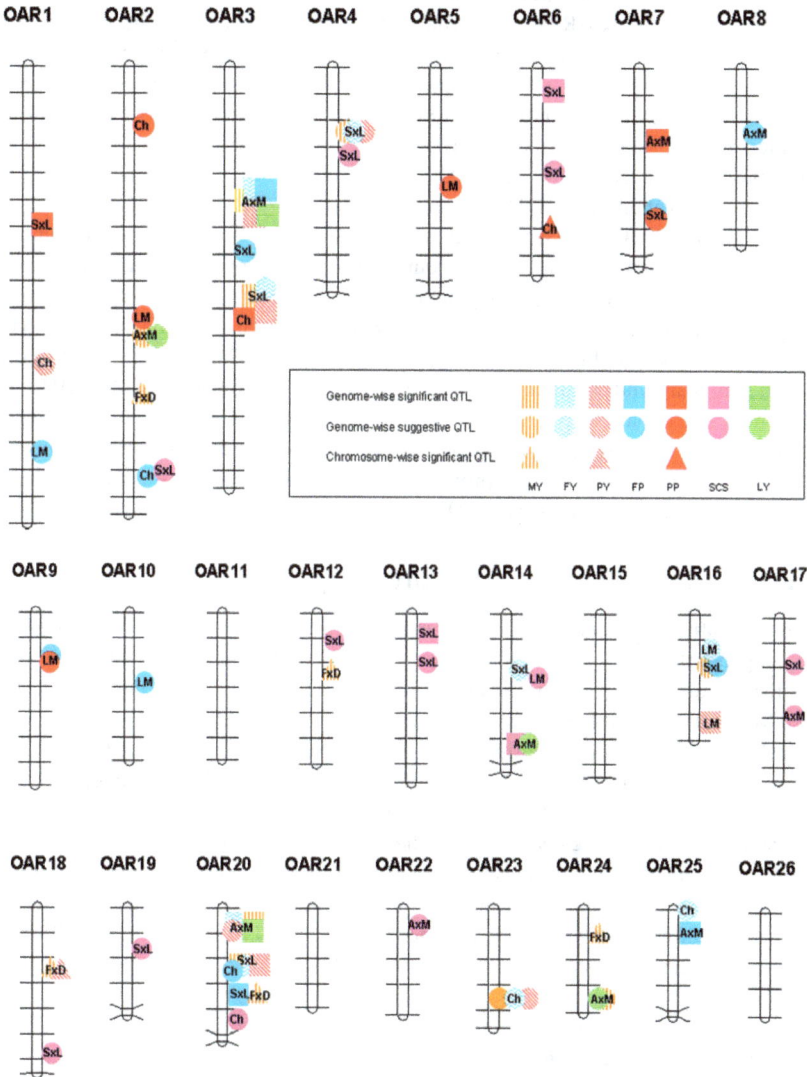

Figure 1. Graphical representation of QTL mapped on the sheep autosome for milk production traits and SCS. Legend abbreviations: MY: Milk yield; FY: Milk fat yield; PY: Milk protein yield; FP: Milk fat percentage; PP: Milk protein percentage; SCS: Somatic cell score; LY: Milk lactose yield.

4.6. Following-up studies on the QTL detected

Few fine-mapping studies have been reported in relation to QTL identified for traits of interest in dairy sheep whereas up to date no QTN explaining a previously identified QTL has been identified for milk traits in sheep. In order to validate the segregation of QTL previously detected on OAR7 in the Sarda x Lacaune population for milk fat and protein contents (Barillet et al., 2006), a new resource population was procreated by mating Sarda rams with BC ewes (Casu et al., 2010). The increased of informative meiosis and a higher marker density allowed to confirm the presence of a chromosome-wise significant QTL affecting milk contents by exploiting a similar approach than the great-grand-daughter design proposed by Coppietters et al. (1999). In Churra sheep, additional analyses have allowed the replication of the QTL detected on OAR20 for FP (García-Gámez et al., 2009) and on OAR3 for PP (García-Gámez et al., 2012) by analyzing additional half-sib families of the Selection Nucleus of Churra sheep commercial population. The increased of marker density reached on the OAR3 QTL region and the use of a combined linkage analysis and linkage disequilibrium analysis (LDLA) has allowed the refinement of the initially described confidence interval for this QTL from 40 to 13 cM (García-Gámez et al., 2012). Further research efforts are under way to perform a high density SNP screening on this QTL region and identify functional and positional candidate genes (García-Gámez et al., under review). High resolution mapping has also been performed on OAR3 and OAR20 in relation to QTL affecting lactation persistency and protein yield segregating in the Australian Awassi X Merino cross population (Singh et al., 2007).

5. Functional candidate genes in dairy sheep

As an alternative approach to QTL mapping, the candidate gene strategy undertakes the study of genes that are supposed to be responsible for a considerable amount of the genetic variation of traits of interest based on their known physiological function (Moioli et al., 2007). More precisely, this is a functional candidate gene, whereas a positional candidate would be any gene mapping within the confidence interval of a described QTL. On this section, we will comment briefly studies performed on functional candidate genes in dairy sheep. In the 1990s, genes coding for milk proteins were studied as potential tools for selection in dairy sheep (see reviews by Barillet et al., 2005, Moioli et al., 2007). For example, for the αs1-casein five polymorphisms were identified (Chianese et al., 1996). Associations of this gene's allelic variants with milk composition traits and renneting properties were identified in different sheep breeds (Piredda et al., 1993; Pirisi et al., 1999). Other milk protein gene extensively studied in dairy sheep is β-lactoglobulin, for which three protein polymorphisms have been described (Erhardt, 1989). However, the lack of consistent results regarding the possible associations of β-lactoglobulin polymorphisms with milk production traits, have discarded this gene as a potential genetic marker.

Based on previously described QTL in the bovine orthologous chromosome (BTA3), Calvo et al. (2004a; 2006) reported a preliminary assessment of genes located on the orthologous ovine chromosome, OAR1, for milk traits in 13 half-sib families of Spanish Manchega sheep.

The genes studied included two α-amylase (*AMY*) genes, annexin A9 (*ANXA9*), solute carrier family 27 member 3 (*SLC27A3*), cingulin (*CGN*) and acid phosphatise 6 lysophophatidic (*ACP6*). However, only within-family associations were detected for *AMY* and *SLC27A*, suggesting the need of larger resource populations to confirm the preliminary results reported. Similar results were found in this population in relation to the heart type *FABP3* gene, studied as a candidate gene for milk fat content (Calvo et al., 2004b). The candidate chromosome approach based on QTL detected in sheep and cattle was also followed by Árnyasi et al. (2009), who looked for any relationship between microsatellite markers localised on OAR6 and milk production traits in a backcross Awassi X Hungarian Merino population. Significant associations were detected for the lactation milk traits studied, with the most significant associations being found for marker *BM143* and lactation milk yield and lactation lactose yield.

A simplistic approach was presented by García-Fernández et al. (2011a) when assessing in Churra dairy sheep the role of genes previously identified to harbour causal mutations or QTN in dairy cattle. Hence, these authors searched for polymorphisms in the genes encoding the acylCoA:diacylglycerol acyltransferase 1 (*DGAT1*; Grisart et al., 2002), the growth hormone receptor (*GHR*; Blott et al., 2003) and the breast cancer resistance protein (*ABCG2*; Cohen-Zinder et al. 2005; Olsen et al., 2007). Also the osteopontin (*SPP1*) gene was considered in this work, as initial candidate gene in dairy cattle (Schnabel et al. 2005) and because its known influence on the expression of milk protein genes (Sheehy et al. 2009). This analysis revealed only significant associations at the nominal level for allelic variants of the *ABCG2* gene, whereas no significant association was found for the other studied genes. These results suggest that milk production traits show a different genetic architecture in sheep and cattle and highlight the need of increasing our knowledge on sheep genomics and not only building up on the advances previously reported in dairy cattle. Scatà et al. (2009) also studied the influence of *DGAT1* polymorphisms on milk traits in three Italian sheep breeds. These authors identified a SNP in the 5`UTR region of the gene showing a significant negative association with milk fat content in the Sarda sheep whereas this allelic variant was rare in Altamurana and Gentile di Puglia breeds, which have a higher milk fat content than Sarda.

Several genes encoding enzymes directly involved in fatty acid metabolism have been studied in the last years. In Churra sheep linkage and association analyses have been performed to assess the influence of the stearoyl-CoA desaturase (*SDS*), acetyl-CoA carboxylase α (*ACACA*) and fatty acid synthase (*FASN*) genes on milk fatty acid composition (García-Fernández et al., 2010a, 2010b, 2011b). Although some significant nominal associations were detected between some of the allelic variants identified in these genes, these analysis did not reveal any major effect of these genes on the fatty acid profile of Churra sheep milk. Crisà et al. (2010) have also studied polymorphisms in genes encoding enzymes putatively involved in the synthesis and metabolism of milk fat in three Italian breeds (Altamurana, Gentile di Puglia and Sarda). These authors identified genes such as α-1-antichymotrypsin-2 (*SERPINA3*), diacylglycerol O-acyltransferase homolog-2 (*DGAT2*), propionyl Coenzyme A carboxylase, β polypeptide (*PCCB*), sulin-like growth factor-1 (*IGF1*) and *FASN* to influence on the variability of the fatty acid profile of sheep milk. Two other genes, *GHR* and zona pellucida glycoprotein-2 (*ZP2*), were found to affect the variability of the total fat content.

6. Short future research and expectations

Despite the moderate number of QTL identified in dairy sheep, the identification of genetic markers to use in marker- or gene- assisted selection programmes is still hampered by the large confidence intervals of the QTL identified by classical linkage analyses based on medium-density microsatellite maps. In the last years, high-density SNP genotyping has become feasible in livestock species, whereas genome sequencing projects for cattle and sheep are now are reality. Whereas the last version of the bovine assembly (btau_4.0; http://www.ensembl.org/Bos_taurus/Info/Index?db=core) is of acceptable quality, the sequencing project of the sheep genome is still ongoing (International Sheep Genomics Consortium et al., 2010). The great research efforts of the International Sheep Genomics Consortium (ISGC; http://www.sheephapmap.org/) to develop public genomic resources to be used by sheep geneticists are providing the research community with the tools required to assist dairy sheep breeding with edge-cutting genomic technology. The Sheep HapMap project allowed the identification of millions of allelic variants in the sheep Genome, part of which have been used for the development of the ovine SNP50 BeadChip. Current efforts of the ISGC are being addressed to progress on the sheep genome sequencing project, with v3.0 of the sheep genome assembly (http://www.livestock genomics.csiro.au/sheep/oar3.0.php) being available, and to develop a high density SNP chip to allow the study of structural variants (J. Kijas, personal communication).

In cattle, several studies can be found already in the literature using high-throughput SNP genotyping platforms to study the genetic basis of milk production traits (Mai et al., 2010; Schopen et al., 2011). In this species, Genomic Selection seems a feasible way of using the advances of the genomic era for a straight-forward implementation in breeding schemes. In dairy sheep, however, the great genetic diversity of the populations makes this approach more difficult to implement in current breeding programmes. In any case, the screening of high density SNP distributed along the sheep genome will, for sure, facilitate the identification of allelic variants directly associated with milk production traits and other traits of interest in dairy sheep. The whole genome association study (GWAS) approach, which directly exploit linkage disequilibrium between markers and causative mutations, together with scans based on the LDLA methodology (Legarra & Fernando, 2009; Druet et al., 2008), are currently under way in some of the resource populations previously used to map classical QTL detection. Several of these genome screenings are included in the framework of the European funded project *Sustainable Solutions for Small Ruminants* (3SR) where the genetic basis of resistance to mastitis, parasite infections and paratuberculosis in different European populations are under dissecting using the last genomic advances. International collaborations with research groups from Australia, China and USA have made of this project a major international effort to increase our knowledge on the field of sheep genomics (see http://www.3srbreeding.eu/ for further details about this project). Another collaborative project performed under the Seventh Framework Programme (FP7) of the European Commission is Quantomics, which aims to deliver a step-change in the availability of cutting edge technologies and tools for the economic exploitation of livestock genomes (http://www.quantomics.eu).

7. Conclusion

Dairy sheep has been the subject of several studies trying to decipher the molecular architecture of milk production traits. Due to dairy sheep production systems characteristics (based on a wide range of local breeds reared under a variety of management systems) and its minor economic importance related to dairy cattle, the results are only modest but provide an initial picture of the genetic basis of milk production in sheep. The development of molecular tools derived from the increasing knowledge on the genome of this species make us envision a promising future where genomic information will assist sheep breeders to analyze their pedigrees and make informed decisions to enhance the improvement of sheep milk production.

Author details

Juan José Arranz and Beatriz Gutiérrez-Gil
Department of Animal Production, Faculty of Veterinary Sciences, University of León, Spain

8. References

Alvarez L.; Gutiérrez-Gil, B.; San Primitivo, F.; de la Fuente, L.F. & Arranz, J.J. (2006). Influence of prion protein genotypes on milk production traits in Spanish Churra sheep. Journal of Dairy Science, Vol.89, No.5, (May 2006), pp. 1784-1791, ISSN 0022-0302

Árnyasi, M.; Komlósi, I.; Lien, S.; Czeglédi, L.; Nagy, S. & Jávor A. (2009). Searching for DNA markers for milk production and composition on chromosome 6 in sheep. Journal of Animal Breeding and Genetics, Vol.126, No.2, (April 2009), pp. 142-147, ISSN 0931-2668

Barillet, F. (1997). Chapter 20: Genetics of milk production, In: The Genetics of Sheep. I. Piper & A. Ruvinsky (Ed.), 539–564., CAB International. ISBN 0 85199 200 5, Oxon, United Kindgom

Barillet, F. (2007). Genetic improvement for dairy production in sheep and goats. Small Ruminant Research, Vol.70, No.1, (June 2007), pp.60–75, ISSN ISSN 0921-4488

Barillet, F. & Boichard, D. (1987). Studies on dairy production of milking ewes. I. Estimation of genetic parameters for total milk composition and yield. Genetics Selection Evolution, Vol.19, No.4, pp. 459–474, 0999-193X

Barillet, F.; Arranz, J.J. & Carta, A. (2005). Mapping quantitative trait loci for milk production and genetic polymorphisms of milk proteins in dairy sheep. Genetics Selection Evolution 37, Suppl 1, S109-123, 0999-193X

Barillet F.; Arranz J.J.; Carta A.; Jacquiet P.; Stear M. & Bishop S. (2006). Final Consolidated Report of the European Union Contract of Acronym GeneSheepSafety, QTLK5-CT-2000-00656

Beh, K.J.; Hulme, D.J.; Callaghan, M.J; Leish, Z.; Lenane, I.; Windon, R.G. & Maddox, J.F. (2002). A genome scan for quantitative trait loci affecting resistance to Trichostrongylus colubriformis in sheep. Animal Genetics, Vol.33, No.2, (April 2002), pp. 97-106, ISSN 0268-9146

Benjamini, Y. & Hochberg, Y. (1995). Controlling the false discovery rate: A practical and powerful approach to multiple testing. Journal of the Royal Statistical Society, Vol.57, No.1, pp. 289-300, ISSN 0035-9246

Blott, S.; Kim, J.J.; Moisio, S.; Schmidt-Küntzel, A.; Cornet, A.; Berzi, P.; Cambisano, N.; Ford, C.; Grisart, B.; Johnson, D.; Karim, L.; Simon, P.; Snell, R.; Spelman, R.; Wong, J.; Vilkki, J.; Georges, M.; Farnir, F. & Coppieters, W. (2003). Molecular dissection of a quantitative trait locus: a phenylalanine-to-tyrosine substitution in the transmembrane domain of the bovine growth hormone receptor is associated with a major effect on milk yield and composition. Genetics, Vol.163, No.1, (January 2003), pp. 253-266, ISSN 0016-6731

Calvo, J.H.; Marcos, S.; Beattie, A.E.; Gonzalez, C.; Jurado, J.J. & Serrano, M. (2004a). Ovine alpha-amylase genes: isolation, linkage mapping and association analysis with milk traits. Animal Genetics, Vol.35, No. 4, (August 2004), pp. 329-332, ISSN 0268-9146

Calvo, J.H; Marcos, S.; Jurado, J.J. & Serrano, M. (2004b). Association of the heart fatty acid-binding protein (FABP3) gene with milk traits in Manchega breed sheep. Animal Genetics, Vol.35, No.4, (August 2004), pp. 347-349, ISSN 0268-9146

Calvo, J.H.; Martínez-Royo, A.; Beattie, A.E.; Dodds, K.G.; Marcos-Carcavilla, A. & Serrano, M. (2006). Fine mapping of genes on sheep chromosome 1 and their association with milk traits. Animal Genetics, Vol.37, No.3, (June 2006), pp. 205-210, ISSN 0268-9146

Carta, A.; Barillet, F.; Allain, D.; Amigues, Y.; Bibé, B.; Bodin, L.; Casu, S.; Cribiu, E.; Elsen, J.M.; Fraghi, A.; Gruner, L.; Jacquiet, J.; Ligios, S.; Marie-Etancelin, C., Mura, L., Piredda, G., Rupp, R., Sanna, S.R., Scala, A., Schibler, L. & Casu, S. (2002). QTL detection with genomic markers in a dairy sheep backcross Sarda x Lacaune resource population, Proceedings of the 7th World Congress on Genetic Applied to Livestock Production 2002, Vol.29, pp. 211-214, ISBN 2738010520, Montpellier, France, 19-23 August, 2002

Carta, A.; Sechi, T.; Usai, M.G.; Addis, M.; Fiori M.; Fraghì A.; Miari S.; Mura L.; Piredda G.; Essen J. M.; Schibler L.; Barrillet F. & Casu S. (2006). Evidence for a QTL affecting the synthesis of linoleic conjugated acid cis-9, trans-11 from 11-c 18:1 acid on ovine chromosome 22, Proceedings 8th World Congress on Genetics Applied to Livestock Production 2006, Communication No.12–03, ISBN 85-60088-01-6, Belo Horizonte, Brazil, 13-18 August, 2006

Carta, A.; Casu, S.; Usai, M. G.; Addis, M.; Fiori, M.; Fraghì, A.; Miari, S.; Mura, L.; Piredda, G.; Schibler; L.; Sechi; T.; Elsen, J.M. & Barrillet F. (2008). Investigating the genetic component of fatty acid content in sheep milk. Small Ruminant Research, Vol.79, No.1, (September 2008), pp. 22-28, ISSN 0921-4488

Carta, A.; Casu, S. & Salaris, S. (2009). Invited review: Current state of genetic improvement in dairy sheep. Journal Dairy Science, Vol.92, No.12, (December 2009), pp. 5814-5833, ISSN 0022-0302

Casu, S. (2004). Recherche de QTL contrôlant la cinétique de l'emission du lait et la morphologie de la mamelle chez les brebis laitières. PhD Thesis. Institut National Agronomique, Paris, France

Casu, S.; Marie-Etancelin C.; Schibler L., Cribiu E., Mura L., Sechi T., Fraghi A., Carta A. & Barillet F. (2003). A genome scan to identify quantitative trait loci affecting udder morphology traits in dairy sheep, Proceedings of the International Workshop on Major Genes and QTL in Sheep and Goat (IWMGQSG), article No.2–19, INRA Castanet-Tolosan, Toulouse, France, 8-11 December

Casu, S.; Sechi, S.; Salaris, S.L. & Carta, A. (2010). Phenotypic and genetic relationships between udder morphology and udder health in dairy ewes. Small Ruminant Research, Vol.88, N. 2-3, (February 2010), pp. 77-83, ISSN 0921-4488

Chamberlain, A.J.; McPartlan, H.C. & Goddard, ME. (2007). The number of loci that affect milk production traits in dairy cattle. Genetics, Vol.177, No.2, (October 2007), pp. 1117-1123, ISSN 0016-6731

Chianese, L.; Garro, G.; Mauriello, R.; Laezza, P.; Ferranti, P. & Addeo, F. (1996). Occurrence of five alpha s1-casein variants in ovine milk. Journal of Dairy Research, Vol.63, No.1, (February 1996), pp. 49-59, ISSN 0022-0299

Churchill G. & Doerge R. (1994). Empirical threshold values for quantitative trait mapping. Genetics, Vol.138, No. 3, pp. 963-971, ISSN 0016-6731

Cohen-Zinder, M.; Seroussi, E.; Larkin, D.M.; Loor, J.J.; Everts-van der Wind, A.; Lee, J.H.; Drackley, J.K.; Band, M.R.; Hernandez, A.G.; Shani, M.; Lewin, H.A.; Weller, J.I. & Ron, M. (2005). Identification of a missense mutation in the bovine ABCG2 gene with a major effect on the QTL on chromosome 6 affecting milk yield and composition in Holstein cattle. Genome Research, Vol.15, No.7, (July 2005), pp. 936-944, ISSN 1088-9051

Coltman, D.W.; Wilson, K.; Pilkington, J.G.; Stear, M.J. & Pemberton, J.M. (2001). A microsatellite polymorphism in the gamma interferon gene is associated with resistance to gastrointestinal nematodes in a naturally-parasitized population of Soay sheep. Parasitology, Vol.122, No.5, (May 2001), pp. 571-582, ISSN 0031-1820

Coppieters, W.; Kvasz, A.; Farnir, F.; Arranz, J.J.; Grisart, B.; Mackinnon, M. & Georges M. (1998). A rank-based nonparametric method for mapping quantitative trait loci in outbred half-sib pedigrees: application to milk production in a granddaughter design. Genetics, Vol.149, No.3, (July 1998), pp 1547-1555, ISSN 0016-6731

Coppieters W.; Kvasz A.; Arranz J.J.; Grisard B.; Farnir F.; Riquet J. & Georges M. (1999). The great-grand-daughter design: a simple strategy to increase the power of a grand-daughter design for QTL mapping. Genetic Research, Vol.74, No.2, (October 1999), pp. 189-199, ISSN 0016-6723

Crawford, A.M.; Montgomery, G.W.; Pierson, C.A.; Brown, T; Dodds, K.G.; Sunden, S.L.; Henry, H.M.; Ede, A.J.; Swarbrick, P.A.; Berryman, T.; Penty, J.M. & Hill D.F. (1994). Sheep linkage mapping: nineteen linkage groups derived from the analysis of paternal half-sib families. Genetics, Vol.137, No.2, (June 1994), pp. 573-579, ISSN 0016-6731

Crisà, A.; Marchitelli C.; Pariset L.; Contarini G.; Signorelli, F.; Napolitano F.; Catillo G.; Valentini A. & Moioli B. (2010). Exploring polymorphisms and effects of candidate genes on milk fat quality in dairy sheep. Journal of Dairy Science, Vol.93, No.8, (August 2010), pp. 3834-3845, ISSN 0022-0302

Davies, G.; Stear, M.J.; Benothman, M.; Abuagob, O.; Kerr, A.; Mitchell, S. & Bishop, S.C. (2006). Quantitative trait loci associated with parasitic infection in Scottish blackface sheep. Heredity, Vol. 96, No.3, (March 2006), pp. 252-258. ISSN 0018-067X de Gortari, M.J.; Freking, B.A.; Cuthbertson, R.P.; Kappes, S.M.; Keele, J.W.; Stone, R.T.; Leymaster, K.A.; Dodds, K.G.; Crawford, A.M. & Beattie C.W. (1998). A second-generation linkage map of the sheep genome. Mammalian Genome, Vol.9, No.3, (March 1998), pp. 204-209, ISSN 0938-8990

de la Fuente L.F.; Fernández G. & San Primitivo F. (1995). Breeding programme for the Spanish Churra sheep breed. Cahiers Options Méditerranéennes, Vol.11, No.11, pp, 165-172, ISSN 1022-1379

de la Fuente, L.F.; Fernández, G. & San Primitivo, F. (1996). A linear evaluation system for udder traits of dairy ewes. Livestock Production Science Vol.45, No. 2-3, (May 1996), pp. 171- 178, ISSN 0301-6226

De Vries, F.; Hamann, H.; Dro¨gemu¨ller, C.; Ganter, M. & Distl, O. (2005). Analysis of associations between the prion protein genotypes and production traits in East Friesian milk sheep. Journal of Dairy Science, Vol.88, No.1, (January 2005), pp. 392–398, ISSN 0022-0302

Dekkers, J.C. (2004). Commercial application of marker- and gene-assisted selection in livestock: strategies and lessons. Journal of Animal Science, Vol.82, E-Suppl, pp. E313-328, Review, ISSN ISSN: 0021-8812

Díez-Tascón C.; Bayón Y.; Arranz J.J.; De La Fuente F. & San Primitivo F. (2001). Mapping quantitative trait loci for milk production traits on ovine chromosome 6. Journal of Dairy Research, Vol.68, No.3, (August 2001), pp. 389-397, ISSN 0022-0299

Dominik, S. (2003). Quantitative trait loci for internal nematode resistance in sheep: a review. Genetics Selection Evolution, Vol.37, Supplement 1, S83, ISSN 0999-193X

Druet, T.; Fritz, S.; Boussaha, M.; Ben-Jemaa, S.; Guillaume, F.; Derbala, D.; Zelenika, D.; Lechner, D.; Charon, C.; Boichard, D.; Gut, I.G.; Eggen, A. & Gautier, M. (2008). Fine mapping of quantitative trait loci affecting female fertility in dairy cattle on BTA03 using a dense single-nucleotide polymorphism map. Genetics, Vol.178, No.4, (April 2008), pp. 2227-2235, ISSN 0016-6731

Eberlein, A.; Takasuga, A.; Setoguchi, K.; Pfuhl, R.; Flisikowski, K.; Fries, R.; Klopp, N.; Fürbass, R.; Weikard, R. & Kühn, C. (2009). Dissection of genetic factors modulating fetal growth in cattle indicates a substantial role of the non-SMC condensin I complex, subunit G (NCAPG) gene. Genetics Vol.183, No.3 (November 2009), pp. 951-964, ISSN 0016-6731

Elsen, J.M.; Mangin, B.; Goffinet, B., Boichard, D. & Le Roy, P. (1999). Alternative models for QTL detection in livestock. I. General introduction. Genetics Selection Evolution, Vol. 31, No.3, (May 1999), pp. 213-224, ISSN 0999-193X

Erhardt, G. (1989). Evidence for a third allele at the β-lactoglobulin (β-Lg) locus of sheep and its occurrence in different breeds. Animal Genetics, Vol.20, No.2, (June 1989), pp 197-204, ISSN 0268-9146

European Commission. (2003). 2003/100/EC Commission decision laying down minimum requirements for the establishment of breeding programs for resistance to transmissible spongiform encephalopathies in sheep. Official Journal L041: 41–45

Galloway, S.M.; Hanrahan, V.; Dodds, K.G.; Potts M.D.; Crawford, A.M. & Hill, D.F. (1996). A linkage map of the ovine X chromosome. Genome Research, Vol.6, No.8, (August 1996), pp. 667-677, ISSN 1088-9051

García-Fernández, M.; Gutiérrez-Gil, B.; García-Gámez, E.; Sánchez, J.P. & Arranz, J.J. (2010a). Detection of quantitative trait loci affecting the milk fatty acid profile on sheep chromosome 22: role of the stearoyl-CoA desaturase gene in Spanish Churra sheep. Journal of Dairy Science, Vol.93, No.1, (January 2010), pp. 348-357, ISSN 0022-0302

García-Fernández, M.; Gutiérrez-Gil, B.; García-Gámez, E.; Sánchez, J.P. & Arranz, J.J. (2010b). The identification of QTL that affect the fatty acid composition of milk on sheep chromosome 11. Animal Genetics, Vol.41, No.3, (June 2010), pp. 324-328. ISSN 0268-9146

García-Fernández, M.; Gutiérrez-Gil, B.; Sánchez, J.P.; Morán, J.A.; García-Gámez, E.; Alvarez, L. & Arranz JJ. (2011a). The role of bovine causal genes underlying dairy traits in Spanish Churra sheep. Animal Genetics, Vol.42, No.4, (August 2011), pp. 415-420, ISSN 0268-9146

García-Fernández, M.; Sánchez, J.P.; Gutiérrez-Gil, B.; García-Gámez, E. & Arranz, J.J. (2011b). Association between Acetyl-CoA Carboylase-α (ACACA) SNPs and milk fatty acid profile in Spanish Churra sheep, Proceedings of the 9th World Congress on Genetics Applied to Livestock Production (ISBN 978-3-00-031608-1), Communication 0196, Leipzig, Germany, 1-6 August 2010

García-Gámez, E.; Gutiérrez-Gil, B.; García-Fernández, M.; Sánchez, J.P.; de la Fuente, F.; San Primitivo, F. & Arranz, J.J. (2009). Confirmation of QTL underlying dairy production traits in Spanish Churra sheep, Proceedings of the XIII Jornadas sobre Producción Animal AIDA 2009, Tomo 1: 30-32

García-Gámez, E.; Gutiérrez-Gil, B.; Sánchez, J.P. & Arranz J.J. (2012). Short communication: replication and refinement of a QTL influencing milk protein percentage on ovine chromosome 3. Animal Genetics, in press (doi: 10.1111/j.1365-2052.2011.02294.x). ISSN 0268-9146

García-Gámez E.; Gutiérrez-Gil B.; Sahana G.; Sánchez JP. Bayón, Y., de La Fuente F. & Arranz JJ. (under review). Genome-Wide Association Analysis for milk production traits in Spanish Churra sheep and genetic support for a quantitative trait nucleotide influencing milk protein percentage in the LALBA gene.

Georges, M. (1998). Mapping genes underlying production traits in livestock, In: Animal Breeding: Technology for the 21st Century, pp: 77-101. (Ed. A. J. Clark). Harwood Academic Publishers. ISBN 9789057022920

Georges, M.; Nielsen, D.; Mackinnon. M.; Mishra, A.; Okimoto, R.; Pasquino, A.T.; Sargeant, L.S.; Sorensen, A.; Steele, M.R.; Zhao X.; Womack, J.E. & Hoeschele I. (1995). Mapping quantitative trait loci controlling milk production in dairy cattle by exploiting progeny testing. Genetics, Vol.139, No.2, (February 1995), pp. 907-920. ISSN 0016-6731

Grisart, B.; Coppieters, W.; Farnir, F.; Karim, L.; Ford, C.; Berzi, P.; Cambisano, N.; Mni, M.; Reid, S.; Simon, P.; Spelman, R.; Georges, M. & Snell R. (2002). Positional candidate cloning of a QTL in dairy cattle: identification of a missense mutation in the bovine DGAT1 gene with major effect on milk yield and composition. Genome research, Vol.12, No.2, (February 2002), pp. 222-231, ISSN 1088-9051

Gutiérrez-Gil, B.; El-Zarei, M.F.; Bayón, Y.; Alvarez, L.; de la Fuente, L.F.; San Primitivo, F. & Arranz J.J. (2007). Short communication: detection of quantitative trait loci influencing somatic cell score in Spanish Churra sheep. Journal of Dairy Science, Vol.91, No.1, (January 2007), pp. 422-426, ISSN 0022-0302

Gutiérrez-Gil, B.; El-Zarei, M.F.; Alvarez, L.; Bayón, Y.; de la Fuente, L.F.; San Primitivo, F. &
    Arranz, J.J. (2008). Quantitative trait loci underlying udder morphology traits in dairy
    sheep. Journal of Dairy Science, Vol.91, Vol.9, (September 2008), pp. 3672-3681, ISSN
    0022-0302
Gutiérrez-Gil, B.; El-Zarei, M.F.; Alvarez, L.; Bayón, Y.; de la Fuente, L.F.; San Primitivo, F. &
    Arranz, J.J. (2009a). Quantitative trait loci underlying milk production traits in sheep.
    Animal Genetics, Vol.40, No.4, (August 2009), pp. 423-434. ISSN 0268-9146
Gutiérrez-Gil, B.; Pérez, J.; Alvarez, L.; Martínez-Valladares, M.; de la Fuente, L.F.; Bayón,
    Y.; Meana, A.; San Primitivo, F.; Rojo-Vázquez, F.A. & Arranz J.J. (2009b). Quantitative
    trait loci for resistance to trichostrongylid infection in Spanish Churra sheep. Genetics
    Selection Evolution, Vol.28, (October 2009), pp. 46-55, ISSN 0999-193X
Gutiérrez-Gil, B.; Williams, J.L.; Homer, D.; Burton, D.; Haley, C.S. & Wiener, P. (2009c).
    Search for quantitative trait loci affecting growth and carcass traits in a cross population
    of beef and dairy cattle. Journal of Animal Science, Vol.87, No.1, (January 2009), pp. 24-
    36
Gutiérrez-Gil, B.; Pérez, J.; de la Fuente, L.F.; Meana, A.; Martínez-Valladares, M.; San
    Primitivo, F.; Rojo-Vázquez, F.A. & Arranz, J.J. (2010). Genetic parameters for resistance
    to trichostrongylid infection in dairy sheep. Animal, Vol.4, No.4, (April 2010), pp. 505-
    512, ISSN: 1751-7311
Gutiérrez-Gil, B.; Alvarez, L.; de la Fuente, L.F.; Sanchez, J.P.; San Primitivo, F. & Arranz J.J.
    (2011). A genome scan for quantitative trait loci affecting body conformation traits in
    Spanish Churra dairy sheep. Journal of Dairy Science, Vol.94. No.8, 2011 (August 2011),
    pp. 4119-4128, ISSN 0022-0302
Haley, C.S. & Knott, S.A. (1992). A simple regression method for mapping quantitative trait
    loci in line crosses using flanking markers. Heredity, Vol.69, No.4, (October 1992), pp.
    315-324, ISSN 0018-067X
Haley, C.S.; Knott, S.A. & Elsen, J.M. (1994). Mapping quantitative trait loci in crosses
    between outbred lines using least squares. Genetics, Vol.136, No.3, (March 1994), pp.
    1195-1207, ISSN 0016-6731
Hamada, H.; Seidman, M.; Howard, B.H. & Gorman, C.M. (1984). Enhanced gene expression
    by the Poly(dTdG)- Poly(dC-dA) sequence. Molecular and Cellular Biology, Vol.4,
    No.12, (December 1984), pp. 2622-2630, ISSN 02707306
Harmegnies, N.; Davin, F.; De Smet, S.; Buys, N.; Georges, M. & Coppieters, W. (2006).
    Results of a whole-genome quantitative trait locus scan for growth, carcass composition
    and meat quality in a porcine four-way cross. Animal Genetics Vol.37, No.6, (December
    2006), pp. 543-553, ISSN 0268-9146
Hunter, N.; Goldmann, W.; Foster, J.D.; Cairns, D. & Smith, G. (1997). Natural scrapie and
    PrP genotype: case-control studies in British sheep. Veterinary Research, Vol.141, No.6,
    (August 2007), pp. 137–140, ISSN 0928-4249
International Sheep Genomics Consortium; Archibald, A.L.; Cockett, N.E.; Dalrymple, B.P.;
    Faraut, T.; Kijas, J.W.; Maddox, J.F.; McEwan, J.C.; Hutton Oddy, V.; Raadsma, H.W.;
    Wade, C.; Wang, J; Wang, W. & Xun, X. (2010). The sheep genome reference sequence: a
    work in progress. Review. Animal Genetics, Vol.41, No.5, (October 2010), pp. 449-553,
    ISSN 0268-9146

Israel, C. & Weller, J.I. (1998). Estimation of candidate gene effects in dairy cattle populations. Journal of Dairy Science, Vol.81, No.6, (June 1998), pp. 1653-166, ISSN 0022-0302

Jaenisch, R. & Bird, A. (2003). Epigenetic regulation of gene expression: how the genome integrates intrinsic and environmental signals. Nature Genetics Vol.33, No.Suppl. (March 2003), pp. 245-254, ISSN 1061-4036

Jonas, E.; Thomson, P.C.; Hall, E.J.; McGill, D.; Lam, M.K. & Raadsma, H.W. (2011). Mapping quantitative trait loci (QTL) in sheep. IV. Analysis of lactation persistency and extended lactation traits in sheep. Genetics Selection Evolution, Vol.43, No.1, (June 2011), pp. 22, ISSN0999-193X

Khatkar, M.S.; Thomson, P.C.; Tammen, I. & Raadsma H.W. (2004). Quantitative trait loci mapping in dairy cattle: review and meta-analysis. Genetics Selection Evolution, Vol.36, No.2, (March 2004), pp. 163-190, ISSN 0999-193X

Knott, S.A. & Haley, C.S. (1992). Maximum likelihood mapping of quantitative trait loci using full-sib families. Genetics, Vol.132, No.4, (December 1992), pp. 1211-1222, ISSN 0016-6731

Knott, S.A.; Elsen, J.M. & Haley, C.S. (1996). Methods for multiple-marker mapping of quantitative trait loci in half-sib populations. Theoretical and applied genetics, Vol.93, No.1, (January 1996), pp. 71-80, ISSN 0040-5752

Labussière, J. (1988). Review of physiological and anatomical factors influencing the milking ability of ewes and the organization of milking. Livestock Production Science, Vol. 18, No.3, (June 1998), pp. 253–274, ISSN 0301-6226

Lander, E. & Kruglyak, L. (1995). Genetic dissection of complex traits: guidelines for interpreting and reporting linkage results. Nature Genetics, Vol.11, No.3, (November 1995), pp. 241-247, ISSN 1061-4036

Lander, E.S. & Botstein, D. (1989). Mapping mendelian factors underlying quantitative traits using RFLP linkage maps. Genetics, Vol.121, No.1, (January 1989), pp. 185-199, ISSN 0016-6731

Legarra, A. & Fernando, R.L. (2009). Linear models for joint association and linkage QTL mapping. Genetics Selection Evolution, Vol.41, (September 2009), pp.43, ISSN 0999-193X

Legarra, A. & Ugarte. E. (2001). Genetic parameters of milk traits in Latxa dairy sheep. Animal Science, Vol.73, No.3, (June 2001), pp. 407-412, ISSN 1357-7298

Legarra, A. & Ugarte, E. (2005). Genetic parameters of udder traits, somatic cell score, and MY in Latxa sheep. Journal of Dairy Science, Vol.88, No.6, (June 2005), pp. 2238-2245, ISSN 0022-0302

Lynch, M. & Walsh, B. (1998). Genetics and Analysis of Quantitative Traits. Sinauer Associates Inc. Publishers. Sunderlan, Massachusset, USA

Maddox, J.F.; Davies, K.P.; Crawford, A.M.; Hulme, D.J.; Vaiman, D.; Cribiu, E.P.; Freking, B.A.; Beh, K.J.; Cockett, N.E.; Kang, N.; Riffkin, C.D.; Drinkwater, R.; Moore, S.S.; Dodds, K.G.; Lumsden, J.M.; van Stijn, T.C.; Phua, S.H.; Adelson, D.L.; Burkin, H.R.; Broom, J.E.; Buitkamp, J.; Cambridge, L.; Cushwa, W.T.; Gerard, E.; Galloway, S.M.; Harrison, B.; Hawken, R.J.; Hiendleder, S.; Henry, H.M.; Medrano, J.F.; Paterson, K.A.; Schibler, L.; Stone, R.T. & van Hest, B. (2001). An enhanced linkage map of the sheep genome comprising more than 1000 loci. Genome Research, Vol.11, No.7, (July 2001), pp. 125-1289, ISSN 1088-9051

Maddox, J.F. (2005). A presentation of the differences between sheep and goats. Genetics Selection Evolution, Vol.37, Suppl.1 (2005), pp. S1-S10, ISSN 0999-193X

Mai, M.D.; Sahana, G.; Christiansen, F.B. & Guldbrandtsen, B. (2010). A genome-wide association study for milk production traits in Danish Jersey cattle using a 50K single nucleotide polymorphism chip. Journal of Animal Science, Vol.88, No.11, (November 2010), pp. 3522-3528, ISSN 0021-8812

Mateescu R,G. & Thonney, M,L. (2010). Genetic mapping of quantitative trait loci for milk production in sheep. Animal Genetics, Vol.41, No.5, (October 2010), pp. 460-466, ISSN 0268-9146

Moioli, B.; D'Andrea, M. & Pilla F. (2007). Candidate genes affecting sheep and goat milk quality. Small Ruminant Research, Vol.68, No.1 (March 2007), pp.179–192, ISSN 0921-4488

Moreno, C.R.; Gruner, L.; Scala, A.; Mura, L.; Schibler, L.; Amigues, Y.; Sechi, T.; Jacquiet, P.; François, D.; Sechi, S.; Roig, A.; Casu, S.; Barillet, F.; Brunel, J.C.; Bouix, J., Carta, A., Rupp, R. (2006). QTL for resistance to internal parasites in two designs based on natural and experimental conditions of infection, Proceedings of the 8th World Congress on Genetics Applied to Livestock Production, Communication 15-05, 2006-08-13-2006-08-1, Belo Horizonte, Brasil, August 13-18, 2006

Moreno, C.R.; Cosseddu, G.M.; Schibler, L.; Roig, A.; Moazami-Goudarzi, K.; Andreoletti, O.; Eychenne, F.; Lajous, D.; Schelcher, F.; Cribiu, E.P.; Laurent, P.; Vaiman, D.; & Elsen, J.M. (2008). Identification of new quantitative trait Loci (other than the PRNP gene) modulating the scrapie incubation period in sheep. Genetics, Vol.179, No.1, (May 2008), pp.723-726, ISSN 0016-6731

Moreno, C.R.; Moazami-Goudarzi, K.; Briand, S.; Robert-Granié, C.; Weisbecker, J.L.; Laurent, P.; Cribiu, E.P.; Haley, C.S.; Andréoletti, O., Bishop, S.C. & Pong-Wong, R. (2010). Mapping of quantitative trait loci affecting classical scrapie incubation time in a population comprising several generations of scrapie-infected sheep. Journal of General Virology, Vol.91, No.2, (February 2010), pp. 575-579, ISSN 0022-1317

Mullis, K.; Faloona, F.; Scharf, S.; Saiki, R.; Horn, G. & Erlich, H. (1986). Specific enzymatic amplification of DNA in vitro: the polymerase chain reaction. Cold Spring Harbor Symposium in Quantitative Biology, Vol.51, No.1, (1986), pp. 263-273, ISSN 0091-7451

Neimann-Sorensen, A. & Robertson, A. (1961). The association between blood groups and several production characteristics in three Danish cattle breeds. Acta Agriculturae Scandinavica, Vol.11, No.2, (January 1961), pp. 163-196, ISSN 0906-4702

Olsen, H.G.; Nilsen, H.; Hayes, B.; Berg, P.R.; Svendsen; M.; Lien, S. & Meuwissen T. (2007). Genetic support for a quantitative trait nucleotide in the ABCG2 gene affecting milk composition of dairy cattle. BMC Genetics Vol.21, (June 2007), pp. 32, ISSN 1471-2156

Ostergard, H.; Kristensen, B. & Andersen, S. (1989). Investigation in farm animals of associations between the MHC system and disease resistance and fertility. Livestock Production Science, Vol.22, No.1, (May 1989), pp. 49–67, ISSN 1871-1413

Othmane, M.H.; De La Fuente, L.F.; Carriedo J.A. & San Primitivo F. (2002). Heritability and genetic correlations of test day milk yield and composition, individual laboratory cheese yield, and somatic cell count for dairy ewes. Journal of Dairy Science Vol.85, No.10, (October 2002), pp. 2692-2698, ISSN 0022-0302

Piredda, G.; Papoff, C.M.; Sanna S.R. & Campus, R.L. (1993). Influence of ovine αs1- casein genotype on the physicochemical and lactodynamographic characteristics of milk. Scienza e Tecnica Lattiero-Casearia, Vol.44, (1993), pp 135–143, ISSN 0390-6361

Pirisi, A.; Piredda, G.; Papoff, C.M.; Di Salvo, R.; Pintus, S.; Garro, G.; Ferranti, P. & Chianese, L. (1999). Effects of sheep αs1-casein CC, CD and DD genotypes on milk composition and cheesemaking properties, Journal of Dairy Research, Vol.66, No.3, (August 1999), pp. 409–419, ISSN 0022-0299

Raadsma, H.W.; Nicholas, F. W.; Jones, M.; Attard, G.; Palmer, D.; Abbott, K. & Grant, T. (1999). QTL mapping for production traits in a multi-stage Awassi x Merino back-inter-cross design, Proceedings of the Association for the Advancement of Animal Breeding and Genetics, Vol.13, pp. 361-364, Bunbury, Western Australia

Raadsma, H.W., Jonas, E.; McGill, D.; Hobbs, M.; Lam, M.K. & Thomson, PC. (2009). Mapping quantitative trait loci (QTL) in sheep. II. Meta-assembly and identification of novel QTL for milk production traits in sheep. Genetics Selection Evolution, Vol.41, No.1, (October 2009), pp.45, ISNN 0999-193X

Rendo, F.; Ugarte, E.; Lipkin, E. & Estonba, A. (2003). Detection of QTLs influencing milk produciton in OAR6 of the Latxa breed, Proceedings of the International Workshop on Major Genes and QTL in Sheep and Goat, CD-ROM Communication nº2-22. Toulouse, France, 8-11 December 2003

Rupp, R.; Lagriffoul, G.; Astruc, J.M. & Barillet, F. (2003). Genetic parameters for milk somatic cell scores and relationships with production traits in French Lacaune dairy sheep. Journal of Dairy Science, Vol.86, No.4, (April 2003), pp. 1476-1481, ISSN 0022-0302

Sanna, S.R.; Carta A. & Casu S. (1997). (Co)variance component estimates for milk composition traits in Sarda dairy sheep using a bivariate animal model. Small Ruminant Research, Vol.25, No.1, (May 1997), pp. 77–82, ISSN 0921-4488

Sax, K. (1923). The association of size differences with seed-coat pattern and pigmentation in Phaseolus vulgaris. Genetics, Vol.8, No.6, (November 1923), pp. 552-560, ISSN 0016-6731

Scatà, M.C.; Napolitano, F.; Casu, S.; Carta, A.; De Matteis, G.; Signorelli, F.; Annicchiarico, G., Catillo, G. & Moioli, B. (2009). Ovine acyl CoA:diacylglycerol acyltransferase 1-molecular characterization, polymorphisms and association with milk traits. Animal Genetics, Vol.40, No.5 (October 2009), pp. 737-742, ISSN 0268-9146

Schibler, L.; Roig, A.; Neau, A.; Amigues, Y.; Boscher, M.Y.; Cribiu, E.P.; Boichard, D.; Rupp, R. & Barillet F. (2002). Detection of QTL affecting milk production or somatic cell score in three French dairy sheep breeds by partial genotyping on seven chromosomes, Proceedings of the 7th World Congress on Genetics Applied to Livestock Production, Vol.29, pp. 215–218, ISBN 2738010520, Montpellier, France, 19-23 August, 2002

Schnabel, R.D.; Kim, J.J.; Ashwell, M.S.; Sonstegard, T.S.; Van Tassell, C.P.; Connor, E.E. & Taylor, J.F. (2005). Fine-mapping milk production quantitative trait loci on BTA6: analysis of the bovine osteopontin gene. Proceedings of the National Academy of Sciences of the United States of America, Vol.102, No.19, (May 2005), pp. 6896-6901, ISSN 0027-8424

Schopen, G.C.; Visker, M.H.; Koks, P.D.; Mullaart, E.; van Arendonk, J.A. & Bovenhuis, H. (2011). Whole-genome association study for milk protein composition in dairy cattle. Journal of Dairy Science, Vol.94, No.6, (June 2011), pp. 3148-3158, ISSN 0022-0302

Seaton, G.; Haley, C.S.; Knott, S.A.; Kearsey, M. & Visscher, P.M. (2002). QTL EXPRESS: mapping quantitative trait loci in simple and complex pedigrees. Bioinformatics Vol.18, No.2, (February 2002), pp. 339–340, ISSN 1460-2059

Seaton, G.; Hernandez, J.; Grunchec, J.A.; White, I.; Allen, J.; De Koning, D.J.; Wei, W.; Berry, D.; Haley, C. & Knott S. (2006). GridQTL: A Grid Portal for QTL Mapping of Compute Intensive Datasets, Proceedings of the 8th World Congress on Genetics Applied to Livestock Production, Belo Horizonte, Brazil, August 13-18, 2006

Serrano, M.; Pérez-Guzmán M. D.; Montoro, V. & Jurado, J. J. (1996). Genetic parameters estimation and selection progress for milk yield in Manchega sheep. Small Ruminant Research, Vol.23, No.1 (November 1996), pp. 51–57, ISSN 0921-4488

Shaw, R. (1987). Maximum likelihood approaches applied to quantitative genetics of natural populations. Evolution, Vol.41, No.4, (April 1987), pp. 812–826, ISSN 0014-3820

Sheehy, P.A.; Riley, L.G.; Raadsma, H.W.; Williamson, P. & Wynn, PC. (2009). A functional genomics approach to evaluate candidate genes located in a QTL interval for milk production traits on BTA6. Animal Genetics, Vol.40, No.4, (August 2009), pp. 492-498, ISSN 0268-9146

Singh, M.; Lam, M.; McGill, D.; Thomson, P.C.; Cavanagh, J.A.; Zenger, K.R. & Raadsma, H.W. (2007). High resolution mapping of quantitative trait loci on ovine chromosome 3 and 20 affecting protein yield and lactation persistency. Proceedings of the Seventeenth Conference of the Association for the Advancement of Animal Breeding and Genetics, Armidale, New South Wales, Australia, 2007, pp. 565-568

Smaragdov, M.G.; Prinzenberg, E.M. & Zwierzchowski, L. (2006). QTL mapping in cattle: theoretical and empirical approach. Animal Science Papers and Reports Vol.24, No.2, (April 2006), pp. 69-110, ISSN 08604037

Soller, M. & Brody, T. (1976). On the power of experimental designs for the detection of linkage marker loci and quantitative loci in crosses between inbred lines. Theoretical and Applied Genetics, Vol.1, No.47, (January 1976), pp. 35-59, ISSN 0040-5752

Vanraden P.M. & Wiggans G.R. (1991). Derivation, calculation, and use of national animal model information. Journal of Dairy Science Vol.74, No.8, (August 1991), pp. 2737-2746, ISSN 0022-0302

Visscher P.M.; Thompson R. & Haley C.S. (1996). Confidence intervals in QTL mapping by bootstrapping. Genetics, Vol.143, No.2, (June 1996), pp.1013-1020, ISSN 0016-6731

Vukasinovic, N.; Moll J. & Künzi, N. (1995). Genetic relationships among longevity, milk production, and type traits in Swiss Brown cattle. Livestock Production Science, Vol.41, No.1, (January 1995), pp. 11–18, ISSN 0301-6226

Weber, J.L. & May, P.E. (1989). Abundant class of Human DNA polymorphisms which can be typed using the polymerase chain reaction. American Journal of Human Genetics, Vol.44, No.2, (March 1989), pp. 388-396, ISSN 0002-9297

Weller J.I.; Kashi Y. & Soller M. (1990). Power of daughter and granddaughter designs for determining linkage between marker loci and quantitative trait loci in dairy cattle. Journal of Dairy Science, Vol.73, No.9, (September 1990), pp. 2525-2537, ISSN 0022-0302

Chapter 6

Milk Protein Genotype Associations with Milk Coagulation and Quality Traits

Elli Pärna, Tanel Kaart, Heli Kiiman, Tanel Bulitko and Haldja Viinalass

Additional information is available at the end of the chapter

1. Introduction

Cheese production is of substantial economic importance in most European countries, where an increasing amount of the produced milk is used for manufacturing cheese (Eurostat 2010a). In Estonia 60% (Statistics Estonia, 2010), in Italy more than 75% (De Marchi et al., 2008), and in Scandinavian countries 33% (Wedholm et al., 2006) of milk is used for cheese production. Recent trends indicate that per capita consumption of cheese is also increasing (Eurostat 2010b).

Milk quality is an essential factor to the dairy industry due to its economic impact. Milk coagulation ability is one of the most important factors affecting cheese yield and quality and has been reviewed (Jakob & Puhan, 1992; Johnson et al., 2001), and therefore is becoming more important as an increasing percentage of milk is used for cheese manufacturing. Milk coagulation properties (MCP) are commonly defined by milk coagulation time (RCT) and curd firmness (A_{30}). It is feasible to design raw milk according to its specific technological use.

Improving cheese yield and quality, through the direct selection of breeding animals on the basis of milk coagulation property traits, is an option due to genetic variation of MCP traits (Ojala et al., 2005). MCP are heritable, quantitative traits; up to 40% of the variation among animals is caused by genetic factors (Ikonen et al., 2004). Estimates of heritability for MCP traits are from 0.30 to 0.40 (Bittante et al., 2002; Ikonen et al., 1999a), and from 0.25 for RCT , and 0.15 for A_{30} (Cassandro et al., 2008) to 0.28 for RCT to 0.41 for A_{30} (Vallas et al., 2010). Predictions of MCP provided by mid-infrared spectroscopy (MIR) techniques have been proposed as indicator traits for the genetic enhancement of MCP (Cecchinato et al., 2009; De Marchi et al., 2009). The expected response of RCT and A_{30} ensured by the selection using MIR predictions as indicator traits was equal to, or slightly less than, the response achievable through a single measurement of these traits. Accordingly, breeding strategies for the enhancement of MCP based on MIR predictions as indicator traits could be easily and immediately implemented for dairy cattle populations where the routine acquisition of spectra from individual milk samples is already measured (Cecchinato et al., 2009). Nevertheless, MCP traits analyzed with different methodologies have significantly different values, due to the diversity of the instruments used and the coagulant activity (Pretto et al., 2011). The type of coagulant could also have an effect, since different coagulants have been used. The method proposed for the prediction of non-coagulation probability of milk samples showed that non coagulating samples from one methodology were highly predictable based on the rennet coagulation time measured with another methodology (Pretto et al., 2011). A standard definition of MCP traits analysis is needed to enable reliable comparisons between MCP traits recorded in different laboratories, and in different animal populations and breeds.

More than 95% of the proteins contained in ruminant milk are coded by six structural genes (Martin et al., 2002). The four casein genes (*CSN1S1, CSN2, CSN1S2,* and *CSN3*) are linked in a cluster, referred to as the CN locus, mapped on chromosome 6 and encode α_{s1}-CN, β-CN, α_{s2}-CN, and κ-CN, respectively, as previously reviewed (Caroli et al., 2009). κ-CN is a key element in renneting, but interactions with the other milk protein systems have to be taken into account, in particular, β-CN and β-LG. Monitoring milk protein variation in different breeds of cattle avoids an increase of alleles with unfavourable effects on cheesemaking (Caroli et al., 2000; Comin et al., 2008; Erhardt et al., 1997; Ikonen et al., 1994, 1999a; Lodes et al., 1996). Therefore, another option for enchancing cheese yield and quality is indirect selection against, or for, some milk protein alleles. Selection against κ-CN E-allele would be a good means to indirectly improve milk quality for cheese production because the E-allele is unfavourably associated with non-coagulating milk, which is common (10%) in Finnish Ayrshire cows (Elo et al., 2007). Likewise the κ-CN G-allele in the Pinzgauer breed (Erhardt et al., 1997) and the κ-CN E-allele in the Italian Friesian breed (Caroli et al., 2000) were both associated with unfavourable coagulation properties. As for a positive association, it is well known that the B variants of β- and k-casein and β-lactoglobulin (β-LG) are favourable for milk coagulation and cheese-making (Dovc & Buchberger, 2000; Losi et al., 1973; Patil et al., 2003; Schaar et al., 1985; Walsh et al., 1998). k-CN allelic variants have been associated with variation in total casein and k-CN concentrations in milk (Hallén et al., 2008; Van den Berg et al., 1992), variation in casein micelle size (Di Stasio & Mariani, 2000; Walsh et al., 1998) and differences in coagulating properties and the cheesemaking quality of milk (Di Stasio & Mariani, 2000). Genetic variants of β-LG have been shown to have an indirect effect on cheese yield through their effect on the ratio of casein to total protein (Coulon et al., 1998; Lundén et al., 1997; McLean, 1986).

In the Dutch Holstein-Friesian population selection for *CSN2-CSN3* haplotype *A*₂*-B*, together with *LGB* B, would result in cows that produce milk more suitable for cheesemaking (Heck et al., 2009). It has also been argued that haplotypes have similar effects in the different breeds and the CN genes themselves were responsible for the haplotype effects observed, rather than genes physically linked to the CN complex (Boettcher et al., 2004).

The objective of this study was to estimate the contribution of the aggregate β-κ-CN and β-LG genotypes on first lactation milk coagulation and quality traits of Estonian Holstein cows. A parallel objective was to identify the variation in genetic polymorphism of milk proteins with the aim to improve the protein composition in milk by selecting for variants of specific genes.

2. Material and methods

2.1. Performance of Estonian cattle populations

In Estonia there are three breeds of dairy cattle – Estonian Red (ER), Estonian Holstein (EHF) and Estonian Native (EN). Distribution of breeds has come to favour EHF (Fig.1). In Estonia, 93.0% of cows are enrolled in an official milk recording programme (Results of Animal Recording in Estonia 2010, 2011) (Fig. 2). Since 1995, average milk yield in Estonia has risen 3947 kg (48%, Fig. 3). The 2010 average actual production for Estonian Holstein herd that were enrolled in producing-testing programs and eligible for genetic evaluation was 7778 kg milk, 317 kg of fat and 260 kg of protein per year. Holstein dairy cattle dominate in Estonian milk production industry because of their excellent production and greater income.

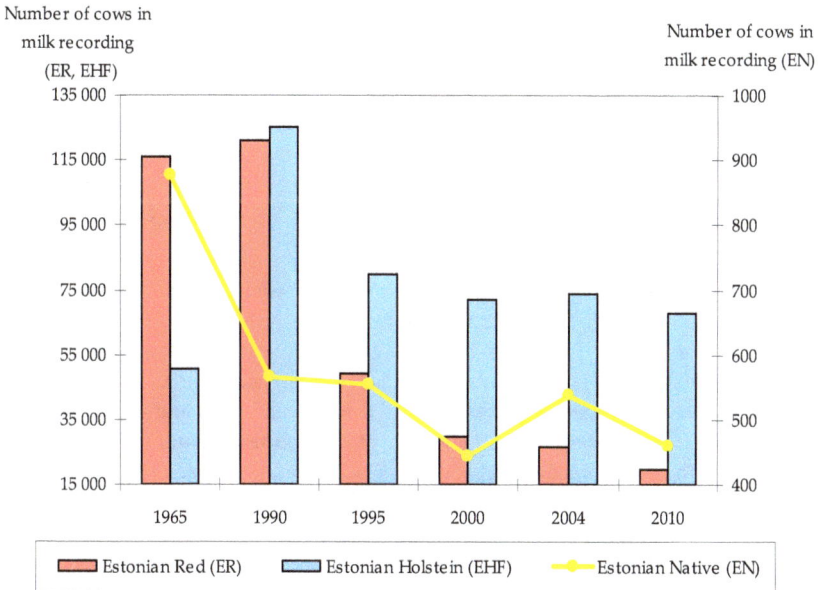

Figure 1. Changes over time in number of cows of each indicated breed on a milk recording programme

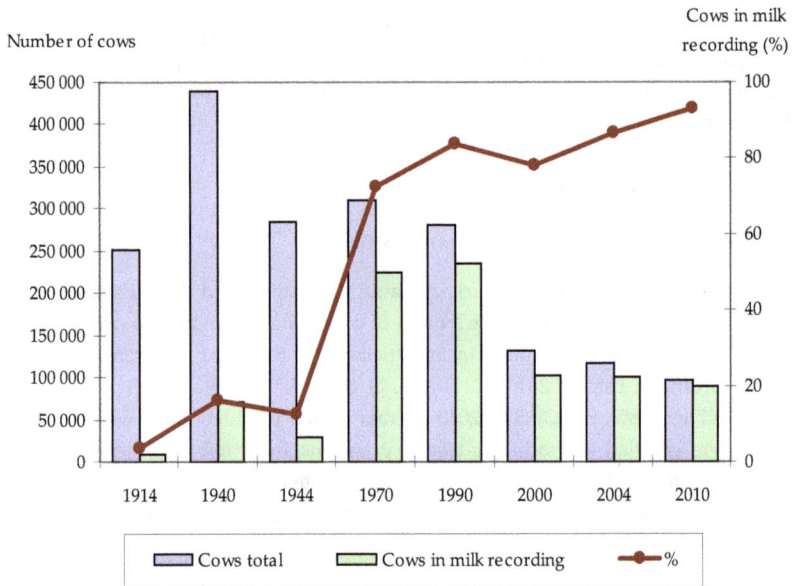

Figure 2. Changes over time in number of Estonian dairy cows and the proportion on a milk recording programme

In the evaluation programme for young bulls of Estonian Holstein breed, ca 25 bulls are tested each year, parallel testing is carried out on 10-12 foreign bulls. The selection of bulls is made from imported American and Canadian embryos, the Estonian Holstein best bull dams and imported young bulls. Often the sons of imported cows are used whose sires are world-known top bulls. The bulls come mainly from the USA, Canada, Germany and the Netherlands.

Currently, estimated breeding values (EBV) for production, conformation and udder health traits for bulls and cows in Estonia are computed by the Animal Recording Centre four times per year (Pentjärv & Uba, 2004). Breeding value estimation is carried out separately for the EHF and the Estonian Red breed (ER), using the best linear unbiased predictor (BLUP) test day animal model for production and udder health traits and the BLUP animal model for conformation traits. The EBV for each production trait – milk (kg), fat (kg) and protein (kg) – is the mean breeding value of the first, second and third lactations, adjusted by the mean average breeding value of the cows born in a defined base year (currently, 1995).

The milk production index (SPAV) is expressed as relative breeding value (RBV) with a mean of 100 and a standard deviation of 12 points for base animals, combining breeding values for milk, fat and protein yield weighted by relative economic values of 0:1:4 for EHF and 0:1:6 for ER (Pentjärv & Uba, 2004).

Milk production (kg)

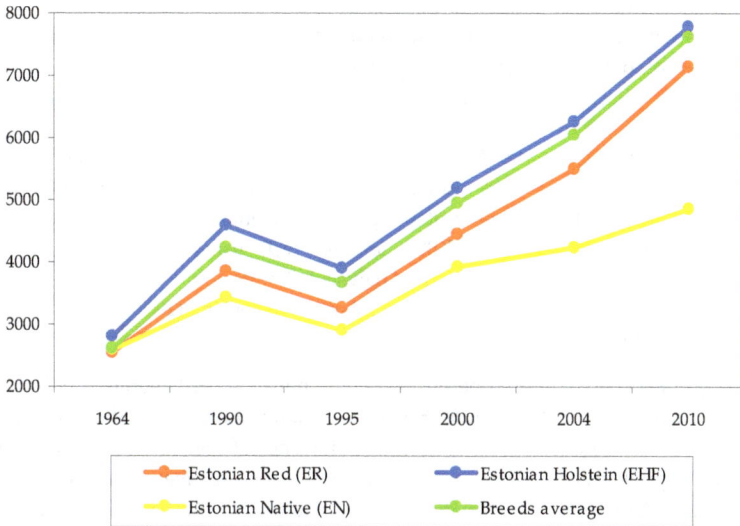

Figure 3. Changes over time in annual milk yield per cow of each indicated breed

The information source for breeding value estimation of udder health traits is somatic cell count (SCC) in one millilitre of milk, transformed into a somatic cell score (SCS) using the formula $SCS = \log_2 (SCC/100000) + 3$ (Pentjärv & Uba, 2004).

The udder health index SSAV is calculated as the sum of EBVs of the first, second and third lactations with index weights 0.26, 0.37 and 0.37, respectively, and is expressed as RBV for genetic evaluation of conformation traits. Data from first lactation cows are used to compute RBVs for 16 linear traits for EHF and 14 linear traits for ER, as well as for three general traits. The conformation index SVAV is expressed as RBV, combining relative breeding values for type, udder and feet by relative economic weights of 0.3:0.5:0.2 for ER and 0.3:0.4:0.3 for EHF.

2.2. Data collection and laboratory milk analysis

First lactation milk samples were collected during routine milk recording as part of a development project for the Bio-Competence Centre of Healthy Dairy Products in Estonia during the period April 2005 – June 2008. The herds had twice-a-day or thrice-a-day milkings. The individual milk samples collected from the cows were either bulked test-day milkings or separate samples from each of the milkings on the test-day. Milk samples were immediately preserved after collection with Bronopol (2-bromo-2-nitropropane-1,3-diol, Knoll Pharmaceuticals, Nottingham, UK) and stored at 4°C during transportation and analyzing periods. Milk samples with a pH lower than 6.5, indicative of colostrum, and non-coagulated milk samples (n=33) were excluded from the analysis. Furthermore, farms with less than 10 cows, and cows with fewer than three test-day records, were removed. The final dataset used for analyses consist of 11,437 test-day records from 2,769 Estonian Holstein cows which were located in 66 herds across the country and were daughters of 169 sires. The number of daughters per sire ranged from 1 to 267. Each cow had from 3 – 6 measurements collected during the different stages (7 – 305 DIM) of the first lactation. Information about the cows, herds and pedigree was obtained from the Estonian Animal Recording Centre (EARC), the Animal Breeders' Association of Estonia and a database, COAGEN®, was produced. The test-day milk yield was recorded and individual milk samples were analyzed for fat percentage, protein percentage and urea content using the MilkoScan 4000 and MilcoScan FT6000, and for SCC using the Fossomatic 4000 and Fossomatic 5000 cell counter at the Milk Analysis Laboratory of the EARC, using methods suggested by the International Committee for Animal Recording (2009). Values of SCC were log-transformed to SCS.

The pH and milk coagulation properties were determined at the Laboratory of Milk Quality of the Estonian University of Life Sciences, usually three days after sampling. The proportion of milk samples with a maximum storage age of seven days was very small, less than 1%. The pH level of the milk was determined using a pH meter (Seven Multi; Mettler Toledo GmbH, Greifensee, Switzerland) at a temperature of 20°C before analyzing the milk coagulation properties. The latter were milk coagulation time in minutes and firmness of curd in volts. Prior to the assessment of the milk coagulation properties, milk samples were heated to the renneting temperature (35°C). The rennet (Milase MRS 750 IMCU/ml; CSK Food Enrichment B.V., The Netherlands) used in the analyses was diluted 1:100 (v/v) with distilled water and 0.2 ml of the solution was added to 10 ml milk. The milk coagulation properties were determined using the Optigraph (Ysebaert, Frepillon, France), which was developed by YDD (Ysebaert Dairy Division) in partnership with the INRA (LGMPA, lab. G. CORRIEU) to define coagulation characteristics in the laboratory, specifically to answer the needs of cheese makers (Ysebaert Dairy Division, 2009).

Measurements made with the Optigraph are not based on a rheological method but on an optical signal in the near-infrared spectrum. During a coagulation test, the light transmitted through the milk gradually weakens because of changes in the micellar structure of casein. The Optigraph then calculates the coagulation parameters (coagulation time, curd firmness, speed of aggregation) by means of particular feature points extracted from the optical information acquired in real time (Optigraph User's Manual).

2.3. Data collection and laboratory blood analysis

Blood samples were collected as part of a development project for the Bio-Competence Centre of Healthy Dairy Products in Estonia during the period of June 2005 to December 2007. Blood samples (n=2,959) were stored in tubes containing K_3EDTA. DNA was extracted from whole blood according to the method described by (Miller et al., 1998) or by using a commercial Puregene Gentra Blood kit (Minneapolis, USA). The quantity of template DNA was approximately 40 to 100 ng for Allele-specific oligonucleotide (ASO) PCR and PCR-RFLP, respectively. Polymorphisms of five milk protein genes were analyzed, four from the casein cluster *(CSN1S1, csn1s2, CSN2, CSN3)* and *LGB*. The list of single nucleotide polymorphisms (SNP) previously reported by (Chessa et al., 2007) was considered to distinguish genetic variants of milk proteins of the sampled cows. ASO primers were designed for the detection of polymorphisms in the *CSN2* (primer sequences in (Värv et al., 2009)) and *CSN1S1* (present study). *CSN1S1* genotyping included amplification of a 155-bp sequence of the gene at exon 17. In accordance with the SNP a26181g in *CSN1S1*, two specific forward oligos were designed to distinguish B and C alleles paired with one reverse primer. An extra mismatch was added to both forward primers at position 2 at the 3' end. Restriction analysis was carried out to genotype *CSN1S2* (Ibeagha-Awemu et al., 2007), *CSN3* (Velmala et al., 1993) and *LGB* (Medrano & Aquilar-Cordova, 1990). In this study, amplified regions of *CSN1S2* with a digestion site to discriminate the protein variants from A-allele, were 330 bp (114 and 216 bp after restriction with *Mbo*II to detect B-allele), 354 bp *(Nla*IV restriction fragments 211 and 143bp to detect C-allele) and 356 bp (*Mnl*I restriction fragments 160 and 196 bp for non-D allele). Products of allele-specific PCR and digestion fragments were separated on agarose gel. Sequencing to verify the studied DNA regions was performed with a BigDye Terminator v3.1 Cycle Sequencing Kit (Applied Biosystems, USA) and analyzed using an ABI Prism 3130 Genetic Analyzer (Applied Biosystems, USA).

2.4. Statistical analyses

Preliminary analyses for testing the significance of fixed effects and single genotype effects were carried out on the SAS System® (SAS, Cary, NC, USA) using the MIXED procedure. Aggregate β-κ-CN genotypes were formed for further analysis. The genotypes with relative frequencies of less than 1% were grouped together into rare β-κ-CN genotype (A$_1$A$_1$-BB, A$_1$A$_1$-BE, A$_1$A$_1$-EE, A$_1$B-BB, A$_1$B-BE, A$_2$A$_2$-AE, A$_2$A$_2$-BE, A$_2$A$_3$-AA, A$_2$B-AA, A$_2$B-BB, BB-AB, BB-BB). Further statistical analysis was carried out using ASReml (VSN International Ltd., Hemel Hempstead, UK), using the following univariate repeatability animal model:

$$y = Xb + Cg + Za + e,$$

where **y** – vector of observations of the dependent variable (log-transformed RCT, A$_{30}$, milk yield, milk protein percentage, fat percentage, SCS); **b** – vector of fixed effects (quadratic polynomial of day in milk, calving age, sample age, sampling year-season, calving year-season); **g** – fixed effects of the β-κ-CN genotypes or β-LG genotypes; **a** – vector of random effects (herd $N(0, I\sigma_h^2)$, additive genetic $N(0, A\sigma_a^2)$ and permanent environmental effect $N(0, I\sigma_{pe}^2)$); **e** – vector of residual random error effects $N(0, I\sigma_e^2)$; **X, C, Z** – known incidence matrices for fixed, genotype and random effects, respectively.

Sample age was included as a covariate in the model only for milk coagulation traits. Sampling year-season and calving year-season were grouped into 3-month classes, 14 classes from April 2005 to June 2008 and 11 classes from December 2004 to August 2007, respectively. Three generations of ancestors with a total number of 17,185 animals in the relationship matrix were included in the analysis.

The class with the largest number of observations, genotype A2A2-AA was used as the class of comparison. It is also homozygous for both loci. Accordingly, the standard errors of the genotype effects are standard errors of the differences between each genotype and the most frequent A2A2-AA genotype (Comin et al., 2008).

3. Results and discussion

3.1. Genetic structure of milk proteins

Allelic variants of casein and *LGB* loci and genotypes are presented in Tables 1 and 2. Four allelic variants of *CSN2*, three allelic variants of *CSN3* and two variants of the *LGB* gene in Estonian Holstein population were detected. *CSN2* occured at a significantly higher frequency for the A2 allele. The B-allele and A3 allele at *CSN2* were rare. The further scan of β-CN, to discriminate protein variant I from variant A2 performed in the Dutch Holstein Friesian population revealed frequency of I allele of 0.14 for proven bulls, 0.27 in young bulls and 0.192 in cows and demonstrated that it is actually one of the common variants for the Holstein population (Visker et al., 2011). This frequency of the β-CN protein variant in Dutch Holstein Friesian population is relatively high compared with the frequencies of other cattle breeds. The Italian Holstein's frequency of the I allele was 0.12 (Jann et al., 2002), while a survey of 30 cattle breeds yielded frequencies up to 0.14 (Jann et al., 2004). As pointed out (Visker et al., 2011), the associations of β-CN protein variant A2 should still interpreted with care, because they may have been combined associations of protein variants A2 and I, which are very different in some traits. Therefore further investigation is needed for Estonian Holstein β-CN I allele. The mean frequency of the β-CN A2 allele in Estonian Holstein was 0.647 (Table 1), ranging in farms in Estonia from 0.35 to 0.73. The mean frequency of the β-CN A1 allele was 0.320, ranging in farms in of Estonia from 0.19 to 0.50. At the *CSN2* locus the A2 allele was highly predominant over other alleles. The heterozygous genotype A1A2 and the homozygous genotype A2A2 at β-CN were represented at almost the same frequencies (0.434 and 0.411, respectively). The most common β-CN genotypes A1A2 and A2A2 frequencies comprised jointly 84%. The occurrence of milk protein genetic variants revealed somewhat higher frequencies of the β-CN A2 allele and the homozygotic genotype A2A2 (Table 1 and 2) in the Estonian Holstein population in comparison with the common European cattle breeds (European Food Safety Authority, 2004). In some large Estonian farms the β-CN A2 allele mean frequency was about 0.70 and homozygotic genotype mean frequency about 0.50.

Protein	Allele	n	Frequency	Protein	Allele	n	Frequency
κ-CN	A	4,356	0.737	α_{s1}-CN	B	5,446	0.983
	B	1,163	0.197		C	92	0.017
	E	391	0.066	α_{s2}-CN	A	5,329	0.962
β-CN	A1	1,888	0.320		B	164	0.030
	A2	3,818	0.647		D	45	0.008
	A3	5	0.001	β-LG	A	2,947	0.498
	B	191	0.032		B	2,967	0.502

Table 1. Allele frequencies of κ-CN, β-CN, α_{s1}-CN, α_{s2}-CN and β-LG in cows of the Estonian Holstein breed

This finding, of a high frequency of the A_2 allele, confirms the advantage of the Estonian Holstein breed that their milk naturally might lower health risks associated with the occurence of the β-CN A_1 allele. The other advantage of the β-CN A_2 allele is its positive association with protein yield (Olenski et al., 2010). The positive effect of the rare A_2A_2-BB genotype, that is the A_2-B haplotype, on milk and protein yields has been reported in previous studies on Californian Holstein (Ojala et al., 1997), Finnish Ayrshire (Ikonen et al., 2001) and Dutch Holstein-Friesian cows (Heck et al., 2009).

Protein	Genotype	n	Frequency	Protein	Genotype	n	Frequency
κ-CN	AA	1,606	0.544	$α_{s1}$-CN	BB	2,877	0.967
	AB	850	0.288		BC	92	0.033
	AE	294	0.100	$α_{s2}$-CN	AA	2,764	0.926
	BB	116	0.039		AB	159	0.057
	BE	81	0.027		AD	42	0.015
	EE	8	0.003		BB	1	0.001
β-CN	A_1A_1	270	0.092		BD	3	0.001
	A_1A_2	1,282	0.434	β-LG	AA	631	0.214
	A^1B	66	0.022		AB	1,685	0.570
	A_2A_2	1,212	0.411		BB	641	0.216
	A_2A_3	5	0.002				
	A_2B	107	0.036				
	BB	9	0.003				

Table 2. Genotype frequencies of κ-CN, β-CN, $α_{s1}$-CN, $α_{s2}$-CN and β-LG in cows of the Estonian Holstein breed

CSN3 had A, B and E allelic variants (Table 1 and 2). κ-CN shows a prevalence of the A allele at a mean frequency of 0.737, followed by the B allele at frequency of 0.197 and the E allele at frequency of 0.066. The most frequent κ-CN genotype of all genotyped Estonian Holstein cows was AA, which was found in slightly more than half of the cows (54.4%), followed by AB (28.8%) and AE (10%). A favourable genetic marker for protein yield, MCP and cheese production, κ-CN B, was rare (3.9% overall) in the homozygous state. The unfavourable κ-CN E allele was also very rare, only eight of the 2,954 sampled cows had the EE genotype.

CSN1S1 had B and C allelic variants (Table 1 and 2), showing the prevalence of the B allele (0.983) and CSN1S2 revealed three allelic variants: A, B, and D, showing the prevalence of the A allele (0.962). Because $α_{s1}$-CN and $α_{s2}$-CN were almost monomorphic (Table 1), they were excluded from the aggregate casein genotypes.

β-LG was represented with two allelic variants A (0.498) and B (0.502). Comparing the results of the genetic structure of milk proteins of the Estonian Holstein breed of Jõudu, 2008 and Värv et al., 2009 to those of this study, the quantity of detected alleles, and their frequencies, are somewhat different. The reason for this could be attributed to the sampling capacity (n=42) in (Värv et al., 2009) and the sampling procedure. Sampling was carried out in this study on a large population (n=2,954, 42 farms) across the whole breed. With a small sampling size the rare allele was not exposed and in one previous paper (Jõudu, 2008) the sampling was not performed across the whole breed, causing some different conclusions about allele frequencies of Estonian cattle breeds in their investigations compared to those of this study.

Expected frequencies of the β-κ-CN genotypes were calculated by multiplying the expected frequencies of the β-CN and κ-CN genotypes. Some alleles at one locus were associated only with certain alleles at the other locus, causing distinct differences between observed and expected frequencies of certain β-κ-CN genotypes (Table 3).

κ-CN genotype	β-CN genotype						
	A_1A_1	A_1A_2	A_1B	A_2A_2	A_2A_3	A_2B	BB
AA	4.0 (117)	23.1 (683)	0.0	27.1 (801)	0.2 (5)	0.1 (2)	0.0
	5.0	23.6	1.2	22.4	0.1	2.0	0.2
AB	1.6 (48)	10.2 (301)	1.5 (45)	12.5 (369)		2.8 (82)	0.1 (3)
	2.6	12.5	0.6	11.8		1.0	0.1
AE	2.5 (73)	7.2 (214)	0.0	0.2 (7)		0.0	0.0
	0.9	4.3	0.2	4.1		0.4	0.0
BB	0.1 (3)	1.2 (34)	0.4 (12)	1.2 (36)		0.8 (24)	0.2 (6)
	0.4	1.7	0.1	1.6		0.1	0.0
BE	0.7 (20)	1.7 (51)	0.3 (9)	0.0 (1)		0.0	
	0.2	1.2	0.1	1.1		0.1	
EE	0.3 (8)	0.0		0.0			
	0.0	0.1		0.1			

Table 3. Observed and expected frequencies (upper and lower line respectively, each given as percentage) of the aggregate β-κ-CN genotypes (numbers of cows in the brackets) in 2,954 Estonian Holstein cows

Some genotypes were observed two to fourfold more frequently than expected (A_1A_1-BE, A_1B-BE, A_1B-BB etc) and A_2B-BB at eightfold more frequently than expected. All cows carrying the κ-CN EE genotype had association only with the β-CN A_1A_1 genotype as has also been reported for Finnish Ayrshire cows (Ikonen et al., 1999a). Some genotypes were less frequent than expected (A_1A_1-BB a quarter and A_2A_2-AE one-twentieth of the expected frequency). Linkage disequilibrium in the casein loci has been observed in dairy cattle population differing in breed and geographical location, leading to unbalanced data (Bovenhuis et al., 1992; Ikonen et al., 1999b; Van Eenennaam & Medrano, 1991). The probable reason for unbalanced data in the Estonian Holstein population could be also the frequent use of few sires carrying, and transmitting, specific casein haplotypes to their offspring. Disequilibrium in the CSN2 and CSN3 loci can also be produced and maintained by selection favouring one combination of alleles over another (Falconer & Mackay, 1996).

3.2. Associations between milk protein genotypes and milk coagulation and quality traits

The associations of β-CN and κ-CN genotypes with milk coagulation (RCT, A_{30}), quality traits (SCS, fat and protein contents), and milk yield was investigated (Table 4).

3.2.1. Milk coagulation traits

Milk coagulation traits (RCT and A_{30}) were affected by aggregate β-κ-CN genotypes (p<0.001, Table 4). The most favourable β-κ-CN genotypes for RCT included the B allele at both loci, as has also been reported elsewhere (Comin et al., 2008) for Italian Holstein cows. Favourable aggregate genotypes for RCT were A_1B-AB and A_2B-AB. The best aggregate genotypes for A_{30} had two B alleles κ-CN, A_1A_2-BB, and the second best had the genotype A_2A_2-BB. κ-CN B was the most favourable for MCP, in every combination with β-CN, as has also been reported by (Comin et al., 2008).

Genotype	N	RCT* (min) Est.	SE	A_{30} (V) Est.	SE	MILK (kg) Est.	SE	PROTEIN (%) Est.	SE	FAT (%) Est.	SE	SCS*** Est.	SE
β-κ-CN		p<0.001		p<0.001		p=0.015		p<0.001		p=0.007		p=0.127	
A_1A_1-AA	110	-0.037	0.014	-0.506	0.255	-1.237	0.424	0.008	0.019	0.041	0.047	0.141	0.139
A_1A_1-AB	42	-0.039	0.021	2.164	0.388	-0.703	0.645	0.073	0.029	0.008	0.071	-0.095	0.213
A_1A_1-AE	70	-0.033	0.016	-0.894	0.305	-1.132	0.508	0.053	0.023	0.035	0.056	0.311	0.168
A_1A_2-AA	633	-0.019	0.007	-0.237	0.130	-0.639	0.217	0.019	0.010	0.046	0.024	-0.049	0.072
A_1A_2-AB	271	-0.056	0.009	2.401	0.173	-0.708	0.288	0.054	0.013	0.064	0.032	0.072	0.095
A_1A_2-AE	207	-0.009	0.010	-0.683	0.190	0.028	0.317	-0.004	0.014	-0.035	0.035	0.051	0.105
A_1A_2-BB	34	-0.073	0.023	4.357	0.422	-2.013	0.705	0.130	0.031	0.246	0.077	-0.560	0.235
A_1A_2-BE	49	-0.052	0.019	2.214	0.355	-0.963	0.595	0.039	0.026	-0.008	0.065	0.084	0.197
A_1B-AB	42	-0.137	0.021	2.390	0.390	0.057	0.648	0.030	0.029	0.009	0.071	-0.081	0.214
A_2A_2-AA	768	0		0		0		0		0		0	
A_2A_2-AB	337	-0.017	0.009	2.157	0.161	-0.383	0.268	0.057	0.012	0.036	0.029	-0.144	0.088
A_2A_2-BB	34	-0.072	0.023	3.865	0.418	-0.460	0.699	0.076	0.031	-0.042	0.077	-0.237	0.233
A_2B-AB	75	-0.092	0.016	1.966	0.295	-0.443	0.493	0.006	0.022	-0.121	0.054	-0.056	0.163
Rare**	93	-0.084	0.014	2.018	0.267	-0.453	0.445	0.029	0.020	-0.051	0.049	-0.168	0.148
β-LG		p<0.001		p<0.001		p=0.462		p=0.648		p=0.356		p=0.571	
AA	589	-0.015	0.006	-0.426	0.127	0.181	0.194	0.000	0.009	-0.016	0.021	-0.047	0.065
AB	1,569	0		0		0		0		0		0	
BB	609	0.033	0.006	0.367	0.131	0.200	0.198	-0.008	0.009	0.023	0.022	0.037	0.065

* Log-transformed
** Genotypes with an occurrence of less than 1%
*** SCS = log2(SCC/100,000) + 3

Table 4. Statistical significance of milk protein genotypes (p), the number of Estonian Holstein cows (n) per β-κ-CN aggregate genotype and β-LG genotype, estimated genotype effects (Est.) with their standard errors (SE) on milk coagulation time (RCT), curd firmness (A_{30}) and milk production and composition

As for the impact of *CSN2* locus, the β-CN A$_2$ allele was more favourable for A$_{30}$ than A$_1$. The A$_1$ allele did not show, in this investigation, the most favourable effect (superior to β-CN A$_2$) which was described by (Comin et al., 2008). The composite β-κ-CN genotypes, including the κ-CN B allele, were also associated with the best MCP in Finnish Ayrshire cattle (Ikonen et al., 1999a; Ojala et al., 2005), but the *CSN2* locus in that sampled population did not include the β-CN B allele. Comparing both casein loci, it seems that *CSN3* affected milk coagulation traits more than *CSN2*. Since the discovery of the micelle-stabilizing protein κ-casein, in 1956, it became evident that κ-CN had an important effect on the stabilization of casein micelles (Waugh & Von Hippel, 1956). In our study, the most frequent β-κ-CN genotype, A$_2$A$_2$-AA, and genotype A$_1$A$_2$-AE were associated with poor milk coagulation time, which is consistent with (Comin et al., 2008). Also, the rare E allele of the κ-CN in the β-κ-CN aggregate genotype had an unfavourable effect on milk coagulation properties. The association of this rare allele of κ-CN with poor milk coagulation has been previously reported (Ikonen et al., 1999a) and (Comin et al., 2008).

3.2.2. Milk yield and protein and fat percentage, SCS

Milk yield and protein and fat contents were affected by aggregate β-κ-CN genotype (p<0.05, Table 4). β-κ-CN genotype A$_1$B-AB is favourable for milk yield. Similarly to A$_{30}$, the most favourable β-κ-CN genotypes for milk protein content were homozygous for the B allele for κ-CN, A$_1$A$_2$-BB and A$_2$A$_2$-BB. The most favourable aggregate genotype for fat content was also A$_1$A$_2$-BB and unfavourable genotype for fat percentage, containing E allele in κ-CN locus, A$_1$A$_2$-AE, but also genotype A$_2$B-AB.

The most favourable for protein content was BB for κ-CN and A$_1$A$_2$ for β-CN (the second best was A$_2$A$_2$, where the A$_2$A$_2$ genotype of β-CN had a slight advantage over the A$_1$A$_1$ genotype). These results were in agreement with those previously reported (Heck, 2009), that the κ-CN genotype was associated with protein content (B>A). Milk with the aggregate genotype A$_1$A$_2$-BB had the best firmness of curd and also the best protein and fat contents. This is in agreement with another investigation (Vallas et al., 2010), where curd firmness had the highest genetic correlation with milk protein percentage (0.48), suggesting that a high protein percentage results in a favourable curd firmness. It has been reported (Cassandro et al., 2008) that there is a correlation coefficient of 0.44 between curd firmness and protein percentage, which is in agreement with the results found in this experiment. The genetic correlations of −0.24 and −0.07 reported (Ikonen et al., 1999a, 2004) for the same traits, however, are different. These inconsistencies indicate that different methodologies used for the investigations may influence the results (Pretto et al., 2011). Curd firmness showed a weak positive genetic correlation (Vallas et al., 2010) with milk fat percentage (0.25) and a weak negative genetic correlation with milk yield (−0.29). Therefore, selection for improved curd firmness may be associated with a somewhat higher protein and fat percentage and slightly reduced milk yield. Genetic correlations for curd firmness with milk yield and fat percentage were negligible in previous studies (Cassandro et al., 2008; Ikonen et al., 1999a). As for the impact of the *CSN2* A$_1$ and A$_2$ alleles on milk production, the A$_2$ allele seems to have slight advantage over A$_1$ in the aggregate β-κ-CN genotype. Genotypes of β-LG were associated with both milk coagulation traits (p<0.001), but had no a significant effect on either milk yield (p=0.462), protein percentage (p=0.648), nor fat percentage (p=0.356) and SCS (p=0.571).

4. Conclusion

The β-κ-CN locus had a strong effect on protein and fat content and milk coagulation properties. Milk with the β-κ-CN aggregate genotype A₁A₂-BB had the best firmness of curd and also the best protein and fat contents. The aggregate genotype A₂A₂-BB, haplotype A₂-B, was also favourable for milk coagulation property traits and protein content. The β-LG locus had no impact on SCS, milk production nor protein and fat contents. The β-LG BB genotype had better curd firmness and AA better milk coagulation time. Linkage disequilibrium in the *CSN2* and *CSN3* loci, which most probably led to unbalanced data, provided justification for the use of aggregate β-κ-CN and β-LG in selection for better milk technological and quality traits.

Author details

Elli Pärna*, Tanel Kaart, Heli Kiiman and Haldja Viinalass
Estonian University of Life Sciences, Tartu, Estonia
Bio-Competence Centre of Healthy Dairy Products, Tartu, Estonia

Tanel Bulitko
Animal Breeders' Association of Estonia, Keava, Rapla County, Estonia

Acknowledgement

The research leading to these results is co-financed by the European Community's Regional Development Fund in the framework of the Competence Centre Programme of the Enterprise Estonia under project No EU22868; EU27789; EU28662; EU30002 of Bio-Competence Centre Of Healthy Dairy Products (Tervisliku Piima Biotehnoloogiate Arenduskeskus OÜ) and by the Targeted Finance Project 1080045s07.

5. References

Bittante, G.; Marusi, M.; Cesarini, F.; Povinelli, M. & Cassandro, M. (2002). Genetic analysis on milk rennet-coagulation ability in Italian Holstein cows, *Proceedings of the 7th World Congress on Genetics Applied to Livestock Production* (WCGALP), CD-ROM Communication No 09-03, ISBN 9782738010520, Montpellier, France, August 19-23, 2002

---

*Corresponding Author

Boettcher, P.J.; Caroli, A.; Stella, A.; Chessa, S.; Budelli, E.; Canavesi, F.; Ghiroldi, S. & Pagnacco, G. (2004). Effects of casein haplotypes on production traits in Italian Holstein and Brown cattle, *Journal of Dairy Science*, Vol.87, No.12 (December 2004), pp. 4311-4317, ISSN 0022-0302

Bovenhuis, H.; Van Arendonk, J.A.M. & Korver, S. (1992). Associations between milk protein polymorphisms and milk production traits, *Journal of Dairy Science*, Vol.73, No.9 (September 1992), pp. 2549-2559, ISSN 0022-0302

Caroli, A.; Bolla, P.; Budelli, E.; Barbieri, G. & Leone, P. (2000). Effect of k-casein E allele on clotting aptitude of Italian Friesian milk, *Zootecnica e Nutrizione Animale*, Vol.26, No.3 (June 2000), pp. 127-130, ISSN 0390-0487

Caroli, A.M.; Chessa, S. & Erhardt, G.J. (2009). *Invited review*: Milk protein polymorphisms in cattle: Effect on animal breeding and human nutrition, *Journal of Dairy Science*, Vol.92, No.11 (November 2009), pp. 5335-5352, ISSN 0022-0302

Cassandro, M.; Comin, A.; Ojala, M.; Dal Zotto, R.; De Marchi, M.; Gallo, L.; Carnier, P. & Bittante, G. (2008). Genetic parameters of milk coagulation properties and their relationships with milk yield and quality traits in Italian Holstein cows, *Journal of Dairy Science*, Vol.91, No.1 (January 2008), pp. 371-376, ISSN 0022-0302

Cecchinato, A.; De Marchi, M.; Gallo, L.; Bittante G. & Carnier, P. (2009). Mid-infrared spectroscopy predictions as indicator traits in breeding programs for enhanced coagulation properties of milk, *Journal of Dairy Science*, Vol.92, No.10 (October 2009), pp. 5304-5313, ISSN 0022-0302

Chessa, S.; Chiatti, F.; Ceriotti, G.; Caroli, A.; Consolandi, C.; Pagnacco, G. & Castiglioni, B. (2007). Development of a Single Nucleotide Polymorphism Genotyping Microarray Platform for the Identification of Bovine Milk Protein Genetic Polymorphisms, *Journal of Dairy Science*, Vol.90, No.1 (January 2007), pp. 451-464, ISSN 0022-0302

Comin, A.; Cassandro, M.; Chessa, S.; Ojala, M.; Dal Zotto, R.; De Marchi, M.; Carnier, P.; Gallo, L.; Pagnacco, G. & Bittante, G. (2008). Effects of composite β- and κ-casein genotypes on milk coagulation, quality, and yield traits in Italian Holstein cows, *Journal of Dairy Science*, Vol.91, No.10 (October 2008), pp. 4022-4027, ISSN 0022-0302

Coulon, J.-B.; Verdier, I.; Pradel, P. & Almena, M. (1998). Effect of lactation stage on the cheesemaking properties of milk and the quality of Saint-Nectaire-type cheese, *Journal of Dairy Research*, Vol.65, No.2 (May 1998), pp. 295-305, ISSN 0022-0299

De Marchi, M.; Bittante, G.; Dal Zotto, R.; Dalvit, C. & Cassandro, M. (2008). Effect of Holstein Friesian and Brown Swiss Breeds on Quality of Milk and Cheese, *Journal of Dairy Science*, Vol.91, No.10 (October 2008), pp. 4092-4102, ISSN 0022-0302

De Marchi, M.; Fagan, C.C.; O'Donnell, C.P.; Cecchinato, A.; Dal Zotto, R.; Cassandro, M.; Penasa, M. & Bittante, G. (2009). Prediction of coagulation properties, titratable acidity, and pH of bovine milk using mid-infrared spectroscopy, *Journal of Dairy Science*, Vol.92, No.1 (January 2009), pp. 423-432, ISSN 0022-0302

Di Stasio, L. & Mariani, P. (2000). The role of protein polymorphism in the genetic improvement of milk production, *Zootecnica e Nutrizione Animale*, Vol.26, No.3 (June 2000), pp. 69-90, ISSN 0390-0487

Dovc, P. & Buchberger, J. (2000). Lactoprotein genetic variants in cattle and cheese-making ability, *Food Technology and Biotechnology*, Vol.38, No.2 (April-June 2000), pp. 91-98, ISSN 1330-9862

Elo, K.; Tyrisevä, A.-M.; Anttila, P.; Vilva, V. & Ojala, M. (2007). Genomic mapping of non-coagulation of milk in the Finnish Ayrshire, *Journal of Animal and Feed Sciences*, Vol.16, No.: suppl. 1, pp. 195-199, ISSN 1230-1388

Erhardt, G.; Prinzenberg, E.M.; Buchberger, J.; Krick-Saleck, H.; Krause, I. & Miller, M. (1997). Bovine k-casein G detection, occurrence, molecular genetic characterization, genotyping and coagulation properties. *Proceedings of the IDF Milk Protein Polymorphism Seminar* II , International Dairy Federation, Brussels, pp. 328-329, ISBN 92-9098-026-9, Palmerston North, New Zealand, February 1997

European Food Safety Authority [EFSA]. (2004). Opinion of the Scientific Panel on Dietetic Products, Nutrition and Allergies on a request from the Commission relating to the evaluation of goats' milk protein as a protein source for infant formulae and follow-on formulae, *The EFSA Journal*, Vol.30, pp. 1-15, ISSN 1831-4732

Eurostat 2010a. Production and external trade of foodstuffs: Dairy products and eggs, 27.06.2011, Available from
http://appsso.eurostat.ec.europa.eu/nui/ show.do?dataset=food_pd_prod4&lang=en

Eurostat 2010b. Gross human apparent consumption of main food items per capita. Cheese, 27.06.2011, Available from
http://appsso.eurostat.ec.europa.eu/nui/ show.do?dataset=food_ch_concap&lang=en

Falconer, D.S. & Mackay, T.F.C. (1996). Introduction to Quantitative Genetics, 4th ed. Longman Group Ltd., Essex, UK, ISBN 9780582243026

Hallén, E.; Wedholm, A.; Andrén, A. & Lundén, A. (2008). Effect of β-casein, κ-casein and β-lactoglobulin genotypes on concentration of milk protein variants, *Journal of Animal Breeding and Genetics*, Vol.125, Issue 2 (April 2008), pp. 119-129, ISSN 0931-2668

Heck, J.M.L. (2009). Milk genomics, opportunities to improve the protein and fatty acid composition in raw milk, *PhD thesis*, Wageningen University, The Netherlands, 32 pp., ISBN 9789085853329

Heck, J.M.L.; Schennink, A.; Van Valenberg, H.J.F.; Bovenhuis, H.; Visker, M.H.P.W.; Van Arendonk, J.A.M. & Van Hooijdonk, A.C.M. (2009). Effects of milk protein variants on the protein composition of bovine milk, *Journal of Dairy Science*, Vol.92, No.3 (March 2009), pp. 1192-1202, ISSN 0022-0302

Ibeagha-Awemu, E.; Prinzenberg, E.; Jann, O.; Luhken, G.; Ibeagha, A.; Zhao, X. & Erhardt, G. (2007). Molecular characterization of bovine CSN1S2*B and extensive distribution of zebu-specific milk protein alleles in European cattle, *Journal of Dairy Science*, Vol.90, No.7 (July 2007), pp. 3522-3529, ISSN 0022-0302

Ikonen, T.; Ahlfors, K.; Kempe, R.; Ojala, M. & Ruottinen, O. (1999a). Genetic parameters for the milk coagulation properties and prevalence of noncoagulating milk in Finnish dairy cows, *Journal of Dairy Science*, Vol.82, No.1 (January 1999), pp. 205-214, ISSN 0022-0302

Ikonen, T.; Bovenhuis, H.; Ojala, M.; Ruottinen, O. & Georges, M. (2001). Associations Between Casein Haplotypes and First Lactation Milk Production Traits in Finnish

Ayrshire Cows, *Journal of Dairy Science*, Vol.84, No.2 (February 2001), pp. 507-514, ISSN 0022-0302

Ikonen, T.; Morri, S.; Tyrisevä, A.-M.; Ruottinen, O. & Ojala, M. (2004). Genetic and phenotypic correlations between milk coagulation properties, milk production traits, somatic cell count, casein content, and pH of milk, *Journal of Dairy Science*, Vol.87, No.2 (February 2004), pp. 458-467, ISSN 0022-0302

Ikonen, T.; Ojala, M. & Ruottinen, O. (1999b). Associations between milk protein polymorphism and first lactation milk production traits in Finnish Ayrshire cows, *Journal of Dairy Science*, Vol.82, No.5 (May 1999), pp. 1026-1033, ISSN 0022-0302

Ikonen, T.; Ojala, M. & Syväoja, E.L. (1997). Effects of composite casein and β-lactoglobulin genotypes on renneting properties and composition of bovine milk by assuming an animal model, *Agricultural and Food Science in Finland*, Vol.6, No.4, pp. 283-294, ISSN 1239-0992

International Committee for Animal Recording [ICAR]. (2009). International agreement of recording practices. Guidelines approved by the General Assembly held June 18, 2008, 13.05.2010, Available from http://www.icar.org/Documents/Rules%20and%20regulations/Guidelines/Guidelines_2 009.pdf

Jakob, E. & Puhan, Z. (1992). Technological properties of milk as influenced by genetic polymorphism of milk proteins – A review, *International Dairy Journal*, Vol.2, No.3, pp. 157-178, ISSN 0958-6946

Jann, O.; Ceriotti, G.; Caroli, A. & Erhardt, G. (2002). A new variant in exon VII of bovine β-CN gene (CSN2) and its distribution among European cattle breeds, *Journal of Animal Breeding and Genetics*, Vol.119, No.1 (February 2002), pp. 65-68, ISSN 0931-2668

Jann, O.C.; Ibeagha-Awemu, E.M.; Özbeyaz, C.; Zaragoza, P.; Williams, J.L.; Ajmone-Marsan, P.; Lenstra, J.A.; Moazami-Goudarzi, K. & Erhardt, G. (2004). Geographic distribution of haplotype diversity at the bovine casein locus, *Genetics Selection Evolution*, Vol.36, No.2 (March-April 2004), pp. 243-257, ISSN 0999-193X

Johnson, M.E.; Chen, C.M. & Jaeggi, J.J. (2001). Effect of Rennet Coagulation Time on Composition, Yield, and Quality of Reduced-Fat Cheddar Cheese, *Journal of Dairy Science*, Vol.84, No.5 (May 2001), pp. 1027-1033, ISSN 0022-0302

Jõudu, I. (2008). Effect of milk protein composition and genetic polymorphism on milk rennet coagulation properties, *PhD thesis*, Estonian University of Life Sciences, Tartu, 114 pp., ISBN 9789949426546

Lodes, A.; Buchberger, J.; Krause, I.; Aumann, J. & Klostermeyer, H. (1996). The influence of genetic variants of milk proteins on the compositional and technological properties of milk. 2. Rennet coagulation time and firmness of the rennet curd, *Milchwissenschaft*, Vol.51, No.10 (December 1996), pp. 543-548, ISSN 0026-3788

Losi, G.; Castagnetti, G.B.; Gracia, L.; Zambonetti, C.; Mariani, P. & Russo, V. (1973). Influenza delle varianti genetiche della caseina κ sulla formazione e sulle caratteristiche della cagliata, *Scienza e Tecnologia degli Alimenti*, Vol.3, pp. 373-374, ISSN 0304-0410

Lundén, A.; Nilsson, M. & Janson, L. (1997). Marked effect of β-lactoglobulin polymorphism on the ratio of casein to total protein in milk, *Journal of Dairy Science*, Vol.80, No.11 (November 1997), pp. 2996-3005, ISSN 0022-0302

Martin, P.; Szymanowska, M.; Zwierzchowski, L. & Leroux, C. (2002). The impact of genetic polymorphisms on the protein composition of ruminants milks, *Reproduction Nutrition Development*, Vol.42, No.5 (September-October 2002), pp. 433-459, ISSN 0926-5287

McLean, D.M. (1986). Influence of milk protein genetic variants on milk composition, yield and cheese making properties, *Animal Genetics*, Vol.18, suppl. 1, pp. 100-102, ISSN 0268-9146

Medrano, J. & Aquilar-Cordova, E. (1990). Polymerase chain reaction amplification of bovine β-lactoglobulin genomic sequences and identification of genetic variants by RFLP analysis, *Animal Biotechnology*, Vol.1, No.1, pp. 73-77, ISSN 1049-5398

Miller, S.A.; Dykes, D.D. & Polesky, H.F. (1998). A simple salting procedure for extracting DNA from human nucleated cells, *Nucleic Acids Research*, Vol.16, No.3 (February 1988), p. 1215, ISSN 0305-1048

Ojala, M.; Famula, T.R. & Medrano, J.F. (1997). Effects of milk protein genotypes on the variation for milk production traits of Holstein and Jersey cows in California, *Journal of Dairy Science*, Vol.80, No.8 (August 1997), pp. 1776-1785. ISSN 0022-0302

Ojala, M.; Tyrisevä, A.-M. & Ikonen, T. (2005). Genetic improvement of milk quality traits for cheese production, In *Indicators of Milk and Beef Quality*, Ed. Hocquette, J.F. & Gigli, S.: Wageningen Academic Publishers, Wageningen, The Netherlands, pp. 307-311, ISSN 0071-2477

Olenski, K.; Kamiński, S.; Szyda, J. & Cieslinska, A. (2010). Polymorphism of the beta-casein gene and its associations with breeding value for production traits of Holstein–Friesian bulls, *Livestock Science*, Vol.131, No.1 (June 2010), pp. 137-140, ISSN 1871-1413

Patil, M.R.; Borkhatriya, V.N.; Boghra, V.R. & Sharma, R.S. (2003). Effect of bovine milk k-casein genetic polymorphs on curd characteristics during Cheddar cheese manufacture, *Journal of Food Science and Technology*, Vol.40, Issue 6 (November 2003), pp. 582-586, ISSN 0022-1155

Pentjärv, A. & Uba, M. (2004). 95 years of animal recording, In: Viinalass, H. (ed) *Animal breeding in Estonia*, pp. 14-16, OÜ Paar, ISBN 9949106311,Tartu

Pretto, D.; Kaart, T.; Vallas, M.; Jõudu, I.; Henno, M.; Ancilotto, L.; Cassandro, M. & Pärna, E. (2011). Relationships between milk coagulation property traits analyzed with different methodologies, *Journal of Dairy Science*, Vol.94, No.9 (September 2011), pp. 4336-4346, ISSN 0022-0302

Results of Animal Recording in Estonia 2010. (2011). Estonian Animal Recording Centre, 52 pp., *Elmatar, ISSN 1406-734X, Tartu*

Schaar, J.; Hansson, B. & Pettersson, H.-E. (1985). Effects of genetic variants of k-casein and β-lactoglobulin on cheesemaking, *Journal of Dairy Research*, Vol.52, No.3, pp. 429-437, ISSN 0022-0299

Statistics Estonia. (2010). Agriculture statistical database, Production of milk products (months), 27.06.2011, Available from http://pub.stat.ee/px-web.2001/I_Databas/ Economy/01Agriculture/02Agricultural_production/04Livestock_production/04Livestoc k_production.asp

Vallas, M.; Bovenhuis, H.; Kaart, T.; Pärna, K.; Kiiman, H. & Pärna, E. (2010). Genetic parameters for milk coagulation properties in Estonian Holstein cows, *Journal of Dairy Science*, Vol.93, No.8 (August 2010), pp. 3789-3796, ISSN 0022-0302

Van den Berg, G.; Escher, J.T.M.; de Koning, P.J. & Bovenhuis, H. (1992). Genetic polymorphism of k-casein and b-lactoglobulin in relation to milk composition and processing properties, *Netherland Milk and Dairy Journal*, Vol.46, No.3-4, pp. 145-168, ISSN 0028-209X

Van Eenennaam, A.L. & Medrano, J.F. (1991). Milk protein polymorphisms in California dairy cattle, *Journal of Dairy Science*, Vol.74, No.5 (May 1991), pp. 1730-1742, ISSN 0022-0302

Värv, S.; Belousova, A.; Sild, E. & Viinalass, H. (2009). Genetic diversity in milk proteins among Estonian dairy cattle, *Veterinarija ir Zootechnika*, Vol.48, No.70 (December 2009), pp. 93-98, ISSN 1392-2130

Velmala, R.; Mäntysaari, E.A. & Mäki-Tanila, A. (1993). Molecular genetic polymorphism at the k-casein and b-lactoglobulin loci in Finnish dairy bulls, *Journal of Agricultural Sciences in Finland*, Vol.2, No.5, pp. 431-435, ISSN 0782-4386

Visker, M.H.P.W.; Dibbits, B.W.; Kinders, S.M.; Van Valenberg, H.J.F.; Van Arendonk J.A.M. & Bovenhuis, H. (2011). Association of bovine β-casein protein variant I with milk production and milk protein composition, *Animal Genetics*, Vol.42, No.2 (April 2011), pp. 212-218, ISSN 0268-9146

Walsh, C.D.; Guinee, T.P.; Reville, W.D.; Harrington, D.; Murphy, J.J.; O'Kennedy, B.T. & FitzGerald, R.J. (1998). Influence of k-casein genetic variant on rennet gel microstructure, Cheddar cheesemaking properties and casein micelle size, *International Dairy Journal*, Vol.8, Issue 8 (August 1998), pp. 707-714, ISSN 0958-6946

Waugh, D.F. & Von Hippel, P.H. (1956). κ-Casein and stabilization of casein micelles, *Journal of American Chemical Society*, Vol.78, Issue 18 (September 1956), pp. 4576-4582, ISSN 0002-7863

Wedholm, A.; Larsen, L.B.; Lindmark-Månsson, H.; Karlsson, A.H. & Andrén, A. (2006). Effect of protein composition on the cheese making properties of milk from individual dairy cows, *Journal of Dairy Science*, Vol.89, No.9 (September 2006), pp. 3296-3305, ISSN 0022-0302

Ysebaert Dairy Division, Optigraph, 10.12.2009, Available from http://www.ysebaert.com/ysebaert/optigraphuk.htm

Control of Mammary Function During Lactation in Crossbred Dairy Cattle in the Tropics

Narongsak Chaiyabutr

Additional information is available at the end of the chapter

1. Introduction

The major problems for the dairy practices in the tropical country are low milk yield and short lactation period of either pure exotic or crossbred dairy cattle. Many factors affect milk production in dairy cattle in tropical areas including high environmental temperature and humidity, lower genetic potential for milk production in indigenous cattle and inadequate supply of food during the dry and hot summer season. Crossbreeding of indigenous and exotic cattle for tropical use has been exploited as an efficient tool for blending the adaptability of native cattle with the high milking potential of exotic breeds resulting in increased milk production. In crossbred cattle, mechanisms of milk secretion are known to be inherited and are thought to be among the causes of differences in metabolic parameters. Milk secretion is a continuous process and requires a continuous supply of substrates for milk productions. Short persistency of lactation is the low ability of cow to continue producing milk at a high level after the peak of lactation. Improved persistency of lactation can contribute to decreasing the cost of the production system because lactation persistency is associated with feeding and health costs including reproductive performance. (Sölkner & Fuchs,1987). There are many different ways to improve persistency of lactation, e.g. during summer, different types of crop residues have been used to feed animals as roughage in attempting to improve dairy productivity during period of scarcity (Jayasuriya & Perera, 1982; Promma et al., 1994). However, there is considerable evidence that understanding the mechanism acting within the body in relation to the control of milk secretion may improve of milk production in crossbred dairy cattle in tropics.

It is well known that mammary growth during pregnancy is a prerequisite for satisfactory lactation in all mammals. During pregnancy, the mammary gland is competing with many other organs for nutrition to sustain growth. Mammary secretory activity in dairy cattle is initiated from pregnancy although it is at a low level. Experimental studies in both periods of pregnancy and lactation will give some clue as to the nature of the developed mechanism of the mammary gland. During pregnancy, maternal bodily functions are altered, e.g. cardiac output and heart rate, while during lactation, many bodily functions are also altered, e.g. general circulation and body fluid (Hanwell & Peaker, 1977). Therefore, knowledge of extra-mammary factors influencing milk production is essential to study in both pregnancy and period of lactation which can provide quantitative information in crossbred Holstein cattle

The lactating mammary gland is dependent upon its blood supply to provide substrates at appropriate rates to sustain milk synthesis. The rate of substrates supplying to the mammary gland is determined by substrate concentration in the plasma and mammary blood flow. There is evidence that substrate supply to the mammary gland is often inadequate to maintain the maximum rate of milk synthesis. This raises the question: do changes in bodily function (water balance, general circulation and mammary circulation) alone have an effect on milk secretion or is the effect solely due to inadequate of the utilization of substrates in the mammary gland in crossbred dairy cattle? As glucose is the principal precursor of lactose, the decrease in milk lactose can be explained by a change of the mammary utilization of glucose (Faulkner & Peaker, 1987). Lactose is a highly osmotic component, which allows the drainage of water from blood to the alveolar compartment. As such, it is the principal milk component regulating the volume of milk production. Glucose is known to play an important role not only in lactose synthesis but also providing the reducing equivalent required for the synthesis of fatty acid *de novo* in the mammary gland (Chaiyabutr et al., 1980). Very few data are available regarding the dynamics and the regulation of glucose metabolism in whole body and the mammary gland of different types of crossbred cattle. Insight into the study in the utilization of glucose metabolism for synthesis of milk components in different metabolic pathways in the mammary gland, have improved understanding of what factors influencing the effect of low milk yield of crossbred dairy cattle in the tropics.

The role of endocrine regulation in initiation and maintenance of lactation is known to occur in many species. However, hormonal requirement among mammalian species differ considerably, for example, in rabbit prolactin alone can maintain lactation, while in cows prolactin is not a rate limiting hormone in established lactation in place growth hormone become relatively more important (Hart, 1973; Mepham, 1993). Very few data are available in the study of circulating hormones during lactation in crossbred dairy cattle. The circulating concentrations of some hormones will be expected to change and relate to the mechanism responsible for the control of milk secretion in different types of crossbred Holstein cattle. Investigation of the plasma levels of various hormones (thyroid hormone, prolactin, cortisol, growth hormone, insulin, glucagon, progesterone and estradiol) involving milk secretion in late pregnancy and different stages of lactation may give some information on the control mechanism of milk secretion either directly or indirectly on the function of mammary gland in different types of crossbred Holstein cattle.

Thus, this chapter is intended to review the highlight the regulatory mechanisms underlying marked low milk yields, by providing an updated summary of the results obtained concerning the physiological changes in both extra-mammary factors and intra-mammary factors of crossbred Holstein cows between 50HF:50Red Sindhi(50%HF) and 87.5HF:12.5RS(87.5%HF) cattle in the tropics

2. Water metabolism and mammary circulation in different stages of lactation of crossbred cattle

It is known that lactating dairy cows metabolize large amounts of water and are affected rapidly by water deprivation (Murphy 1992). In the tropics, dairy cattle utilize body water to maintain homeostasis in the mechanisms of thermoregulation and lactation. The effect of excessive heat load in high environment temperature leading to increase in heat dissipation (Hahn et al., 1999), which may lead to a loss body water and decrease in milk yield in lactating cattle. However, an increase in water intake during lactation closely match to increase in water secrete in milk, since milk composition has about 87% of water. During lactation, many bodily functions are altered; for example, blood volume and cardiac output are increased (Hanwell & Peaker 1977). These changes may effectively alter body fluid and thus circulatory distribution including the blood supply to the mammary gland.

2.1. Water metabolism

2.1.1. Measurement of water metabolism

In this section, a brief description of methods for measuring water metabolism is presented. According to the study of Chaiyabutr et al., (1997) in measurements of water metabolism, the water turnover rate (WTO), total body water (TBW) in dairy cattle were performed using tritiated water dilution techniques. Briefly, the dairy cattle was injected intravenously with carrier free tritiated water in normal saline at a single dose of 3,000 μCi per animal. The equilibration time was determined by taking serial venous blood samples for 3 days after the injection. Blood samples were collected at 20, 30, 40, 50, 60 min and 4, 8, 20, 26, 32, 44, 50, 56, 68 and 74 h subsequent to the injection. Preparation of samples for counting was achieved by the internal standardization technique. The corrected activity of samples, in disintegrate per minute (d.p.m.), were plotted on semi-logarithmic paper against time. The dilution curve of tritiated water in plasma was described by an exponential equation using a one compartment model, which is

$$Yi = Ae^{-klt},$$

where Y is concentration of tritium in plasma at time t (nci/ml); A is plasma concentration intercept 1 in nci/ml.

The extrapolated activity at theoretical zero time of complete mixing of radio-isotope was used to determine the total body water space (TOH). The TOH space was calculated:

TOH space (ml) = [standard count (dis/min) x dose (ml)] / [radio activity counts at zero time (dis/min)].

The biological half-life of tritium-labelled water (T 1/2) was determined from the slope of the linear regression line obtained from plot on semi-logarithmic paper of the activity of the samples taken over the period of 3 days against time. The water turnover rate was calculated from the equation:

$$WTO \, (l/day) = 0.693 \times TOH \, space \, / \, T \, 1/2.$$

Total body water (TBW) was calculated by using the corrected factor (1 - fraction of plasma solids) x TOH space (Chaiyabutr et al. 1997).

Work on water metabolism has produced some interesting results. The regulation of body fluids and mammary circulation in different types of crossbred Holstein Friesians (HF) cattle showed some differences between 50%HF and 87.5%HF cattle during late pregnancy and lactation (Chaiyabutr et al., 1997). A short persistency of milk yield during the transition period from early to mid lactation has been observed in 87.5%HF cattle. In contrast to 50%HF cattle, persistent lactation seemed to be apparent throughout periods of lactation (Chaiyabutr et al. 2000a). The different control mechanisms were at play in the regulation of milk production as lactation advances in 50%HF and 87.5%HF cattle. The 87.5%HF cattle had lower efficiency in water retention mechanism and poor adaptation to tropical environment in comparison to 50%HF (Chaiyabutr et al., 1997; 2000a). The high genetic similarity of 87.5%HF cattle to the exotic bos taurus breed may lead to poor adjustment to the tropical environment, while yielding high milk production in early lactation. However, the milk yield of 87.5%HF cattle has been shown to be higher in early lactation than 50%HF cattle while the ratio of total dry matter intake to milk yield was lower in 87.5%HF as compared to 50%HF cattle. The energy output in milk and for maintenance for 87.5%HF cattle was greater than the energy consumed in the food in early lactation. The 50%HF cattle was approximately in energy equilibrium, there being no change in the ratio of dry matter intake to milk yield among periods of lactation.

Water is known to play a prominent role in the processes involved in the formation of milk, since milk is isoosmotic with plasma. In intra-mammary process, the net transfer of water from plasma to the alveolar lumen is directly proportional to the transfer of solute. Water is drawn into secretory vesicles because of the presence of osmotically active lactose (Kaufmann & Hagemeister,1987; Linzell & Peaker, 1971),which is formed in the golgi apparatus of the lactating cells. Thus, the amount of water secreted into milk is regulated by the rate of lactose synthesis. However, cellular water metabolism in lactating cell is not a final story of the processes of milk formation. The extra-mammary factor about body water content will be a prerequisite for satisfactory lactation, which may influence on lactation persistency in crossbred cattle.

Measurements of water turnover rate and total body water using tritiated water as a marker in different stages of lactation of crossbred Holsteins cattle is shown in Fig.1 (Chaiyabutr et al.,1997; 2000a). Chaiyabutr et al (2000a) have noted that the total body water as a percentage of body weight of 87.5%HF cattle was lower than 50%HF cattle in all periods of lactation. This may be attributed to a relatively lower efficiency in the water retention mechanism, although the water intake was higher in 87.5%HF animals during high milk yield. Low water content may be related to the poor adaptation of 87.5%HF to the tropical environment. Poorer lactation persistency in higher yielding cows has also been noted (Chase, 1993; Coulon et al., 1995). The studies in both 87.5%HF and 50%HF housing in the same shed under the same environment by Chaiyabutr et al., (2000a) have shown that a lower water turnover rate was apparent especially in late lactation of the 87.5%HF cattle in comparison with 50%HF cattle.

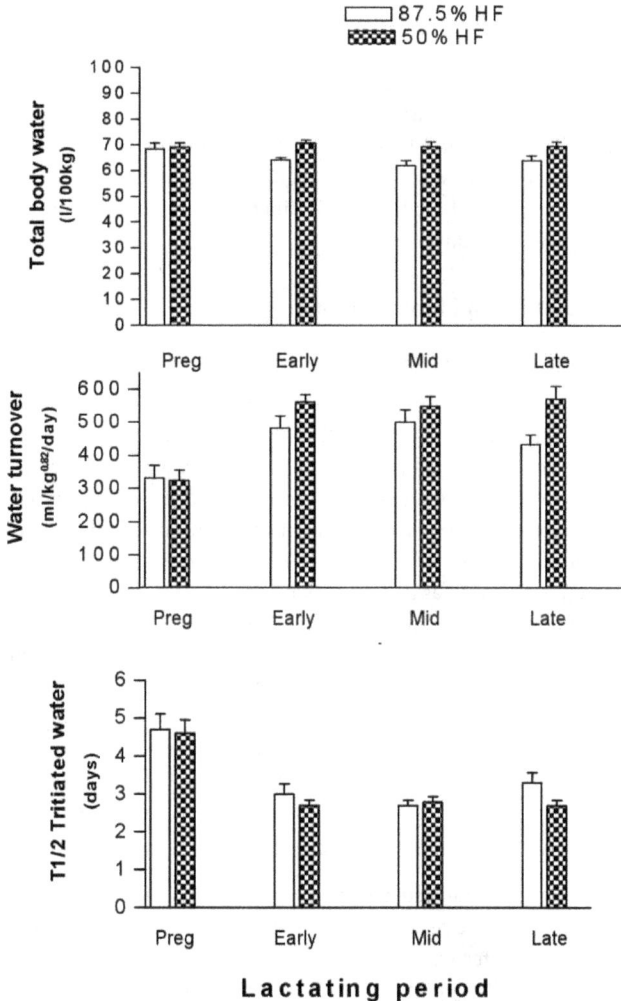

Figure 1. Water turnover rate, total body water and half-life of tritiated of water during late pregnancy and lactation of 87.5%HF and 50%HFcattle in tropical environment. (Data from Chaiyabutr et al., 1997; 2000a).

The water turnover rate of both types of crossbred cattle was not influenced by environmental conditions, although marked differences of water turnover rate and half-life of body water in animals have been reported during the winter and summer (Ranjhan et al., 1982). A higher water reserve in 50%HF cattle would not only provide a higher reservoir of soluble metabolites for biosynthesis of milk but was also useful in slowing down the elevation in body temperature of this breed during lactation in hot conditions (Nakamura et al.,1993). The 87.5%HF cattle did not require greater amounts of water, while it could restore their body fluids to equilibrium in all lactating periods. The differences between crossbred and purebred animals in body composition and water turnover rate have also been reported (Macfarlane & Howard 1970). The study of the regulation of body fluid and mammary circulation during late pregnancy and early lactation of crossbred Holstein cattle by Chaiyabutr et al., (1997) have found that the water turnover rate was significantly higher approximately 45 % in lactating cattle than in pregnant cattle in either 87.5%HF or 50%HF cattle. During the lactating period the half-life of tritiated water (2.9 days) was significantly lower than pregnant period (4.7 days).

2.2. Mammary circulation and blood volume

2.2.1. Measurement of udder blood flow

According to the study of Chaiyabutr et al., (1997), the measurement of udder blood flow in dairy cattle was described. Briefly, blood flow through half of the udder was determined by measuring the dilution of dye T-1824 (Evans blue) by a short term continuous infusion in the milk vein of dairy cattle. A dye (T-1824) was dissolved in sterile normal saline and diluted to a concentration of 100 mg/l. The solution was infused by a peristaltic pump at a constant rate of 80 ml/min into the milk vein for 1-2 min. Before infusion, blood was drawn from downstream in the milk vein as a pre-infusion sample. About 10 seconds after starting infusion, 10 ml of blood was drawn from downstream in the milk vein at a constant rate into a heparinized tube. Two consecutive plasma samples were taken during dye infusion. Blood flow of half of the udder was calculated from the concentration of dye in plasma samples using the equation derived by Thompson & Thomson (1977). In lactating cows, quarter milking showed that the yields of the two halves of the udder were similar. Udder blood flow was therefore calculated by doubling the flow measured in one milk vein.

2.2.2. Measurement of plasma volume and blood volume

In the studies of plasma volume and blood volume, plasma volume was measured by dilution of Evan's blue (T-1824) dye. The injection of 20 ml of the dye (0.5 g/100 ml normal saline) into the ear vein catheter was performed and venous blood samples were collected from the jugular vein at 30, 40 and 50 min after dye injection. Dilution of dye at zero time was determined by extrapolation. Blood volume was calculated from the plasma volume and packed cell volume (PCV). Plasma osmolality was measured using the freezing point depression method.

The results of study by Chaiyabutr et al (2000a) have noted that the mammary circulation of crossbred dairy cattle between 87.5%HF and 50%HF varied during different periods of lactation. Mammary blood flow is known to be a major determinant controlling milk production in a way to carry milk precursors to the mammary gland at the process of milk synthesis. A decrease in blood flow to the mammary gland coinciding with a short persistency of milk yield during the transition period from early to mid lactation occurred in 87.5%HF cattle (Chaiyabutr et al. 2000a). The milk yield of 87.5%HF cattle significantly declined in mid-lactation from early lactation. In contrast to 50%HF cattle, short persistent lactation seemed to be apparent throughout periods of lactation (Fig. 3). The study for alteration of lactation persistency in different crossbred animals will throw some light on a useful index for studying adaptability in crossbred cattle, which will provide information on choosing suitable crossbred dairy cattle in the tropics. A higher mammary blood flow during early lactation compared to mid-lactation in 87.5%HF cattle could not be attributed to a change in blood volume and plasma volume, which remained nearly constant (Fig.2). Plasma osmolality remained unchanged during the course of lactation in both 87.5%HF and 50%HF cattle indicating that homeostasis was being maintained throughout all periods of lactation. Such differences of the mammary circulation between 87.5%HF and 50%HF animals can be attributable to disparities in breed. The ratio of mammary blood flow to the rate of milk yield did not change during the course of lactation in 50%HF animals. The marked decrease in the mammary blood flow of 87.5%HF at mid-lactation would correlate with the decrease of milk yield. The decline in milk yield after the peak will be due primarily to a decreased availability of substrates to the mammary gland. The marked increase in the ratio of mammary blood flow to the rate of milk yield during lactation advance to mid-lactation (548:1) and late lactation (668:1) in 87.5%HF cattle indicated a more decreased secretory activity of mammary tissue. The question then arises as to whether mammary metabolism influences mammary blood flow or mammary blood flow influences mammary metabolism.

Many studies on mechanisms concerned with regulation of mammary blood flow, local and extra-mammary production of vasoactive agents and activity of mammary sympathetic nerves have been reviewed comprehensively (Linzell, 1974). The study on mammary circulation in relation to the general circulation of crossbred cattle at different stages of lactation is shown in Fig. 3. The mean values of mammary resistance (the ratio of mean arterial pressure to mammary blood flow) did not significant change in different periods of lactation in either 87.5%HF or 50%HF cattle. It is clear that the local changes for vasoconstriction in the udder were not apparent in different periods of lactation for both types of crossbred cattle. This proposal was indicated that humoral factors were responsible, especially the level of plasma growth hormone in the regulation of mammary circulation and milk production between 87.5%HF and 50%HF cattle. A number of studies have demonstrated that similar proportion increases in milk secretion and mammary blood flow occurred during growth hormone treatment in goats and cows (Hart et al., 1980; Chaiyabutr et al 2007). A marked decline in blood flow to the mammary gland during the transitional period from early lactation to mid-lactation in 87.5%HF cattle with the constancy of the plasma volume and blood volume were apparent (Chaiyabutr et al 2000a). Regulation of mammary blood flow from the early lactation to mid-lactation in 87.5%HF cattle may be regarded a major homeorhetic principle (Bauman & Currie, 1980).

Mammary growth during pregnancy has been known to be a prerequisite for satisfactory lactation. The degree of local vasoconstriction was greater during pregnancy and less in lactation for both types of crossbred cattle. In late pregnancy, the mammary circulation of

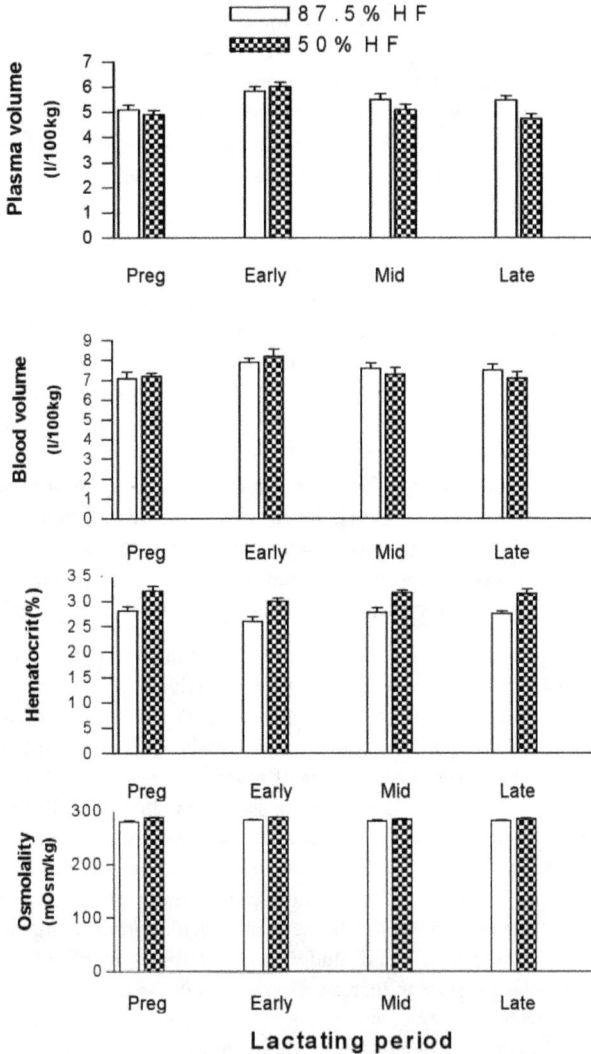

Figure 2. Plasma volume, blood volume, hematocrit and plasma osmolality during late pregnancy and lactation of 87.5%HF and 50%HF cattle in tropical environment. (Data from Chaiyabutr et al., 1997; 2000a).

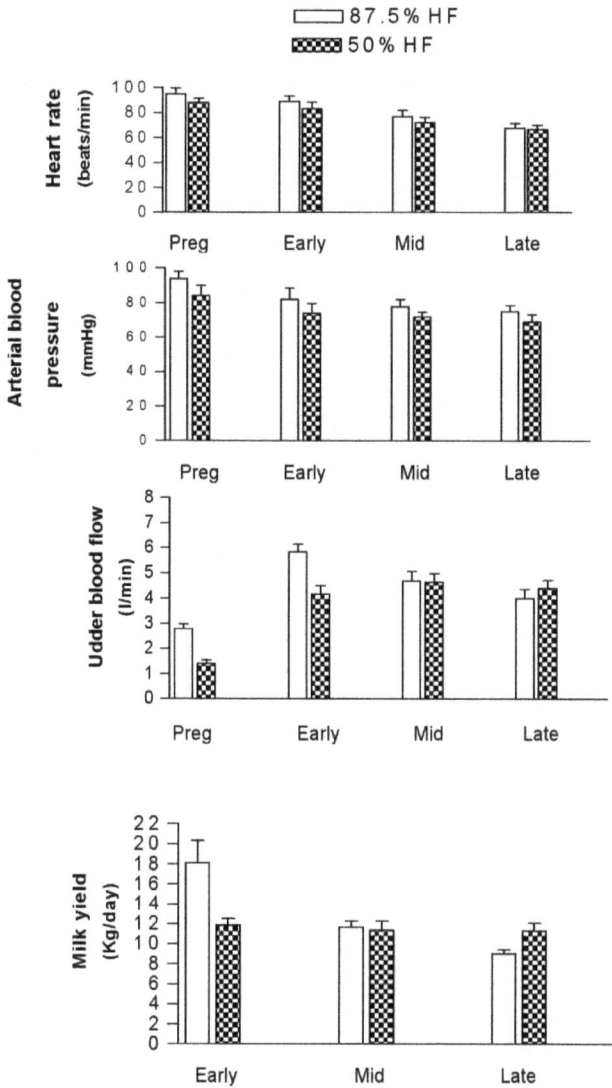

Figure 3. The mean arterial pressure, heart rate, udder blood flow and milk yield during late pregnancy and lactation of 87.5%HF and 50%HF cattle in tropical environment. (Data from Chaiyabutr et al., 1997; 2000a).

50%HF was less than 87.5%HF cattle which could be due to variations in the developments of mammary blood vessels and mammary cells. The high genetic blood level closing to the exotic *bos taurus* breed of 87.5%HF animals may cause a rapid rate in the development of the secretory activity of the mammary cell in the late pregnancy. Local vasodilatation produced by the active cells (Hanwell & Peaker, 1977; Lacasse et al., 1996) would decrease in the resistance of the vascular bed and a higher mammary blood flow of 87.5%HF in comparison with 50%HF cattle (Chaiyabutr et al., 2000a). These physiological changes would accompany with increases in plasma volume and blood volume in the lactating period when compared with the pregnant period in either 87.5%HF or 50%HF cattle. These changes are agree to other reports that blood volume, plasma volume and water turnover are markedly higher in lactating animals than in pregnant animals (Hanwell & Peaker, 1977; MacFarlane & Howard, 1970). The packed cell volume significantly decreased in the lactating period of 87.5%HF animals while it did not significantly change in 50%HF animals. The packed cell volume of 50%HF animals was significantly higher than that of 87.5%HF animals.

There were no significant differences in heart rate, arterial blood pressure and plasma osmolality between the periods of late pregnancy and early stage of lactation in either 87.5%HF or 50%HF cattle. Udder blood flow was nearly three times higher in lactating period than in late pregnant period in either 87.5%HF or 50%HF cattle. The udder blood flow of 87.5%HF was significantly higher while mammary resistance was significantly lower than in 50%HF animals. Milk secretion in early lactation of 87.5%HF animals was higher than that of 50%HF animals. The ratio of DM intake to milk production for 87.5%HF animals was lower than that of 50%HF animals. From these results it can be concluded that the difference between breeds was found in 87.5%HF animals which had a higher milk yield but a lower adjustment for the regulation of body fluids during pregnancy and lactation in comparison to 50%HF animals.

3. Role of substrates supply for milk production

It is known that the rate of substrates supplying to mammary gland is determined by substrate concentration in the plasma and mammary blood flow. The sustainable milk synthesis in the mammary gland is dependent upon its blood supply to provide substrates at appropriate rates. There is evident that substrate supply to the mammary gland is often inadequate to maintain the maximum rate of milk synthesis (Linzell & Mepham 1974). The mammary gland may be producing milk at a rate below its potential. The rate of milk production depends on function of number of secretory cells and their metabolic activity. However, the mechanisms can be devided into three main levels of regulation, which are the arterial flow of substrates in the mammary gland, the substrate extraction by the mammary gland and the metabolic and secretion activities of the mammary epithelial cell.

3.1. Changes of substrates supply

3.1.1. Measurement of mammary uptake of substrates

According to the study of Chaiyabutr et al., (2002), the measurement of udder blood flow in dairy cattle was described. Briefly, The uptake of substrate by the udder (Us), expressed as $\mu mol/min$, was calculated from the equation:

$$Us= MPFx (P_A-P_V)$$

The substrate extraction by the mammary (Es):

$$Es = (P_A-P_V)/ P_A$$

Where:

MPF = Mammary plasma flow (ml/min)

P_A = Concentration of substrate in coccygeal arterial plasma (μmol/ml)

P_V = Concentration of substrate of plasma from milk vein (μmol/ml)

The results of study by Chaiyabutr et al (2002) in both 87.5%HF and 50%HF cattle have shown that the mammary uptake of many substrates between late pregnancy and lactating periods in crossbred HF cattle was not based on changes in arteriovenous concentration differences and extraction ratio. Mammary growth during pregnancy has been known to be a prerequisite for satisfactory lactation. The pattern of mammary growth varied during pregnancy and early lactation for both 87.5%HF and 50%HF cattle. Changes of substrates across the udder will account for changes in mammary blood flow. Both 87.5%HF and 50%HF cattle showed the low values of both arteriovenous concentration differences of substrates across the mammary gland and mammary extraction ratio during late pregnancy, which differ from the lactating period with a marked increase of mammary uptake of substrates for milk synthesis.

Volatile fatty acid in form of acetate is known to be the major of source energy of normal fed ruminants. No differences in mammary arteriovenous concentration differences and extraction ratio of acetate have been shown between late pregnancy and lactation in both 87.5%HF and 50%HF cattle. The low rate of mammary uptakes for acetate during pregnancy was dependent upon rate of blood flow. During late pregnancy, circulating β-hydroxybutyrate level which arise mainly from rumen butyrate in the normal fed animal (Leng & West, 1969). There were no obvious of arteriovenous concentration differences and extraction ratio of β-hydroxybutyrate across the mammary gland of both 50%HF and 87.5%HFcattle.

The concentrations of arterial plasma free fatty acid and triacylglycerol increased in the late pregnancy of both 50%HF and 87.5%HFcattle and were more sensitive to alteration than other blood substrates; this phenomenon has been proposed as an indication of under-nutrition (Reid & Hinks, 1962). The mobilization of body fat in late pregnancy occurs in response to hormonal secretion (Lindsay, 1973). It would be a physiological phenomenon, not a consequence of under-nutrition. The measurement of arteriovenous differences of FFA across the mammary gland together with mammary blood flow did not provide a quantitative estimation of their total uptake by mammary tissue in either 50%HF and 87.5%HF cattle, since the release of FFA into venous blood due to triacylglycerol hydrolysis during the uptake of plasma triacylglycerol as in lactation has been noted (West et al., 1967). During the development of the mammary gland in normal late pregnancy, an occurring the utilization of FFA may cause changes in arteriovenous differences, the extraction ratio and net uptake of FFA. The net uptake of triacylglycerol by the mammary gland significantly increased in lactating period in comparison to the late pregnancy in both 87.5%HF and 50%HF cattle. It is possible that a change for releasing of FFA is a result of changes of enzymatic activity of lipoprotein lipase in the mammary tissue. This enzyme activity has been reported to be higher in lactating bovine mammary tissue relative to pregnant tissue (Shirley et al., 1973).

3.2. Glucose metabolism and glucose utilization in the mammary gland

It is known that an increase of milk yield can be achieved only by increasing the rate of lactose synthesis. The lactating udder utilizes most of the glucose entering the circulation of ruminants and irreversible glucose loss from plasma is highly correlated with lactose output (Bickerstaffe et al., 1974; Horsfield et al., 1974). The role of glucose in regulating milk secretion has also been demonstrated in the isolated perfused udder (Hardwick et al., 1961). The regulation of the milk yield of animals is therefore mainly based on the mechanisms governing the quantity of glucose extracted by the udder and converted into lactose. The metabolic fate of glucose metabolism will consider the utilization in the whole body relating to the utilization in the mammary gland in either pregnancy or lactation.

3.2.1. Measurements of Glucose kinetic and glucose utilization in the mammary gland

According to Chaiyabutr et al.,(2000c), body glucose metabolism and intramammary glucose metabolism has been described. In brief, glucose kinetic studies at different stages of lactation were carried out by using continuous infusion of both [U-^{14}C]-glucose and [3-^{3}H]-glucose solution. A priming dose of radioactive glucose in 20 ml of normal saline containing 30 µCi [3-^{3}H]-glucose and 15 µCi[U-^{14}C]-glucose was administered intravenously via the ear vein catheter and followed by a continuous infusion(using Peristatic pump) of 1 ml/min of normal saline solution (0.9%) containing 0.7 µCi/ ml of [U-^{14}C]-glucose and 1.5 µCi/ml of [3-^{3}H]-glucose for 3 h. During the last 1 h of continuous infusion, three sets of blood samples were collected at 20 min intervals.

The glucose turnover rate in the whole animal (T), expressed as µmol/min, was calculated from the equation:

$$T = I/G_A$$

where, I = rate of infusion of [U-^{14}C]glucose or [3-^{3}H]glucose (µCi/min) and G_A = specific activity of ^{14}C- or ^{3}H-glucose in arterial plasma at equilibrium (µCi/µmol).

Recycling of glucose carbon in the whole animal, expressed as % glucose turnover, was calculated from the equation:

$$Recycling = (T_3 - T_{14}) \times 100/T_3$$

Where:

T_3 = Reversible turnover of glucose calculated from [3-^{3}H]glucose
T_{14} = Irreversible turnover of glucose calculated from [U-^{14}C]glucose

Uptake of glucose by the udder (U_G), expressed as µmol/min, was calculated from the equation:

$$U_G = MPFx \ (P_A P_V)$$

Where:

MPF = Mammary plasma flow (ml/min)
P_A = Concentration of glucose in coccygeal arterial plasma (µmol/ml)
P_V = Concentration of glucose of plasma from milk vein(µmol/ml)

The milk component output (MO), expressed as µmol/min and was calculated from the equation:

$$MO = Ms \ x \ Cc/1000$$

Where:

Ms = Milk secretion rate (ml/min)
Cc = Concentration of components in milk (µmol/ L)

Incorporation (A) of radioactivity from glucose into milk components was calculated from the equation:

$$A = M_A/G_A \ x \ t$$

Where:

A = Incorporation of radioactivity from glucose into milk components (µmol/min)
M_A = Total activity of ^{3}H or ^{14}C in the milk components (µCi)
G_A = Specific activity of ^{14}C-or ^{3}H-glucose in arterial plasma at equilibrium (µCi/µmol)
t = Time of infusion (min)

On the glucose kinetics studies using 3-[^{3}H] glucose and [U-^{14}C] glucose infusion, the total glucose entry rate (the reversible turnover rate of 3-[^{3}H] glucose) and utilization rates of glucose (the irreversible turnover rate of [U-^{14}C] glucose) increased significantly during early lactation in comparison to late pregnancy for both 50%HF and 87.5%HF cattle (Fig.4). Recycling of glucose-C was approximately 20% in both crossbred cattle which was unaffected by the stage of late pregnancy or early lactation. During late lactation, glucose carbon recycling and plasma glucose clearance increased in both 50%HF and 87.5%HF cattle. Comparing 50%HF and 87.5%HF cattle, arterial plasma glucose concentrations were slightly higher during pregnant period but significantly higher in lactating periods in 50%HF cattle (Chaiyabutr et al.,1998). This indicates that steady state conditions between the rate of utilization of glucose and the rate of gluconeogenesis differ in the body pool of glucose. The uptake, arteriovenous differences and extraction ratio for glucose across the udder increased in the lactating period in either 87.5%HF or 50%HF cattle. Glucose uptake by the udder accounted for 65% of the total glucose turnover rate in 87.5%HF cattle and 46% in lactating 50%HF cattle, although both crossbred cattle exhibited the same body glucose turnover rate (Chaiyabutr et al., 2000b). The 87.5%HF cattle had high milk yield, body glucose metabolism and udder glucose metabolisms as compared with 50%HF cattle in early lactation.

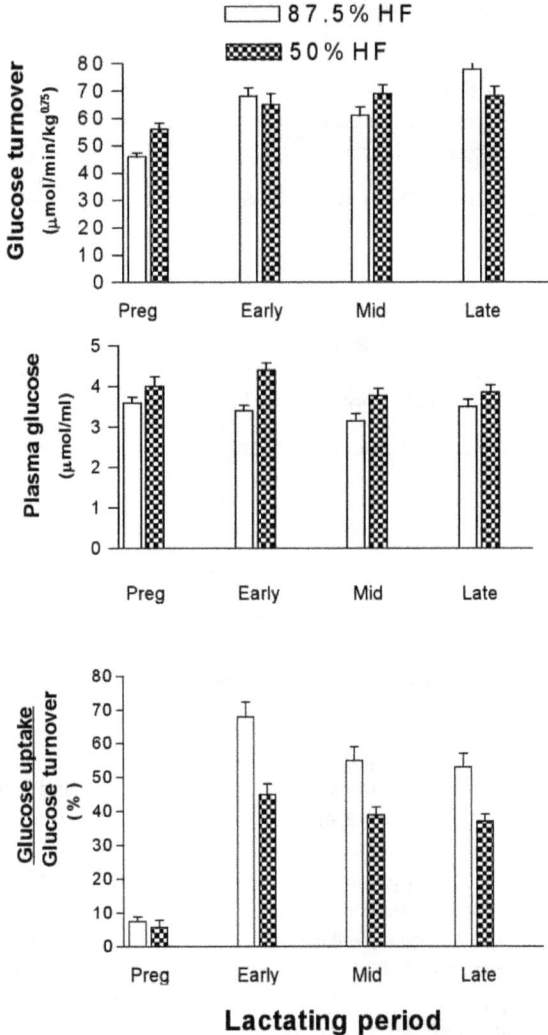

Figure 4. Glucose turnover, plasma glucose concentration and percentage of glucose uptake to glucose turnover during late pregnancy and lactation of 87.5%HF and 50%HF cattle in tropical environment. (Data from Chaiyabutr et al., 1998; 2000b).

In early lactation, the high milk yields and lactose secretion of 87.5%HF cattle were related to glucose uptake by the udder and udder blood flow as compared with 50%HF cattle. An increase in udder blood flow during early lactation was higher in 87.5%HF cattle than in 50%HF cattle. The marked decreases in udder blood flow, glucose uptake, lactose secretion and milk yield of 87.5%HF cattle were apparent in mid- and late lactation. In contrast to 50%HF cattle, no alterations for udder blood flow, udder glucose uptake, lactose secretion and milk yields were apparent throughout the course of lactation. The 87.5%HF cattle had a higher milk yield but a shorter peak yield and poorer persistency in comparison with 50%HF cattle (Chaiyabutr et al., 2000a). Changes in the utilization of glucose by the mammary gland for milk production in crossbred cattle would be dependent on changes in intramammary factors.

In general the rate of milk secretion is thought to be determined primarily by the secretion of lactose and consequent osmotic water movements follows lactose which create milk secretion as whole (Rook & Wheelock, 1967; Linzell & Peaker, 1971). An increase in milk yield can be attributed to an increase in the rate of lactose synthesis. Glucose is essential for milk secretion (Hardwick, Linzell & Price, 1961) and glucose moiety of lactose arises directly from plasma glucose (Ebner & Schanbacher, 1974). No differences of mammary arteriovenous differences of glucose in different periods of lactation in either 87.5%HF or 50%HF cattle (Chaiyabutr et al., 2000b, 2002) suggested that glucose uptake was determined mainly by mammary blood flow(Linzell, 1973). Mammary blood flow, milk yield and lactose yield were not affected by alterations of plasma glucose concentrations in both 87.5%HF and 50%HF cattle (Chaiyabutr et al. 2000b). There is evidence to confirm that the plasma glucose concentration is not a major factor in determining lactose output. The major change would occur in the metabolism of the mammary epithelial cell in different periods of lactation. It might be reasonable to believe that the reduction of glucose uptake by the mammary gland during mid- and late lactation of 87.5%HF cattle would not be a result of the decrease in the rate of phosphorylation of glucose by hexokinase. The decrease in mammary glucose uptake has been shown to be insulin independent. The plasma insulin concentrations of both 87.5%HF and 50%HF cattle showed similar ranges in different periods of lactation (Chaiyabutr et al., 2000d). Transport of glucose in the mammary gland is known to be a rate-limiting step for milk synthesis which relates to glucose transporters in the acinar cell (Burnol et al., 1990). Thus, the extraction efficiency of glucose by the mammary gland would depend mainly on the capacity for transmembrane transport and/or intracellular metabolism of glucose. However, in mammary epithelial cells of ruminants, the GLUT1 transporter protein predominating in the bovine mammary gland under normal physiological conditions (Komatsu et al, 2005; Zhao et al., 1996) is probably never saturated. The regulation of glucose extraction by bovine mammary epithelial cells may occur at another level than transmembrane transport (Xiao & Cant, 2003).

According to Chaiyabutr et al. (2000c), the studies of glucose utilization in the mammary cell using 3-[^{3}H] glucose and [U-^{14}C] glucose as markers were performed in 87.5%HF and 50%HF cattle. Then, incorporation of radioactive glucose from blood into milk components, the uptake of glucose including other substrates and milk components output were calculated and expressed as μmol/min. in the metabolic pathway. As schemes shown in Fig. 5 & 6, the quantitative utilization of the glucose taken up by the mammary gland for the synthesis of lactose, metabolize via the pentose phosphate pathway, Embden-Meyerhof pathway and the tricarboxylic acid cycle showed differences between 87.5%HF and 50%HF cattle. The calculated percentage of metabolism of glucose 6-phosphate pool to galactose moiety of lactose during early lactation was about 92% in 87.5%HF cattle and 80% in 50%HF cattle (Chaiyabutr et al 2000c). The percentages of these values were consistent with the results reported in normally fed goats and bos Taurus cows (Wood et al., 1965; Linzell, 1968). During lactation advance to late lactation, the percentages of these values decreased markedly in 87.5%HF cattle. The synthesis of lactose usually involves combination of glucose and UDP-galactose; the UDP-galactose originates from glucose-6-phosphate (Ebner & Schanbacher, 1974). The availability of cytosolic glucose 6-phosphate in the cell is usually sufficient to account for the cytosolic lactose synthesis in all periods of lactation. In mid- and late lactation of 87.5%HF cattle, a decrease in the process of lactose biosynthesis might result of it impaired synthesis. The mammary gland cannot synthesize its own glucose, since the activity of glucose 6-phosphatase is low or absent in the mammary tissue (Davis & Bauman, 1974), reducing the chance for reversibility of the glucose 6-phosphate to glucose reaction. It is possible that a decrease in the process of lactose biosynthesis may occur at the last step catalyzed by lactose synthetase (Jones, 1978), and at the same time there is an increase in cytosolic concentration of glucose 6-phosphate. The reduction of glucose uptake during lactation advance in 87.5%HF cattle might be due to the inhibition of the enzyme hexokinase by high cytosolic concentration of glucose 6-phosphate pool. However, lactose biosynthesis is a complex process (Kuhn et al., 1980), there is still a need for more information to elucidate the changes in enzymatic activity in this particular system in different stages of lactation and in different breeds of cattle.

During the reduction in the lactose synthesis in mid- and late lactation in 87.5%HF cattle, there were no significant changes in the rate of fatty acid synthesis *de novo* in the mammary gland (Chaiyabutr et al 2000c). In contrast to 87.5%HF cattle, there was a low in the flux of glucose 6-phosphate through the pentose phosphate pathway in 50%HF cattle (Fig. 6). The glucose taken up by the mammary gland were metabolized in the pentose phosphate pathway about 10-21% in 87.5%HFcattle and 5-7% in 50%HF cattle in early to mid lactation (Chaiyabutr et al 2000c). Glucose metabolism accounted for the NADPH production from pentose phosphate pathway required for fatty acid synthesis *de novo* in the udder about 21% in 87.5%HF cattle and 15% in 50%HF cattle. The differences in glucose and lipid metabolism between 87.5%HF and 50%HF cattle would be accounted for by the fact of the different breeds of animals with different rates of udder metabolism. The high rate of glucose metabolism accounted for the NADPH production in the pentose phosphate pathway during early and mid-lactation, which also associated with the high transfer of carbon atom of glucose to glycerol for the esterification in milk fat. The percentages of glycerol synthesized from glucose carbon (using U-^{14}C glucose as marker) in vivo, were lower than those reported in studies in vitro by Hardwick and co-worker (1963), in which 23% of triacylglycerol glycerol was synthesized from glucose in the perfused goat udder. The difference can be attributed to the fact that the glucose carbon taken up by the mammary gland in intact conscious animals can be accounted for synthesis of lactose, citrate, CO_2 and lost as venous-plasma lactate. Some glucose may also be converted into amino acids for milk protein synthesis (Linzell & Mepham, 1968). In 87.5%HF cattle, the utilization of glucose carbon by mammary epithelial cells for the synthesis of triacylglycerol(using U-^{14}C glucose infusion) was lower in early lactation and conversely, the values expressed as percentages are higher during mid- and late lactation (Chaiyabutr et al 2000c). An increase in the proportion of ^{14}C-glucose converted to triacylglycerol was consistent with a large proportion of glucose 6-phosphate which was metabolized via the Embden-Meyerhof pathway (EMP) in late lactation in 87.5%HF cattle (Fig.5).

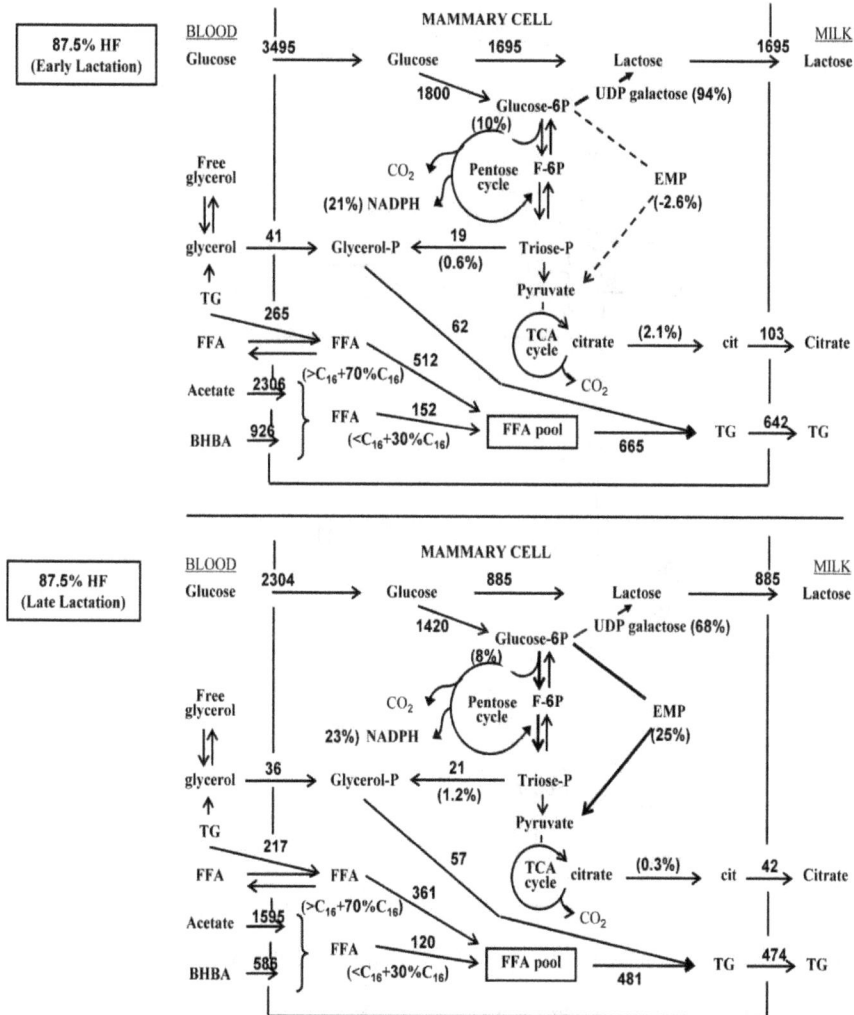

Figure 5. Scheme for calculated metabolic pathways involve intracellular partitioning of precursors of milk in mammary epithelial cells during early and late lactation of 87.5%HF cattle. Values are shown in micromole/min. (Calculated data from Chaiyabutr et al 2000c)

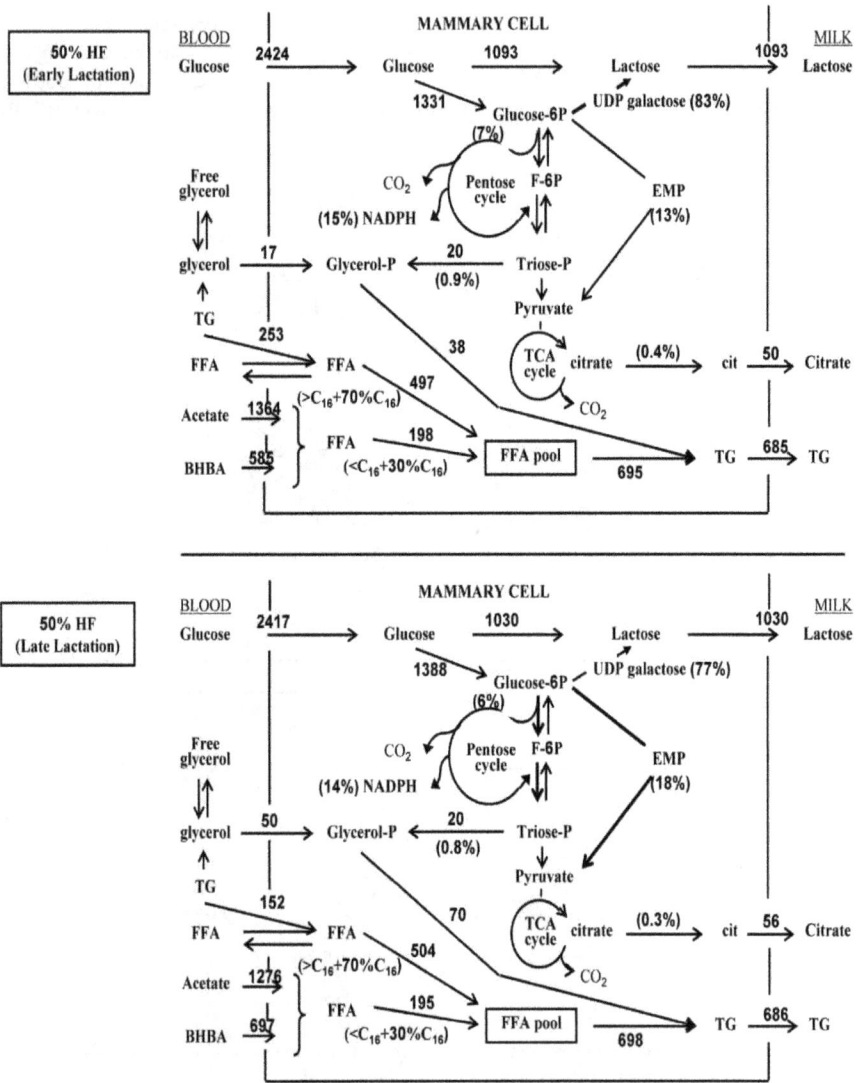

Figure 6. Scheme for calculated metabolic pathways involve intracellular partitioning of precursors of milk in mammary epithelial cells during early and late lactation of 50%HF cattle. Values are shown in micromole/min. (Calculated data from Chaiyabutr et al 2000c)

It is known that acetate, β-hydroxybutyrate, long chain fatty acids of triacylglycrol and free fatty acid fractions of plasma are major precursor of milk fatty acids. The balance data for the utilization of both short chain and long chain fatty acid were performed by calculating their likely contribution to milk free fatty acid knowing its composition and subtracting these values from the measured uptake of the substrates (Fig.5 & 6). Acetate and β–hydroxybutyrate were grouped together because both acetate and β–hydroxybutyrate were contributed to the synthesis of milk fatty acids up to and including C_{16} (Annison et al., 1968; Palmquist et al., 1969). In normally fed cattle, acetate is the main substrates for milk fat synthesis and β–hydroxybutyrate also make a more significant contribution to oxidative metabolism in the udder. There is a negligible oxidation of free fatty acid by the udder in normally fed animals (Bickerstaffe et al., 1974). The insignificant net uptake of free fatty acids by the mammary gland has been shown to be due to the simultaneous uptake of free fatty acids and release into mammary venous blood of fatty acid derived from plasma triacylglycerol which are hydrolysed during their uptake by the mammary gland (West et al., 1972). During mid- and late lactation of 87.5%HF animals there are marked decreases in utilization of both acetate and β–hydroxybutyrate which are due to the decrease in their supplies from the blood stream (Chaiyabutr et al., 2002).

During early lactation, the milk fatty acid concentrations with a chain length of C_6 to C_{18} of 50% HF cattle were higher than those of 87.5% HF cattle feeding on similar diet (Chaiyabutr et al 2000c). During mid- and late lactation, similar milk fatty acid concentrations were maintained as in early lactation of 50% HF animals. There was considerable variation with advanced lactation in the level of milk fatty acid concentration of 87.5% HF cattle. During mid-lactation and late lactation, the milk fatty acid concentration, particularly with a chain length of C_{16} to C_{18}, increased to the same level as that in 50% HF animals.

4. Role of changes of hormone supply

It has been realized that during late pregnancy, lactogenesis occurs concurrently with mammary development and many hormones are needed for maximal stimulation of lactogenesis. There are many factors have been reported to capable of influencing lactation persistency in ruminant. The hormonal control of substrates uptake of the mammary gland and milk yield may be expected to occur in crossbred lactating cattle. However, the way in which hormones act is complex. The studies of Chaiyabutr et al (2000d) in measurements plasma hormone levels of different types of lactating crossbred cattle are shown in Table 1. There were no differences in the mean plasma thyroxine (T4) concentrations during late pregnant period and lactating periods of experimental animals between 87.5%HF and 50%HF cattle which were given the same period and identical rations. However, plasma triiodothyronine (T3) concentrations of both types of crossbred HF cattle were lowered in late pregnancy as compared to lactating periods. The difference of the pattern of changes between T3 and T4 at the onset of lactation may be suggestive of an active and rapid transformation of T4 to T3 (Boonnamsiri et al 1979). The maintenance in high levels of plasma T3 concentrations throughout lactation suggested that T3 might act as an important factor in the regulation of lactation.

Changes in endocrine status during the transition period from late pregnancy to lactation will influence metabolism and the nutritional status. The pattern of differences in insulin concentrations between late pregnancy and early lactation could not be attributed to diurnal variation (Bines et al., 1983) and feeding effect (Bassett 1974). The mean plasma insulin concentration of both 87.5%HF and 50%HF cattle increased during the lactating period as compared to the late pregnant and it remained constant in a higher level throughout the lactating period. An elevation of plasma insulin levels during the onset of lactation might be a factor involved in changes in glucose turnover rate (Chaiyabutr et al., 1998). However, the plasma glucose concentrations in 87.5%HF and 50%HF cattle did not coincide with an increase in the plasma insulin concentration throughout periods of lactation. The plasma glucose concentrations of 50%HF cattle in both pregnant and lactating periods were higher than those of 87.5%HF cattle. The hyperactivity of adrenal cortex to produce the higher levels of plasma cortisol in 50%HF cattle might play a role for a rise of the plasma glucose level.

There is evidence that a lipogenic role will be expected for an elevation of plasma insulin levels during lactation by the documented fall in plasma FFA concentrations which occurred throughout lactation in crossbred HF cattle. During late pregnancy, low levels of the plasma insulin concentration coincided with higher level of plasma FFA concentrations. Several mechanisms can propose to contribute to the changes in lipogenesis, e.g. the movement of energy substrates away from the adipose tissues stores causing an elevation of plasma FFA during low levels of the plasma insulin (Yang & Baldwin, 1973) or the decreases in the sensitivity of adipose tissue to insulin at the onset of lactation (Faulkner & Pollock, 1990). During late pregnancy, mammary growth and foetus development could account for energy deficit relating to the elevation of plasma FFA concentrations. The depressed plasma insulin during late pregnancy might represent a part of the mechanism permitting mobilization of energy store. The higher level of plasma progesterone during pregnancy in both types of crossbred HF might be the other factors that contributed to an increase in plasma FFA concentration (Shevah et al., 1975). A marked elevation of plasma estradiol, primarily estrone, during late pregnancy and dramatic decreases after parturition in both 87.5%HF and 50%HF cattle, would decrease feed intake at the period of prepartum and developing high plasma FFA levels (Grummer et al., 1990).

The higher levels in plasma growth hormone (GH) during late pregnancy in 50%HF cattle could account for an increase in plasma FFA concentrations in comparison to those of 87.5%HF cattle. During lactation advance to mid and late lactation in 87.5%HF cattle, the decrease in the level of GH coincided with the decreases in the plasma FFA concentration and milk yield. The higher level of GH in early lactation of both 87.5%HF animals might influence on mammary blood flow in this period. An increase in mammary blood flow will relate to increase milk yield by contributing to a partitioning of nutrients to the mammary gland (Davis & Collier 1985; Peel et al., 1981). The control mechanism for the mammary function during transition period from pregnancy to lactation probably differed between 50%HF and 87.5%HF cattle. The higher level of GH during late pregnancy comparing to early lactation of 50%HF appeared to have no effect on mammary blood flow. The triggering of mammary blood flow and lactogenesis will involve a complex interaction of hormonal events. In late pregnancy the onset of copious lactation is overcome by the inhibition action of progesterone. Falling concentrations of progesterone after parturition would release the mammary gland from this inhibition and the rate of milk synthesis and mammary blood flow rise to a value that becomes limited by new factors perhaps the actions of GH. Plasma prolactin concentrations vary within narrow limits and did not differ between 50%HF and 87.5%HF cattle either in late pregnancy or during lactation. It has been shown that circulating prolactin in cattle can be raised or lowered by day length (Bourne &Tucker 1975) or ambient temperature (Wetteman and Tucker, 1974).

Hormones	Cattle	Late pregnant	Early lactation	Mid lactation	Late lactation
Triiodothyronine	87.5%HF	87.8	100.5	114.8	114.8
(ng/100 ml)	50%HF	87.5	132.2	131.5	126.4
Thyroxine (T4)	87.5%HF	3.66	3.67	3.94	3.69
(µg/100 ml)	50%HF	3.78	3.99	3.69	3.71
Prolactin (ng/ml)	87.5%HF	4.38	5.31	6.10	8.51
	50%HF	6.72	6.69	6.74	7.79
Growth hormone	87.5%HF	8.45	12.05	9.10	7.91
(ng/ml)	50%HF	15.16	9.03	8.76	9.76
Cortisol (ng/ml)	87.5%HF	8.9	14.2	5.7	6.2
	50%HF	6.8	29.2	25.3	35.9
Insulin	87.5%HF	12.35	18.65	23.71	21.06
(µ U/ml)	50%HF	16.55	23.03	22.91	22.92
Glucagon	87.5%HF	41.8	54.7	82.7	65.5
(pg/ml)	50%HF	57.0	67.2	118.6	112.1
Progesterone	87.5%HF	3.85	0.38	0.72	2.27
(ng/ml)	50%HF	4.45	0.12	1.42	2.16
Estradiol	87.5%HF	122.4	14.0	11.9	14.2
(pg/ml)	50%HF	175.6	18.3	22.1	25.9

Table 1. Concentrations of triiodothyronine, thyroxine, prolactin, growth hormone, Cortisol, insulin, glucagon, progesterone and estradiol in plasma of crossbred HF animals during late pregnancy and different stages of lactation. (From Chaiyabutr et al., 2000d)

5. Other physiological responses in different stages of lactation in crossbred Holstein cattle

The mechanisms that limit the rate of milk yield and shorter lactation persistency as lactation advances in crossbred dairy cattle in tropics are not only animal genetics that have to be considered but other factors, for example, high environmental temperatures and hormonal factors can influence milk production of cows (Collier et al., 1982). High environmental temperatures are known to affect milk secretion at various levels of mechanisms in dairy cattle both directly and indirectly. Thermal effect with heat stress will affect bodily functions of dairy cattle. Many technologies are required to improve milk production of dairy cattle in the tropics. Many studies have been done in attempting to improve dairy productivity by management strategies in high environmental temperatures. Environmental modification is the most common to reduce the impacts of high temperature for increase milk production, for example water spray with fans (Fike et al., 2002), or evaporative cooling system (Chan et al., 1997; Chaiyabutr et al., 2008). Bovine growth hormone or somatotropin (bST) is a homeorrhetic hormone connected with growth and lactation in ruminant (Bauman, 1992). The concentration of plasma bovine somatotropin (bST) in 87.5% HF cattle decreased rapidly as lactation progressed to mid and late lactation(Chaiyabutr et al., 2000d). This decrease would accompany with a reduction in both mammary blood flow and milk yield. Many studies have demonstrated the efficacy of bST for improvement in milk yield (Breier et al., 1991 ; Burton et al., 1994). Long term exogenous recombinant bovine somatotropin (rbST) in 87.5% crossbred Holstein cattle increased in milk yield which accompanied with an increase in the rate of mammary blood flow, but the stimulant effect for milk yield was less in late lactation despite a high level of mammary blood flow (Chaiyabutr et al. 2007). It is not known which factors are the cause and which factors are the effects for such reduction.

The additive effects of cooling and supplemental recombinant bovine somatotropin (rbST) in responsible for the short persistency of milk yield in crossbred Holstein cattle has been carried out in 87.5%HF cattle. Long term exogenous recombinant bovine somatotropin (rbST) in 87.5% HF cattle could increase the rate of mammary blood flow, total body water (TBW) and extracellular fluid (ECF) in association with an increase in plasma IGF-I (Chaiyabutr et al. 2005). An increase in mammary blood flow would contribute nutrients partitioning to the mammary gland for milk synthesis. Greater water retention during rbST administration would not only provide a greater reservoir of soluble metabolites for biosynthesis of milk, but it may be useful in slowing down the elevation in body temperature during heat exposure.

The milk yield of 87.5% HF cattle either under misty fan cooling or non-cooling increase after peak during supplemental rbST in early stage of lactation; thereafter milk yields continued to decline as lactation advanced, while mammary blood flow and body fluids volume increased during rbST supplementation in each stage of lactation in both cooled cow under misty fan and non-cooled cow (Sitprija et al.,2010). An increase in ECF by the effects of rbST in both cooled and non-cooled cows lead to an increase in MBF as secondary responses in facilitating increased milk production. However, changes in other bodily functions of crossbred cattle were also involved by effects of mist-fan cooling and supplemental rbST. In crossbred lactating cows, the response to rbST supplementation for milk production could be enhanced under mist-fan cooling. An increase milk production during rbST supplementation was mediated via increase in efficiency of feed utilization without changes in diets digestibility (Chanchai et al., 2010). Alterations of plasma hormones concentration especially IGF-I would increase during rbST supplementation in each stage of lactation in both cooled and non-cooled cows. The effect of exogenous rbST for increase in milk productions required IGF-I as a mediator to increase in mammary blood flow for increasing the availability of substrates to the mammary gland for milk synthesis.

6. Conclusion

The aim of this review is to highlight the regulatory mechanisms underlying marked low milk yields, by providing an updated summary of the results obtained concerning extra-mammary factors and intra-mammary factors of crossbred dairy cows in the tropic. In conclusion, the 87.5%HF animal had the genetic potential for a high milk yield and homeorhetic adaptation for mammary function differed from 50%HF animals during periods of lactation. Altering lactation persistency in 87.5%HF was regulated mainly by chronically acting of growth hormones through the period of lactation. The utilization of glucose in the mammary gland was determined by measuring rates of glucose uptake and the incorporation of glucose into milk components in both groups of 50% HF and 87.5% HF cattle. In early lactation, there were no significant differences of the total glucose entry rate and glucose carbon recycling between crossbred cattle. The percentages and values of non-mammary glucose utilization increased during lactation advance in both 50% HF and 87.5% HF cattle. The percentage of glucose uptake for utilization in the synthesis of milk lactose by the mammary gland of 87.5% HF cattle was higher than 50% HF cattle. Intracellular glucose6-phosphate metabolized via the pentose phosphate pathway accounted for the NADPH (reducing equivalent) of fatty acid synthesis in the mammary gland being higher in 87.5% HF animals during mid- lactation. A large proportion of metabolism of glucose via the Emden-Meyerhof pathway in the mammary gland was more apparent in 50% HF cattle than 87.5% HF cattle during early and mid- lactation, while it markedly increased for 87.5% HF cattle during late lactation. The glucose utilization for biosynthetic pathways in the mammary gland of 50% HF animals was maintained in a similar pattern throughout periods of lactation. A poorer lactation persistency of 87.5% HF animals occurred during lactation advance, which was related to a decrease in the lactose biosynthetic pathway. As advanced lactation, local changes for biosynthetic capacity within the mammary gland would be a factor in identification of the utilization of substrates in the rate of decline in milk yield. The proportion of glucose would be metabolized less for lactose synthesis, but metabolized more via the Embden-Meyerhof pathway and the tricarboxylic acid cycle as lactation advances to late lactation.

Author details

Narongsak Chaiyabutr
Department of Physiology, Faculty of Veterinary Science, Chulalongkorn University, Bangkok, Thailand

7. References

Annison, E.F.; Linzell, J.L. & West, C.E. (1968). Mammary and whole animal metabolism of glucose and fatty acids in fasting lactating goats. *Journal of Physiology*, Vol. 197, pp. 445-459.

Bassett J.M. (1974). Diurnal patterns of plasma insulin, GH, corticosteriod and metabolite concentrations in fed and fasted sheep.Australian Journal of Biological Sciences, Vol. 27, pp. 167-168.

Bauman D.E. & Currie W.B. (1980). Partitioning of nutrients during pregnancy and lactation: a review of mechanisms involving homeostasis and homeorhesis. *Journal of Dairy Science*, Vol. 63, pp. 1514 –1529.

Bauman D.E. (1992). Bovine somatotropin: review of an emerging animal technology. *Journal of Dairy Science*, Vol. 75, No. 3, pp. 3432-3451.

Bickerstaffe,R.; Annison,E.F. & Linzell, J.L. (1974). The metabolism of glucose, acetate, lipids and amino acids in lactating dairy cows. *Journal of Agricultural Science*, Vol82, pp 71-85.

Bines, J.A.; Hart, I.C. & Morant, S.V. (1983) Endocrine control of energy metabolism in the cow: diurnal variations in the concentrations of hormones and metabolites in the blood plasma of beef and dairy cows. *Hormone and metabolic Research*, Vol. 15. pp. 330-334.

Boonnamsiri, V.; Kermode, J.C. & Thompson, B.D. (1979). Prolonged intravenous infusion of labelled iodocompounds in the rat: [125]I thyroxine and (^{125}I) triiodothyronine metabolism and extrathyroidol conversion of thyroxine to triiodothyronine. *Journal of Endocrinology*, Vol. 82, pp. 235-243.

Bourne, R.A. & Tucker, H.A. (1975) Serum prolactin and LH responses to photoperiod in bull calves. *Endocrinology*, Vol. 97, pp. 473-475.

Breier, B.H.; Gluckman, P.D.; McCutcheon, S.N. & Davis, S.R. (1991). Physiological responses to somatotropin in the ruminant. *Journal of Dairy Science*, Vol. 74 (suppl.2), pp. 20-34.

Burnol,A.F.; Leturque, A.; Loizeaau, M.; Postic, C., & Girard, J. (1990). Glucose transporter expression in rat mammary gland. *Biochemical Journal*, Vol. 270, pp. 277-279.

Burton, J.L.; McBride, B.W.; Block, E.; Glimm, D.R. & Kennelly, J.J. (1994).A review of bovine growth hormone. *Canadian Journal of Animal Sci*ence, Vol. **74** pp. 167-201.

Chaiyabutr, N.; Faulkner, A. & Peaker, M. (1980). The utilization of glucose for the synthesis of milk components in the fed and starved lactating goat in vivo. *Biochemical Journal*, Vol. 186 ,pp. 301-308.

Chaiyabutr, N.; Komolvanich, S.; Sawangkoon, S.; Preuksagoon, S. & Chanpongsang, S. (1997). The regulation of body fluids and mammary circulation during late pregnancy and early lactation of crossbred Holstein cattle feeding on different types of roughage. *Journal of Animal Physiology and Animal Nutrition*, Vol. 77, pp. 167-179.

Chaiyabutr, N., Komolvanich, S.; Sawangkoon, S.; Preuksagorn, S.& Chanpongsang, S. (1998). Glucose metabolism in vivo in crossbred Holstein cattle feeding on different types of roughage during late pregnancy and early lactation. *Comparative Biochemistry and Physiology. Part A*, Vol. 119, pp. 905-913.

Chaiyabutr,N.; Preuksagorn,S.; Komolvanich,S. & Chanpongsang,S. (2000a). Comparative study on the regulation of body fluids and mammary circulation at different states of lactation in crossbred Holstein cattle feeding on different types of roughage. *Journal of Animal Physiology and Animal Nutrition*, Vol. 83, pp. 74-84.

Chaiyabutr,N.; Preuksagorn, S.; Komolvanich, S. & Chanpongsang, S.(2000b). Glucose metabolism in crossbred Holstein cattle feeding on two types of roughage at different stages of lactation. *Comparative Biochemistry and Physiology. Part A*, Vol. 125(1), pp.121-130.

Chaiyabutr,N.; Komolvanich, S.; Preuksagorn,S. & Chanpongsang, S. (2000c). Comparative studies on the utilization of glucose in the mammary gland of crossbred Holstein cattle feeding on different types of roughage during different stages of lactation. *Asian-Australasian Journal of Animal Science*, Vol. 13(3), pp. 334-347.

Chaiyabutr,N.; Komolvanich,S.; Preuksagorn, S. & Chanpongsang, S.(2000d). Plasma levels of hormones and metabolites as affected by the forages type in two different types of crossbred Holstein cattle. *Asian-Australasian Journal of Animal Science*, Vol. 13(10), pp. 1359-1366.

Chaiyabutr, N.; Thammacharoen, S.; Komolvanich, S. & Chanpongsang, S. (2002). Studies on the mode of uptake of plasma glucose, acetate, β-hydroxybutyrate triglyceride fatty acids and glycerol by the mammary gland of crossbred Holstein cattle feeding on different types of roughage. *Asian-Australasian Journal of Animal Science*, Vol. 15(10), pp. 1445-1452.

Chaiyabutr, N.; Thammacharoen, S.; Komolvanich, S. & Chanpongsang, S. (2005). Effects of long-term administration of recombinant bovine somatotropin on milk production and insulin like growth factor-I and insulin in crossbred Holstein cows. *Journal of Agricultural Science (Cambridge)*, Vol. 143, pp. 311-318.

Chaiyabutr, N.;. Thammacharoen, S.; Komolvanich, S. & Shanpongsang, S. (2007). Effects of long-term exogenous bovine somatotropin on water metabolism and milk yield in crossbred Holstein cattle. *Journal of Agricultural Science (Cambridge)*, Vol. 145, pp. 173-184.

Chaiyabutr, N.; Chanpongsang, S. & Suadsong, S. (2008). Effects of evaporative cooling on the regulation of body water and milk production in crossbred Holstein cattle in a tropical environment. *International Journal of Biometeorology*, Vol. 52, pp. 575-585.

Chan, S.C.; Hubber J.T.; Chen K.H.; Simas J.M. & Wu, Z. (1997). Effects of ruminally inert fat and evaporative cooling on dairy cows in hot environmental temperatures. *Journal of Dairy Science*, Vol. 80, pp. 1172–1178.

Chanchai, W.; Chanpongsang, S. & Chaiyabutr,N. (2010). Effects of cooling and supplemental recombinant bovine somatotropin on diet digestibility, digestion kinetics and milk production of cross-bred Holstein cattle in the tropics. *Journal of Agricultural Science (Cambridge)*, Vol. 148, pp. 233–242.

Chase L.E., (1993). Developing nutrition programs for high producing dairy herds. *Journal of Dairy Science*, Vol. 76, pp. 3287-3293.

Collier, R.J.; Beede, D.K.; Thatcher, W.W.; Israel, L.A. & Wilcox, C.J. (1982). Influences of environment and its modification on dairy animal health and production. *Journal of Dairy Science*, Vol 65, pp.2213–2227.

Coulon, J.B.; Perochon, L. & Lescourret, F. (1995). Modelling the effect of the stage of pregnancy on dairy cow's milk yield. *Journal of Animal Science*, Vol. 60, pp. 401-408.

Davis, C.L. & Bauman, D.E. (1974). General metabolism associated with the synthesis of milk. In: *Lactation Vol. II*, B.L. Larson,& V.R Smith,(Eds.), pp. 3-30, Academic Press, New York and London.

Davis S.R, & Collier R.J. (1985). Mammary blood flow and regulation of substrate supply for milk synthesis. *Journal of Dairy Science*, Vol. 68, pp. 1041–1058.

Ebner, K. E. & Schanbacher, F.L. (1974). Biochemistry of lactose and related carbohydrates. In: *Lactation Vol. II,*. B.L. Larson,& V.R Smith,(Eds.), pp. 77-113, Academic Press, New York and London.

Faulkner, A. & Peaker, M. (1987). Regulation of mammary glucose metabolism in lactation. In: *The Mammary gland: development, regulation and function*, M.C. Neville & Daniel CW (Eds.), pp 535–562,Plenum Press, New York.

Faulkner, A. & Pollock, H.T. (1990). Metabolic responses to euglycaemic hyperinsulinaemia in lactating and non-lactating sheep in vivo. *Journal of Endocrinology*, Vol. 124, pp. 59-66.

Fike, J.H.; Staples, C.R.; Sollenberger, L.E.; Moore, J.E. & Head, H.H. (2002). Southeastern pasture-based dairy systems: housing, posilac, and supplemental silage effects on cow performance. *Journal of Dairy Science*, Vol. 85, pp. 866–878.

Grummer, R.R.; Bertics, S.J.; LaCount, D.W.; Snow, J.A.; Dentine, M.R. & Staauffacher, R.H. (1990). Estrogen induction of fatty liver in dairy cattle. *Journal of Dairy Science*, Vol. 73, pp 1537.

Hahn, G.L.; Mader, T.L.; Gaughan, J.B.; Hu, Q. & Nienaber, J.A. (1999). Heat waves and their impacts on feedlot cattle, *Proceedings of 15th International Society Biometerology congress*, Sydney, Australia, September, pp. 353-357.

Hanwell, A. & Peaker, M. (1977). Physiological effects of lactation on the mother. In: *Comparative Aspects of Lactation*, M. Peaker, (Ed.), pp.297-312, The Zoological Society of London, Academic Press.

Hardwick, D.C.; Linzell, J.L. & Price, S.M. (1961). The effect of glucose and acetate on milk secretion by the perfused goat udder. *Biochemical Journal*, Vol.80, pp.37-45.

Hardwick, D.C.; Linzell, J.L. & Mepham, T.M. (1963). The metabolism of acetate and glucose by the isolated perfused udder 2. The contribution of acetate andglucose to carbondioxide and milk constituents. *Biochemical Journal*, Vol.88, pp.213-220..

Hart, I.C. (1973). Effect of 2-Bromo-2-ergocryptine on milk yield and the level of prolactin and growth hormone in the blood of the goat at milking. *Journal of Endocrinology*. Vol. 57, pp.179-180.

Hart, I.C.; Lawrence, S.E. & Mepham, T.B. (1980). Effect of exogenous growth hormone on mammary blood flow and milk yield in lactating goats. *Journal of Physiology*,Vol. 46, pp. 308.

Horsfield, S.; Infield, J.M. & Annison, E.F. (1974). Compartmental analysis and model building in the study of glucose kinetics in the lactation cow. *Proceedings of the Nutrition Society*, Vol. 33, No.1, pp.9-15.

Jayasuriya, M.C.N. & Perera, H.G.D. (1982). Urea-ammonia treatment of rice straw to improve its nutritive value for ruminants. *Agricultural Wastes*, Vol.4, pp.143-150.

Jones, E.A. (1978). Lactose Biosynthesis. In: *Lactation Vol. II*, B.L. Larson,& V.R Smith,(Eds.), pp. 371-385, Academic Press, New York and London.

Komatsu, T.; Itoh, F.; Kushibiki, S. & Hodate K. (2005). Changes in gene expression of glucose transporters in lactating and nonlactating cows. *Journal of Animal Science*, Vol. 83, pp. 557–564.

Kaufmann, W. & Hagemeister, H. (1987). Composition of milk. In: *Dairy-cattle production*, H.O. Gravert (Ed), pp 107–171, Elsevier Science Publishers BV, Amsterdam.

Kuhn, N.J.; Carrick, D.T. & Wilde, C. J. (1980). Milk synthesis. *Journal of Dairy Science*, Vol. 63, pp 328-336.

Lacasse, P.; Farr, V.C. ; Davis, S.R. & Prossser, C.G. (1996). Local secretion of nitric oxide and the control of mammary blood flow. *Journal of Dairy Science,* Vol 79, pp.1369-1374.

Leng, R. A. and C. E. West. 1969. Contribution of acetate, butyrate, palmitate, stearate and oleate to ketone body synthesis in sheep. *Research in Veterinary Science,* Vol. 10, pp.57-63.

Lindsay, D.B. (1973). Metabolic changes induced by pregnancy in the ewe. In: *Production disease in farm animals,* E.J.M. Payne; K.G. Hibitt & B.F. Sansom, (Eds). pp.107-114, Bailliere, Tindal, London.

Linzell, J.L. & Mepham, T. B. (1968). Mammary synthesis of amino acids in the lactating goat. *Biochemical Journal,* Vol.107, pp.18-19.

Linzell, J.L. & Peaker, M. (1971). Mechanisms of milk secretion. Physiological Review, Vol. 51, pp.564-597.

Linzell, J.L. (1968). The magnitude and mechanisms of the uptake of milk precursors by the mammary gland. *Proceedings of the Nutrition Society,*Vol. 27, pp. 44-52.

Linzell J.L. (1973). The demands of the udder and adaptation to lactation. In: Production disease in farm animals,J.M. Payne; K.G. Hibbitt & B.F. Sansom (Eds.),pp 89-106, Bailliere, Tidal, London.

Linzell, J. L. & Mepham, T. B. (1974). Effect of intramammary arterial infusion of essential amino acids in the lactating goat. *Journal of Dairy Science,* Vol. 41, pp. 101-109.

Linzell, J.L. (1974). Mammary blood flow and methods of identifying and measuring precursors of milk. In: *Lactation Vol. II,* B.L. Larson,& V.R Smith,(Eds.), pp. 143-225, Academic Press, New York and London.

MacFarlane, M.V. a& Howard, B. (1970). Water in the physiological ecology of ruminants. In: *Physiology of digestion and metabolism in Ruminants.* Phillipson, A.T. (Ed.), pp. 362-374, Oriel Press, Newcastle upon Tyne.

Mepham, T. B. (1993). The development of ideas on the role of glucose in regulating milk secretion. *Australian Journal of Agricultural Research,* Vol.44, pp. 508-522.

Murphy, M. R. (1992). Symposium: Nutritional factors affecting animal water and waste quality. *Journal of Dairy Science,* Vol.75, pp 326-333.

Nakamura, R.M.; Araki, C.T. & Chaiyabutr, N. (1993). Temperate dairy cattle for hot climates: Telemetry studies and strategy. In: *Livestock Environment IV, Fourth International Symposium* , pp.16-22, University of Warwick, England.

Palmquist, D.L.; Davis, C.L.; Brown, R.E. & Sachan, D.S.(1969). Availability and metabolism of various substrates in ruminants. V. Entry rate into the body and incorporation into milk fat of D(-)-β- hydroxybutyrate. *Journal of Dairy Science,* Vol. 52, pp. 633-638.

Peel, C.J.D.; Bauman, D.E.; Gorwit, R.C. & Sniffen,C.J.(1981). Effect of exogenous growth hormone on lactational performance in high yielding dairy cows. *Journal of Nutrition,* Vol.111, pp.1662-1671.

Promma, S.; Tasaki, I.; Cheva-Isarakul, B. & Indratula, T. (1994). Digestibility of Neutralized urea-treated rice straw and nitrogen retained in crossbred Holstein streers. *Asian-Australasian Journal of Animal Science,* Vol. 7(4), pp. 487-491.

Ranjhan, S.K.; KalanidhI, A.P.; Gosh, T.K.; Singh, U.B. & Saxena, K.K. (1982). Body composition and water metabolism in tropical ruminants using tritiated water. In: *Use of Tritiated Water in Studies of Production and Adaptation in Ruminants.* pp 117-132. International Atomic Energy Agency, Vienna.

Reid, R. L. & Hinks, N. T. (1962). Studies on the carbohydrate metabolism of sheep, XVIII. The metabolism of glucose, free fatty acid, ketones and amino acids in late pregnancy and lactation. *Australian Journal of Agricultural Research*, Vol. 13, pp.1112-1123.

Rook, J.A.F. & Wheelock, J.V. (1967). The secretion of water and water soluble constituents in milk. *Journal of Dairy Research*. Vol. 34, pp. 273-287.

Shevah, Y.; Black, W.J.M.; Carr, W.R. & Land, R.B. (1975). The effects of nutrition on the reproductive performance of Finnx Dorset ewes. 1. Plasma progesterone and LH concentrations during late pregnancy. *Journal of Reproductive Fertility*, Vol. 45, pp.283-288.

Shirley, J. E.; Emerry, R. S.; Convey, E. M & Oxender, W. D. (1973). Enzymic changes in bovine adipose and mammary tissue, serum and mammary tissue hormonal changes with initiation of lactation. *Journal of Dairy Research*, Vol 56, pp.569-574.

Sitprija, S.; Chanpongsang, S.& Chaiyabutr, N. (2010). Effects of Cooling and Recombinant Bovine Somatotropin Supplementation on Body Fluids, Mammary Blood Flow, and Nutrients Uptake by the Mammary Gland in Different Stages of Lactation of Crossbred Holstein Cattle. *Thai Journal of Veterinary Medicine*, Vol. 40, pp.195-206.

Sölkner, J. & Fuchs, W. (1987). A comparison of different measures of persistency with special respect to variation of test-day milk yields. *Livestock Production Science* Vol.16, pp.305-319.

Thompson, G.E. & Thomson E.M. (1977). Effect of cold exposure on mammary circulation oxygen consumption and milk secretion in the goat. *Journal of Physiology* Vol. 272 (1), pp.187-196.

Tucker, H.A. (1981). Physiological control of mammary growth, lactogenesis and lactation. *Journal of Dairy Science*, Vol. 64, pp.1403-1421.

Tucker, H.A.(1987). Quantitative estimates of mammary growth during various physiological states: a review. *Journal of Dairy Science*, Vol. 70, pp.1958-

West, C.E.; Annison, E.F. & Linzell, J.L. (1967). Plasma free fatty acid uptake and release by the goat mammary gland. *Biochemical Journal*, Vol. 102: 23P.

West, C.E.; Bickerstaffe, R.; Annison, E.F. & Linzell, J.L. (1972). Studies on the mode of uptake of blood triglycerides by the mammary gland of the lactating goat. The uptake and incorporation into milk fat and mammary lymph on labelled glycerol, fatty acids and triglycerides. *Biochemical Journal*, Vol 126, pp. 477-490.

Wetteman, R.P. & Tucker, H.A. (1974). Relationship of ambient temperature to serum prolactin in heifers. *Proceedings Society of Experimental Biology and Medicine*, Vol.146, pp.908-911.

Wood, H.G.; Peeters, G.J.; Verbeke, R.; Lauryssens, M. & Jacobson, B. (1965). Estimation of the pentose cycle in the perfused cow's udder. *Biochemical Journal*, Vol.96, pp.607-615.

Xiao, C.; Cant, J.P. (2003).Glucose transporter in bovine mammary epithelial cells is an asymmetric carrier that exhibits cooperativity and trans-stimulation. *American Journal of Physiology, Cell Physiolology*, Vol.285, pp. C1226-C1234.

Yang, Y.T. & Baldwin, R.L. (1973). Lipolysis in isolated cow adipose cells. *Journal of Dairy Science*, Vol 56, pp.366-374.

Zhao, F.Q.; Moseley, W.M.; Tucker, H.A. & Kennelly, J.J. (1996). Regulation of glucose transporter gene expression in mammary gland, muscle, and fat of lactating cows by administration of bovine growth hormone and bovine growth hormone-releasing factor. *Journal of Animal Science*, Vol.74, pp.183–189.

Chapter 8

Milk Casein Alleles, Haplotypes and QTL Effect on Protein and Fat Content and Milk Yield in Argentinean Criollo and Cross Goats

M.E. Caffaro, C. Suárez, D.L. Roldán, and M.A. Poli

Additional information is available at the end of the chapter

1. Introduction

Globally, most of the goats are in marginal areas and hostile environments, raised by rural poor, smallholders and small producers (Morand-Fehr & Boyazoglu, 1999).

In Argentina, the existence of goats according to data released by SENASA (National Agrarian Health Service) is 4,256,716 animals (existence until March 2011), which are held by not more than 50,000 farmers, mostly the rural poor.

The predominant activity is to produce offsprings, followed in importance of fiber, milk and hides.

The production systems are essentially extensive with grazing on natural degraded grasslands. Most of the population is in the center-north, where production is oriented mainly to meat and milk, while in northern Patagonia are most of the hairs producing goats (mohair). In the north of the province of Buenos Aires there are intensive production systems, which mainly engaged in milk production and less production of meat.

According to data published by the Ministry of Agriculture, Livestock and Fisheries Argentina milk production reaches the 3,200 tons/year from about 10,000 goats with an average production of 300 liters per animal for 210 days lactation. The production structure is divided into two groups: first, business-oriented dairy large-scale productions, which use technologies and breeds of potentially high milk yield (Toggenburg, Pardo Alpina, Saanen and Anglo Nubian) and secondly small and medium producers, who have mostly Criollo goats which produce milk and meat on a seasonal basis.

The dairy goats for cheese making have been an ancient activity in the Argentinean regions of the arid valleys of the Northwest and Cuyo, a craft developed by producers in subsistence conditions.

The Figure 1 shows Northwest Argentina area (Salta, Jujuy, Tucumán, Santiago del Estero and Catamarca provinces).

Figure 1. Map of Argentina. Northwest area highlighted

The production systems are based on the use of natural pastures and less often, irrigated pasture implanted. The herd is managed so that the kids after calving remain part of the day with the goats, thus allowing the animals to milk to obtain milk intended for cheese making. These cheeses are intended for family consumption, barter or sell the local market.

In Argentina, the Criollo goat is a "native" breed which can be considered as like-dairy breed that lives in isolated and very hard areas. Up to now, this breed has not been artificially selected for any specific trait and it has only been subjected to natural selection since brought by the Spanish people nearly 500 years ago (Rodero et al., 1992).

In the Northwest (NW) region of Argentina there are 91,532 goats and it is around of 21.4 % of the country (SENASA, existence until March 2011). Their great adaptation to specific environments allows for the exploitation in low rural areas, providing rural communities with typical products of animal origin. Moreover they constitute a potential genetic resource to be evaluated and preserved before some characteristic can be lost due to the unsystematic crosses with other highest selected breeds. Most breeders from NW region milk their goats to produce and trade a "homemade" cheese which is an important familiar economic income.

2. Milk goat proteins

Due to the molecular and genomics structure and organization of the milk protein are deeply described in a special chapter in this book, here we only give a general overview of milk protein.

In ruminants the four caseins represent about 80% of the protein in milk. It is well known that caseins are encoded by four linked and clustered genes including α_{s1}-casein (*CSN1S1*),

Milk Casein Alleles, Haplotypes and QTL Effect on Protein and Fat Content and Milk Yield in
Argentinean Criollo and Cross Goats

175

β-casein (*CSN2*), α_{s2}-casein (*CSN1S2*) and κ-casein (*CSN3*) genes. In goats, the entire casein gene cluster region spans about 250 kb on chromosome 6 (Hayes et al., 1993).

In the last years, the genetic polymorphism of goat α_{s1}-casein has been intensively studied because of its extensive variability and α_{s1}-casein its direct relationships with milk quality and composition (review by Grasclaude et al., 1994; Martin et al., 2002). So far, 18 alleles associated with different rates of protein synthesis have been identified. On the basis of the milk content of caseins, the variants can be classed into four groups: "high" alleles (A, B1, B2, B3, B4, C, H, L and M) producing almost 3.5 g/L of α_{s1}-casein each; "intermediate" alleles (E and I: 1.1 g/L); "low" alleles (D, F and G: 0.45 g/L) and "null" alleles (O1, O2, O3 and N) producing almost 0 g/L of α_{s1}-casein each (Moioli et al., 2007).

Owing to the importance of κ-casein in the technological properties of milk, the polymorphisms in the κ-casein gene have been extensively studied in ruminants. The κ-casein fraction plays a crucial role in the formation, stabilization, and aggregation of the casein micelles and thus affects the technological and nutritional properties of milk. Cheese-making is based on the cleavage of the κ-casein Phe-Met peptide bond by enzymes or heat.

In goats, the gene *CSN3*, like that of *CSN1S1* is polymorphic, which is being studied intensively. Further polymorphisms in exon 4 of the κ-casein locus were detected. Several studies, using various techniques, have reported polymorphisms in the goat κ-casein gene. Eight polymorphisms sites have been detected (Yahyaoui et al., 2003; Jann et al., 2004).

Both in cattle (Boettcher et al., 2004) and goats (Caroli et al., 2004; Prinzenberg et al., 2005) are discussed the possibility that contributions be found not only the genes specific to each casein, but the haplotypes.

3. Allele and haplotype frequency of α_{s1}-casein and κ-casein

In the last years the *"animal breeding world"* has been facing to the new paradigma quantitative-molecular for improvement. Over the past decades, unequivocal evidence has emerged for the existence of genes with major effects on many performance and fitness characteristics. The opportunity to use molecular screening of individual animals to rapidly increase the frequency of such genes is now a reality. The use of genetic markers to efficiently introgress genes into different genetic backgrounds likewise can potentially change specific genetic characteristics of a population without greatly diluting other established adaptation and product-quality traits. But the first step to use a molecular marker in any breding plan is to know the allele variability and its population frequencies.

3.1 α_{s1}-casein allele frequency and genotypes

In her Master of Science Thesis, Suárez (2003) described the polymorphism at the *CSN1S1* gene in a population of Criollo goats from the NW region. Blood samples were taken in four provinces of Argentina and genomic DNA was extracted from leukocytes following a protocol by Madisen et al. (1987). All samples were used to estimate the allelic frequencies at the *CSN1S1* gene.

The genotyping at the *CSN1S1* locus was carried out by a multiplex amplification using fluorescent primers. The PCR products were analyzed with an ABI PrismTM 310 (Applied Biosystem Inc., USA) automated sequencing system equipped with GenScanTM (version 2.1) and the GenotyperTM software (Babilliot and Amigues, personal communication). This procedure was developed by Labogena (Jouy-en-Josas, France). Also, some DNA samples were amplified by an allele specific PCR for allele E (Amills, 1996).

The *CSN1S1* allelic distribution present in this breed shows the most frequency of "high" alleles (0.68) is predominant over the "intermediate", "low" and "null" alleles (0.19, 0.23 and 0.02, respectively). The "high" alleles, A and B, have the highest frequencies, 0.30 and 0.31, respectively, followed by the E allele with a frequency of 0.19. The F allele and the O allele were found at a frequency of 0.04 and 0.02, respectively. During the genotyping process a pattern that does not belong to any known alleles was found. We checked the inheritance of this "new" variation in a heterozygous buck family with 13 kids and 11 does. We have preliminary designated it as "X allele" which was present at a frequency of 0.07.

The polymorphic information content (PIC) is a measure of informativeness related to expected heterozygosity and likewise is calculated from allele frequencies (Botstein et al., 1980; Hearne et al., 1992).

The PIC value found for the *CSN1S1* locus was 0.73 and the expected hetorozygosity 0.77. From the 28 possible genotypes (7 alleles) 23 (82%) were found. The A/B genotype had the highest frequency (0.19) and the C/X, F/F and O/X genotypes the lowest (0.005). Both A/B and B/B genotypes accounted for 31.7 % of all the genotypes analyzed. The genotypes B/O, C/C, C/O, F/O and O/O were absent. From the 17 goat breeds in which *CSN1S1* frequencies were evaluated (Table 1), the Criollo and Canaria breeds were the most similar for the A allele (0.30 and 0.28, respectively), B allele (0.31 and 0.32, respectively) and E allele (0.19 and 0.20, respectively) but were very different in the O allele (0.02 and 0.20, respectively). Although the Jónica breed seems also similar for the A, B and C allele frequencies to the Criollo breed, the F allele frequency was quite different (0.04 Criollo and 0.28 Jónica). Within "high" casein content alleles the C allele was found only in a few breeds and its frequency was always very low, but in the Criollo breed we found a frequency of 0.07, higher than any other breed.

Taken into account the historical information, that states that: " . . . during the first decade of America colonization European introduced different species of animals and some of them (swine, chicken, sheep, goats, cattle and horses) were boarded in the Canarias Island La Gomera in 1493 by Cristóbal Colón, . . ." (Gonzalez-Stagnaro, C. 1997) and considering also the phenotypic similarities between the Criollo and the Iberian breeds like Murciana-Granadina, Malagueña, Canaria and Payoya, we run a Principal Components Analysis (PCA).

The central idea of PCA is to reduce the dimensionality of a data set consisting of a large number of interrelated variables, while retaining as much as possible of the variation present in the data set (Jolliffe, I. 2002). In the PCA we found that the Criollo breed is away

Milk Casein Alleles, Haplotypes and QTL Effect on Protein and Fat Content and Milk Yield in
Argentinean Criollo and Cross Goats

177

Breed	n	A	B	C	E	F	O
Alpina (Fr)[1]	213	0.14	0.05	0.01	0.34	0.41	0.05
Alpina (It)[1]	80	0.00	0.00	0.00	0.35	0.59	0.06
Saanen (Fr)[1]	159	0.07	0.06	0.00	0.41	0.43	0.03
Saanen (It)[1]	70	0.05	0.00	0.00	0.49	0.46	0.00
Poitevine (It)[1]	209	0.05	0.35	0.00	0.45	0.14	0.00
Corse (Fr)[1]	106	0.06	0.13	0.00	0.14	0.59	0.08
Rove (Fr)[1]	147	0.12	0.05	0.00	0.62	0.10	0.11
Gargánica (It)[2]	38	0.28	0.41	0.03	0.00	0.22	0.08
Maltesa (It)[2]	70	0.41	0.16	0.00	0.06	0.37	0.00
Murciana Granadina(Sp)[1]	77	0.08	0.25	0.00	0.62	0.05	0.00
Malagueña (Sp)[1]	56	0.00	0.25	0.00	0.70	0.05	0.00
Payoya (Sp)[1]	39	0.04	0.14	0.00	0.82	0.00	0.00
Canaria (Sp)[1]	74	0.28	0.32	0.00	0.20	0.00	0.20
Vallesana (It)[2]	83	0.03	0.13	0.00	0.28	0.39	0.17
Roccaverano (It)[2]	77	0.23	0.12	0.00	0.21	0.38	0.04
Jónica (It)[2]	110	0.35	0.30	0.00	0.06	0.28	0.00
Norway Multicolor[3]	147	0.00	0.11	0.00	0.00	0.03	0.86
Criolla (Ar)[4]	214	0.30	0.31	0.07	0.19	0.04	0.02

Table 1. Allelic frequencies at the CSN1S1 locus in different breeds n: Sample size; Fr: French; It: Italy; Sp: Spain; Ar: Argentina. [1] Grosclaude et al. (1994); [2] Sacchi et al. (2003); [3] Adnoy et al. (2003); [4] Suárez (2003).

from the 4 Iberian breeds (Figure 2). The Murciana-Granadina, Malagueña and Payoya were the most important breeds in the first axe (about 41% of the total variability). In axe 2, the Criolla breed account for 20% of the total variability and in axe 3 the Canaria breed also account for 20% of the total variability.

Although the PCA results from α_{s1}-casein show that Criolla breed is quite different from Iberian breeds morphometric measures and zoometric characteristics can help to establish any phylogenetic relationship between the Criollo breed and Iberian goats. Nevertheless 500 years in the America continent under natural selection process and eventual population contraction and expansion periods, one would not expect to observe so much genetic similarities between the Criollo breed and its Iberian ancestors.

In other study, Caffaro (2007), described the polymorphisms at the CSN1S1 and CSN3 genes in a population of Criollo in CriolloxCriollo and CriolloxSaanen crosses.

The allelic variants for the CSN1S1 gene were detected using the PCR-AS and PCR-RFLP technique. The B and C alleles were gathered in B-(C) groups since they are indistinguishable with the methods developed in this study.

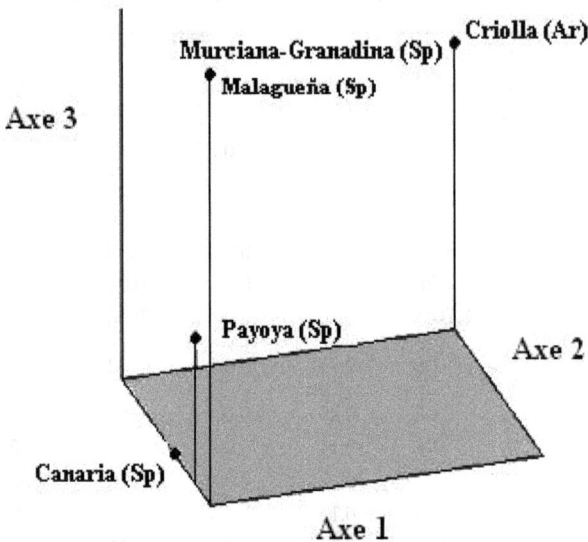

Figure 2. Results of principal component analysis by goat breeds

For *CSN1S1* gene, "high" alleles were found with most frequency (A, B and C) with 0.66. The "intermediate" allele presented by 0.17. The "low" allele was present in 0.03. Finally, the "null" allele was presented at 0.02.

For *CSN1S1* gene allelic frequencies were similar to the reported by Suárez (2003) for Creole goats in the NW. Furthermore, it is showing the X allele (0.10), as found by Suárez (2003), with a frequency slightly higher.

3.2. κ-casein allele frequencies and genotypes

In the same Criollo population described above, Caffaro (2007) using the PCR technique amplified the *CSN3* gene. The PCR products were purified by commercial kits and then the purified product was sequenced in both senses. The consensus sequence of each animal was aligned whit different alleles.

In the *CSN3* locus was found only 3 of 8 allelic variants described. The highest frequency was for the B allele (0.74), then the A allele (0.24) and lower D allele frequency (0.02) and only in heterozygous state. Moreover, the emergence of the state homozygous the allele B is significantly higher than other alleles. The number of individuals homozygous for this locus was approximately three times higher than in heterozygous (116 on 42 animals).

Two animals were also found (dam and daughter) with allele B' (subtype B allele) and an animal with a substitution at nucleotide 91 (for the sequence of allele B) not yet reported in the literature.

Milk Casein Alleles, Haplotypes and QTL Effect on Protein and Fat Content and Milk Yield in Argentinean Criollo and Cross Goats

179

Comparing the allelic frequencies found in Criollo goats with that reported for the different breed from different countries shows that the frequencies of three alleles in Criollo goats have a similar behavior to that reported in most other breed, i.e., most often found for the B allele, followed by the A allele and low frequency D allele was found.

In the Table 2 allele frequencies are presented for 26 goat breeds in different countries. Allele frequencies of this Table were taken from the following works: Chessa et al. (2003); Yahyaoui et al. (2003); Prinzenberg et al. (2005), Sacchi et al. (2005), Caroli et al. (2006 and 2007).

Breed	*n*	A	B	C	D	E	F	G	M
Murciana-Granadina[1]	30	0.37	0.63	-	-	-	-	-	-
Canaria[1]	30	0.58	0.42	-	-	-	-	-	-
Malagueña[1]	11	0.45	0.55	-	-	-	-	-	-
Teramana[3]	28	-	0.70	0.02	0.10	-	0.14	0.04	
Montefalcone[3]	17	-	0.59	-	-	0.41	-	-	-
Girgentana[3]	19	0.34	0.45	-	0.05	-	0.05	0.11	-
Sarda[3]	19	0.31	0.61	-	-	-	0.08	-	-
Alpina[2]	28	0.34	0.66	-	-	-	-	-	-
Saanen[2]	28	0.39	0.48	0.13	-	-	-	-	-
Saanen[3]	22	0.25	0.66	0.09	-	-	-	-	-
Gargania[3]	72	0.10	0.75	0.01	0.13	-	-	-	0.01
Jonica[3]	229	0.13	0.73	0.02	0.11	-	-	-	0.01
Maltesa[3]	105	0.08	0.70	0.01	0.20	-	-	-	0.01
Camosciata[3]	45	0.36	0.63	-	0.01	-	-	-	-
Vallesana[3]	83	0.37	0.60	0.03	-	-	-	-	-
Roccaverano[3]	77	0.25	0.60	0.06	0.07	-	-	-	0.01
Frisa[3]	70	0.33	0.65	0.01	0.01	-	-	-	-
Orobica[3]	66	0.08	0.55	0.02	0.35	-	-	-	-
Verzasca[3]	67	0.28	0.72	-	-	-	-	-	-
Camasciata[3]	88	0.39	0.61	-	-	-	-	-	-
Angora[4]	43	0.15	0.67	0.04	0.11	-	-	0.02	-
Weisse Deutsche Edelziege[5]	32	0.34	0.55	0.06	0.05	-	-	-	-
Borno[5]	66	0.46	0.54	-	-	-	-	-	-
Red Sokoto[5]	88	0.40	0.58	-	-	-	-	-	0.02
West African Dwarf[5]	92	0.36	0.64	-	-	-	-	-	-
Cabra de pelo[5]	50	0.31	0.26	0.04	0.25	-	-	0.14	-
Criolla[6]	137	0.24	0.74	-	0.02	-	-	-	-

Table 2. Allelic frequencies at the *CSN3* locus in different breeds N: Sample size. [1]Spain; [2]French, [3]Italy, [4]Turkey, [5]Germany, [6]Argentina

3.3. Haplotype frequencies

Haplotype means a set of closely linked alleles (genes or DNA polymorphic) inherited as a unit.

To reconstruct the most probable haplotype they use the LSPH program (Baruch et al., 2006).

LSPH program assumes that no recombination between loci. The haplotypes are reconstructed in two stages: first, the haplotypes in the parental generation are determined and the progeny haplotype for more information (no missing genotypes) are assigned. Then, the origin of each haplotype (paternal or maternal) is determined.

Once built haplotype most likely, the relative frequency was estimated by procedure FREQ of program SAS (Institute, Inc., Cary, NC).

Haplotypes frequencies the *CSN1S1-CSN3* cluster are reported in Table 3. A total of 12 possible haplotypes resulted from the alleles combination considered, 6 were highest frequency than 0.05. And the most frequent haplotypes were E-B (n=89) and B(C)-B (n=62).

	Haplotype											
	A-A	A-B	A-D	B(C)-A	B(C)-B	E-A	E-B	F-A	F-B	O-A	O-B	X-B
Frec (%)	0.035	0.116	0.013	0.038	0.200	0.087	0.289	0.025	0.071	0.011	0.025	0.090
NH	11	36	4	12	62	27	89	8	22	3	8	28

Table 3. Haplotype found. Frec (%): relative frequencies. NH: number of animals in each haplotype

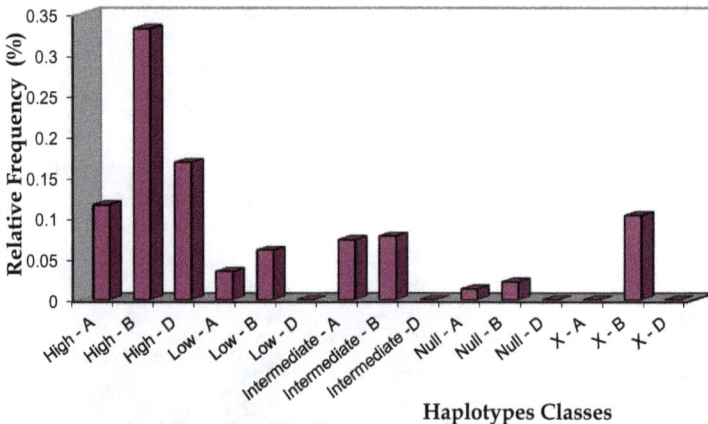

Figure 3. Relative frequency of haplotype grouped into haplotypes classes.

The X allele was only found associated with the B allele of *CSN3* (X-B). Also, the D allele of *CSN3* gene was only associated with *CSN1S1* A (A-D) and the B allele of *CSN3* was found associated with all *CSN1S1* alleles present in this study.

The results of this haplotype grouping in classes is similar those reported by other authors (Sacchi et al., 2005 and Prinzenberg et al., 2005), where they found that the B allele of *CSN3* appeared associated with two of the strongest *CSN1S1* alleles (A and B) (Figure 3).

4. Effect of α$_{s1}$-casein, κ-casein and haplotypes on production traits

Milk yield and milk components are influenced by a number of factors including animal genetic and environmental conditions. An association among physiological state of the animal, breed, parity number, season of kidding and level of production were reported by Gipson & Grossman (1990) and for milk composition by Soryal et al. (2005); Dimassi et al. (2005) and Fekadu et al. (2005).

In our study milk samples were collected from the Criollo goat belonging to the experimental flock (*n*=150) located in the Experimental Farm Leales – INTA (National Institute of Agricultural Technology), located 52 km southeast of the city of San Miguel de Tucumán, 27° 11′ latitude south and 65° 17′ west and at an altitude of 335 m above sea level. The majority of these Criollo goats were rescued from the different private flocks around the NW region and they can be considered the best representative of the breed. A sub set of 84 goats were milked. Milk samples were collected monthly by hand-milking into an individual container (50 ml) then refrigerated and analyzed for milk components (fat %-FP and protein %-PP) by Milkoscan using infrared procedures. Milk yield (MY) was recorded weekly.

To evaluate the effect of the *CSN1S1* genotypes on milk traits (MY, FP and PP) a set of 147 records from 84 milking goats was analyzed using the follow mixed model (1):

$$y = X\beta + Zu + \varepsilon \qquad (1)$$

where y is the vector of phenotype data, X and Z are the design matrix, β is the unknown vector of fixed effects, u is the vector of random effects, and ε is the error vector. Records of MY, PP and FP were transformed to aggregate data by Fleishmann method (Peña Blanco et al., 2005). The influence of four fixed factors was considered: genotype (7 levels), kidding season (autumn or spring), parity number (6 levels) and number of kids (single and double) and every lactation was considered as a repetitive data. The PROC MIXED from SAS package was used (SAS, 2005). We wrote estimable functions and performing contrast to evaluate the differences between the available genotypic groups.

The average values estimated for MY, FP and PP were: 1.214 kg/day, 2.825 % and 2.672 %, respectively.

We did not detect any differences for the four fixed effects on MY or FP from aggregate data (Table 4), but we found a significant effect of season of kidding on PP. Angulo et al. (2002) reported that the season of kidding in Malagueña goats affected both MY and PP.

Fixed effects	MY	FP	PP
Genotype	0.2679	0.5370	0.0170 *
Kidding season	0.3907	0.0600	0.0008 **
Parity number	0.2468	0.6270	0.1974
Number of kids	0.6981	0.2890	0.2630

Table 4. Significance level for fixed effects of MY (milk yield), FP (fat percentage) and PP (protein percentage). * P≤0.05; ** P≤0.01

It has been reported that the difference in casein content between an homozygous animal for "high" alleles, like A/A, and an homozygous animal for "low" alleles, like F/F, is 6 g/l (Grosclaude *et al.*, 1994), moreover the efficient transport of caseins seems to be dependent upon CSN1S1, thus animals with "low" alleles (F, G) would have a reduce solid content in milk (Chanat et al., 1999). In general goat milk with high levels of CSN1S1 have been found to have better milk composition, not only protein content, but also regarding fat, total solids, phosphorous and lower pH than milks with low levels of CSN1S1 (Grosclaude et al., 1987; Barbieri et al., 1995).

According to these results, we did not find any difference between genotypes for MY or FP. This result agree with those reported by Mahé et al. (1993); Vassal et al. (1994); Février et al. (2000) and Adnoy et al. (2003), but are rather uneasy to explain. The milk fat content is modulated by genetics and environmental factors (Barroso et al., 1999), and the regulation of gene expression in the mammary gland is not clear.

Leroux et al. (2003) investigated the association between the polymorphism at CSN1S1 locus and lipid content in caprine milk with and they conducted RT-PCR and macroarrays analyses and concluded that milk fat content is not due to differences in the expression of some enzymes of the lipogenesis pathway but could be related to the expression of some genes encoding protein anchored in the milk fat globule membrane. An alternative hypothesis has been mentioned for a possible physical linkage between the CSN1S1 locus and an unknown QTL that might influenced fat content (Manfredi, 2003), but this has not yet been investigated.

Regarding PP, significant differences exist with respect to the genotype in Criollo goats. This is also in agreement with previous results in French, Italian and Spanish breeds (Serradilla, 2003). To identify which genotype had the main influence for PP, we grouped the records according to the available genotypes and compared the differences between the estimated means (Table 5).

Milk samples having genotype for "high" alleles, A, B, C, had higher protein percentage (1.109 gr/kg) than homozygous samples for the "intermediate" allele E, as shown in Table 5. This is in agreement with the results reported by Angulo et al. (2002) for the Malagueña breed. In the Criollo goat from the NW region, although significative (P<0.05), the difference in protein content between "high" and "intermediate" genotypes was smaller than those reported in the literature.

In addition, we compare animals homozygous for the "X" allele and we did not find any significative difference between animals bearing "high" alleles and X/X goats, but we found

Milk Casein Alleles, Haplotypes and QTL Effect on Protein and Fat Content and Milk Yield in
Argentinean Criollo and Cross Goats

183

a difference with goats homozygous for the intermediate E allele (P=0.0154). This suggests
that this "X" allele might belong to the "high" alleles group.

Genotype	PP Estimated (± sd)	p-value
AA-BB-CC-AC-BC vs EE	1.109 (± 0.430)	0.0140
AA-BB-CC-AC-BC vs XX	-0.915 (± 0.727)	0.2170
EE vs XX	-2.024 (± 0.789)	0.0154

Table 5. Estimated differences among *CSN1S1* genotypes for PP mean and standard deviation and p-value.

To evaluate the effect of the *CSN3* and haplotypes on milk traits (MY, FP and PP), records
from 86 milking goats were analyzed using the follow lineal model (2)

$$y = Xb + e \tag{2}$$

where y is the vector that includes n records in each animal. Records of MY, FP and PP
were transformed to aggregate data by Fleishmann method (Peña Blanco et al., 2005). b is
the unknown vector of fixed effects, and e is the vector of residuals with covariance matrix
$I\sigma_e^2$. X is the incidence matrix relating the vector data with the vector of fixed effects.

The fixed effects considered in the matrix X were: racial biotype (Criollo and different
percentages of crossing with Saanen), age of the animal (from 9 months to 14 years), calving
season (2 levels, autumn and winter), lactation number (6 levels, from first lactation to number
6, assuming that animals born on the premises, had no previous lactations), number of kids
(single and double), α_{s1}-casein genotype (23 levels according to genotypes present) genotypes
κ-casein (4 levels: 1 = AA genotype, 2 = BB genotype, genotype 3 = AB, 4 = genotype AD). It
also adjusted the interactions between genotype α_{s1}-casein and κ-casein genotypes.

The GLM procedure from SAS package was used (SAS, 2005). We wrote estimable functions
and performing contrast to evaluate the differences between the available genotypic groups.

With respect to the *CSN3* gene, there was no significant effect on production variables.
Contrary to these results, Chiatti et al. (2005), who studied the effect of *CSN3* gene on
production variables in four Italian breeds of goats, they found that animals with BB
genotype had a higher percentage of protein and casein in milk, and animals with genotype
AB, had milk with higher fat content. These authors suggest that the results are independent
of gene expression levels *CSN1S1*.

On the other hand, when we assessed the association of the haplotypes present in the
population (α_{s1}-casein/κ-casein) with dairy character, significant effect on protein percentage
was found

We found that animals with haplotypes "high" for *CSN1S1* and *CSN3* B allele were more
frequent but also the presence of allele B was associated with higher protein levels

compared with the A allele when both were associated with alleles "high" for the gene *CSN1S1* (0.040 and 0.036, respectively). The milk from animals that contain the D allele of *CSN3* give values lower protein content than milk from animals with *CSN3* B allele when alleles are also associated with "high" locus *CSN1S1*.

Sacchi et al. (2005) and Prinzenberg et al. (2005) postulated that the B allele of *CSN3* occured most often associated with the "high" alleles *CSN1S1*, levels of casein in milk were higher than the presence of the A allele of *CSN3*. On the other hand, the average values were similar between different alleles found for *CSN3* gene when they were associated with "intermediate" *CSN1S1* gene.

5. QTL detection for milk production traits in goats

The use of quantitative information in livestock breeding programmes has become more sophisticated over time in order to allow breeders to make faster progress in a chosen set of traits. Quantitative information was initially used in mass selection, whereby individuals with better trait values were chosen to be parents of the next generation. This progressed to using information from relatives and multiple traits, by assuming an infinitesimal (and usually additive) model of inheritance.

This approach maybe criticized for the simplicity of this assumption with regard to the actual mechanisms of gene action, but has proven highly successful in generating genetic change in livestock populations.

Most of the economically important traits for breeding animals show a continuous distribution of observations as a result of the action of polygene and the environment (Falconer & Mackay, 1996; Lynch & Walsh, 1998). Turn polygene are the result of a large number of additive gene actions between quantitative trait loci (QTL) scattered throughout the genome. Each QTL shows a small effect on the total variation present in the character. Such genes could be detected as segregation in pedigrees, either by observation (in the case of extreme effects) or by sophisticated statistical analysis. The advent of easily scored genetic markers (i.e., microsatellites and SNPs-Single Nucleotype Polymorphism) spaced across the genome of a species has allowed the inheritance of each position in the genome to be traced from parents to progeny, and consequently has allowed more powerful tests of segregation to be developed. Then, we can define a QTL as a chromosomal region that contains one o more genes that encoded a quantitative trait (Andersson & Georges, 2004).

Linkage-based quantitative trait loci (QTL) mapping is based on the linkage disequilibrium observed within a family (Lynch & Walsh, 1998), exploiting recombination in pedigreed and genotyped generations. In livestock the most usual breeding system is to use one male mated with many females producing paternal half-sib family designs. This is the simplest and common commercial herds desing situation to QTL detection.

For mapping genes, there are basically two strategies: the candidate gene and genomic sweep. In the candidate gene locus identifies a polymorphism which is known to have effect on the phenotype in one species and, potentially, could have similar effect on others.

Genomic scanning strategy explores the association between a phenotype and selected markers across the genome, thus identifying chromosomal regions that include genes partly responsible for the expression of that character.

Figure 4 illustrates the principle of detecting a QTL when using a single marker M (with alleles M1 and M2) located at a distance of θ centiMorgan (cM) of QTL. Nevertheless, today the most frequent methodology used in animals for QTLs detection is the method of "multiple markers," which performs the mapping interval and estimated the position of a QTL located between two markers. This strategy is unbiased only if within the chromosomal segment bounded by the two markers there is a single QTL (Kinghorn & van der Werf, 2000).

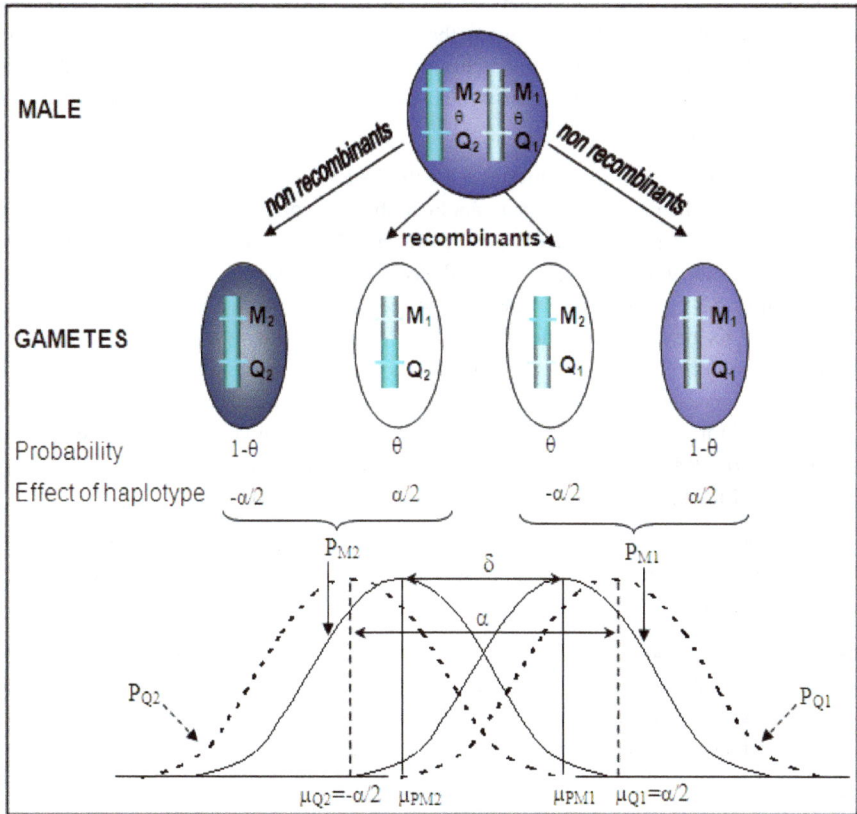

Figure 4. Detection of QTL with a single bi-allele marker. (Adapted to Gautier, 2003)

Most studies of QTL detection for milk production traits have been carried out in cattle, using mostly aggregated data (Zhang et al., 1998; Heyen et al., 1999; Plante et al., 2001; Viitala et al., 2003; Ashwell et al., 2004)). Moreover, Rodriguez-Zas et al. (2002) measured the association between chromosomal regions and the scale and shape that describe the lactation curve in dairy cattle using a two-stage procedure: a random regression model to predict the elements of the lactation function on an animal basis, followed by a regression interval mapping using the predictions obtained in the first stage of the analysis. Work of QTL detection in goats is scarced (for a reviews of QTL mapping in goat see Maddox & Cocket, 2007). Analyses on QTLs affecting milk traits of dairy goats are lacking. However, the only well-documented genetic association with dairy traits in goats is the one related with the highly variable α_{s1}-casein polymorphisms (Grosclaude et al., 1994; Adnoy et al., 2003; Manfredi, 2003; Suárez, 2003; Sacchi et al., 2005; Caffaro, 2007).

In searching for QTLs in goats, one may look at those significant associations found between milk production traits and genetic markers in cattle (reviewed by Khatkar et al., 2004), and take advantage of the homology between the genetic maps of the cow and the goat. For example, Roldán et al. (2008) identified chromosomal regions associated with variation in the lactation function of goats using the two-stage procedure as employed by Rodriguez-Zas et al. (2002).

The goat population used in this study was established in 1998 at the Experimental Farm of Leales – INTA, in the province of Tucumán. Phenotypes recorded were milk yield (MY), fat percentage (FP) and protein percentage (PP) from 212 female goats. Milk samples were collected at the morning milking, at each of the two kidding seasons (fall-winter and spring-summer) during 6 years (1999–2004). Each goat was sampled for FP and PP once a month, either five or six times per lactation. Records of daily MY for a given test day were the averages of an entire week, with sampling taking place every month but on an irregular basis (up to 3 weeks a month). Lactations with three or fewer records were deleted, and the lactation stage ranged from 3 to 321 days. The total number of observations were 897 for MY, and 814 for FP and PP. Averages (standard deviations) across all families were 0.879 kg (SD 0.563) for MY, 4.382% (SD 1.864) for FP and 4.065% (SD 0.757) for PP.

Genotypic data from eight paternal half-sib families composed of 87 young females and 75 older goats were used for QTL detection. Whereas six families were purebred Criollo goats, the two other families consisted of Saanen by Criollo crosses.

Taking account the homologies between the genetic maps of cattle and goats and the significant associations between molecular marker and milk production traits summarized by Khatkar et al. (2004) in *Bos taurus* autosome (BTA) 3, 6, 14 and 20 a set of 37 microsatellite were selected. Details of molecular markers about map positions, distance can be see in the web.

The statistical analysis was conducted in two stages. They first calculated the predictions of the random regression coefficients, and then they tested the effects of the chromosome regions on those predictions. The random regression model used in the first stage included the goat-specific parameters plus the predictions of individual breeding values and permanent environmental effects. The model equation was as follows:

$$y_{ij\ln rs} = c_l + b_n + d_r + p_s + \sum_{m=1}^{5} \kappa_{(ns)m}\ \phi_{ij(ns)m}(t) + \sum_{m=1}^{5} \psi_{(n)m}\ \phi_{ij(n)m}(t) +$$

$$\sum_{m=1}^{5} \tau_{ijm}\ \phi_{ijm}(t) + \sum_{m=1}^{5} \pi_{ijm}\ \phi_{ijm}(t) + \varepsilon_{ij\ln rs} \tag{3}$$

In (3), y_{ijlnrs} is the record of MY, FP, or PP, for the ith animal measured on the jth test-day taken on year l (l = 1999,..., 2004), under lactation number n (n = 1,..., 6), with the rth number of kids at parturition (r = 1, 2), and in the sth season (s = fall-winter, spring-summer). Fixed effects in the model were year (c_l), lactation number (b_n), number of kids at parturition (d_r), and season (p_s). Additionally, the parameters of the regression function of Ali & Schaeffer (1987) were fitted for each combination of lactation and season ($\kappa_{(ns)m}$). The regression coefficients ($\psi_{(n)m}$) in the Ali & Schaeffer (1987) function were also fitted as random variables that are associated with the permanent environmental effects that are common to the nth lactation number. For animal i, the random variables τ_{ijm} and π_{ijm} are the regression coefficients for the breeding values and the permanent environmental effects, respectively, of the lactation curve proposed by Ali & Schaeffer (1987). For animal i measured on day j, the random regression function which mimics the model of Ali & Schaeffer (1987) can be written as

$$\sum_{m=1}^{5} \lambda_{ijm}\ \phi_{ijm}(t) = \alpha_i + \beta_i \left(\frac{t_{ij}}{280}\right) + \gamma_i \left(\frac{t_{ij}}{280}\right)^2 + \delta_i \ln\left(\frac{280}{t_{ij}}\right) + \phi_i \left[\ln\left(\frac{280}{t_{ij}}\right)\right]^2$$

where λ_{ijm} is either τ_{ijm} or π_{ijm}, and t is a vector such that $t' = [1,(t_{ij}/280), (t_{ij}/280)^2, (\ln(280/t_{ij})), (\ln(280/t_{ij}))^2]$, being t_{ij} the jth test-day for animal i. The scale parameter that characterizes the overall level of the trait (MY, FP or PP) for the curve of individual i is α_i. The remaining parameters are responsible for the shape of the curve, and represent the rate of change of the trait at different stages of the lactation: β_i and γ_i are associated with a decreasing slope of the curve, and δ_i and ϕ_i with increasing slope. The lactation length was taken to be 280 days. The variance and covariance functions in the random regression model (3) were estimated by Restricted Maximum Likelihood using the program VCE5 (Kovac & Groeneveld, 2003). The permanent environmental matrix (E) was estimated with a submodel of (3) in which the fifth and sixth terms were left out. To fit (3), E was split into 'across lactations' (E_B) and 'within lactation' (E_W) components, such that E_W = 0.4 E and E_B = (1-0.4) E. The value 0.4 corresponds to the correlation between permanent environmental effects of first and second lactations. The estimates of the fixed effects, as well as of the additive genetic and permanent environmental effects were calculated by solving the mixed model equations using a program written in PROC IML (SAS, 2005).

In the second stage, we performed a QTL analysis using the half-sib regression interval mapping method of Knott et al. (1996), with the software QTL Express (Seaton et al., 2002). The test statistics were computed every centiMorgan (cM) over the mapped chromosome. F-statistic thresholds for chromosome-wise level were calculated from 10,000 permutations

(Churchill & Doerge, 1994). Families that displayed the highest evidence for a QTL at the location in the across-family analysis were taken from the QTL-express output (Knott et al., 1996).

5.1. Genetics parameters

The estimated heritability for MY ranged from 0.142 to 0.593, and the average estimated over the whole lactation period was 0.343. These values were in agreement to those reported in the literature for different goat breeds, and with estimates obtained from either single or multiple trait models.

Our results seem to be slightly lower than those of Weppert & Hayes (2004) for Nubian, Alpine, Saanen and Toggenburg goats, when maternal effects were included in the analysis. The estimated heritability of MY from a model without maternal effects was equal to 0.19 (Weppert & Hayes, 2004). Similar estimates of heritability to the values found in the current study were reported for South African Saanen goats (0.30, Muller et al., 2002), and for Alpine and Saanen females (0.23, Clément et al., 2002).

The average heritability estimate of FP was equal to 0.092 (ranging from 0.093 to 0.141), and the average heritability estimate of PP was 0.160 (ranging from 0.007 to 0.515). These values were lower than those obtained and reviewed by Muller et al. (2002) using data from several goat breeds: the range of FP was 0.160–0.540 (Spain, Alpine, Saanen and Toggenburg breeds) and the range of PP was 0.250–0.620 (Spain, Greece and Saanen goats).

5.2. QTL detection

5.2.1. Milk yield

Seven tests were significant for at least one parameter. In CHI6, we detected a significant effect for all parameters in the interval flanked for the MS BM4621 and BM415 (from 70 to 78 cM).

One chromosomal region was associated with δ and Φ on CHI14. Several studies based on cumulative single records in dairy cattle detected the presence of QTL at the genome-wise and suggestive thresholds on BTA3, in the interval from 16 to 32 cM (Heyen et al., 1999), at 39 cM (Plante et al., 2001), and at 40 cM (Vandervoort & Jansen, 2002).

On the other hand, neither Viitala et al. (2003) nor Ashwell et al. (2004), reported evidence of a QTL for MY on BTA3. Rodriguez-Zas et al. (2002) using test-day milk records reported a significant association between marker MS BL41 (32 cM) and the parameter that describes the shape of the function at the beginning of the lactation.

Additionally, these workers reported significant associations between the scale parameter and two chromosomal regions of BTA3 that were located in the centromere from 0 to 36 cM and in the telomere from 91 to 113 cM (close to MS HUJI177 at 100 cM and MS BR4502 at 113 cM).

An association with increasing slope parameters d and Φ was found at 70 and 71 cM, respectively, on CHI6. Also at 75 cM, the scale parameter (α) and the descriptor of the shape

Milk Casein Alleles, Haplotypes and QTL Effect on Protein and Fat Content and Milk Yield in
Argentinean Criollo and Cross Goats

189

at the end of the lactation (γ) were found to be significant. Another association with the remained descriptor of the decreasing slope (β) was detected at 78 cM. The correlation between the estimates of α, γ, δ and Φ was high (0.922 – 0.998).

Similarly, the correlation between β and the other parameters ranged from 0.652 to 0.758. Such correlations may suggest a QTL with pleiotropic effects on these parameters. Rodriguez-Zas et al. (2002) detected a putative QTL between 0 and 21 cM in dairy cattle that affected the scale parameter for MY. They also reported the finding of another QTL affecting the shape parameters for MY, which is located in the region from 108 to 129 cM. In dairy cattle, Zhang et al. (1998) reported a putative QTL in the interval between 30 to 50 cM on BTA6 (between MS BM1329 and BM143).

Remember, in goat, the casein gene cluster has been mapped to the distal region of CHI6 (Grosclaude et al., 1994), and is composed of four genes (αs1- casein, αs2-casein, β-casein, and κ-casein). Moioli et al. (2007) reviewed the several associations between casein genes, and dairy traits of goats have been reported in the literature.

This is especially so for the *CSN1S1* gene, which displays a higher level of polymorphism than the one observed in the bovine, and has been related to fat and protein contents (Grosclaude et al., 1994; Adnoy et al., 2003; Manfredi, 2003; Sacchi et al., 2005). However, the effect of αs1-casein alleles on MY does not seem to be important (Moioli et al., 2007). In the current research, they used αs1-casein gene as a marker gene (MS CSN) and did not find association to MY. These results and previous described by Suarez (2003) and Caffaro (2007) suggest that the effect of *CSN1S1* and *CSN3* gen or the haplotypes are not very important for MY but they could have an effect on lactation curve shape.

Moving to CHI14, there was evidence for a QTL at 14 cM (the interval flanked by ILSTS011 and RM011) related to δ and Φ. This result suggests that there may be a QTL with pleiotropic effects on both parameters. It may also be the case that, due to the high correlation (0.996) between the estimates of those parameters, one of the associations observed may be a false positive.

Similarly, Rodriguez-Zas et al. (2002) reported an association between the chromosomal region at about 13 cM on BTA14 (marker CSSM66), and the shape parameters that describe the changes in milk yield during mid and late lactation. However, in dairy cattle a QTL affecting MY was seemingly associated with BTA14 by Khatkar et al. (2004).

Several studies in cattle reported QTLs for MY on BTA20. The chromosomal regions in the bovine were at 21 cM (Plante et al., 2001), 82 cM (Viitala et al., 2003) and 68 cM (Ashwell et al., 2004). In the current study, no microsatellite marker on CHI20 was associated with some lactation descriptors.

5.2.2. Fat percentage

Although several studies in dairy cattle reported a putative QTL on BTA3 and BTA6 affecting FP (Khatkar et al., 2004), we did not find evidence for the effects of a chromosomal region in either CHI3 or CHI6, at the chromosomal-wise threshold. However, for FP they detected significant effects from CHI14 and CHI20.

When analyzing CHI14, one chromosomal region at 63 cM (between CSSM66 and CSSM36) associated with variation of the parameter β was found significant.

Conversely, analyses with dairy cattle found strong evidence for a putative QTL affecting FP near the centromere of BTA14, being the MS CSSM66 the nearest marker (Coppieters et al., 1998; Zhang et al., 1998; Heyen et al., 1999; Ashwell et al., 2004).

A QTL proximal to the centromere on BTA14 with an effect on FP has consistently been reported (Grisart et al., 2002 and Winter et al., 2002), and the mutation underlying this QTL has been identified (Winter et al., 2002) as the K232A substitution in exon VIII of acylCoA / diacylglycerol acyltransferase 1 enzyme (DGAT1).

This enzyme is considered to be of importance in controlling the synthesis rate of triglycerides in adipocytes. Nevertheless, no associations with several microsatellite markers on BTA14 were found by Rodriguez-Zas et al. (2002) using a longitudinal mapping model. Comparing the goat and bovine linkage maps, CHI14 shows a partial homology with BTA14 from MS CSSM66 to the telomere.

The centromeric region of CHI14 flanked between MS ETH225 and MS BM757, corresponds to the same region of chromosome 9 in cattle. Therefore, it is likely that an association with FP could be found near the centromeric end of CHI9.

At 72 cM on CHI20, where the nearest marker is BMS1719, we detected an association with parameter α ($p < 0.00002$). Many additional QTL with significant effects on FP and FY has been reported for chromosome 20 in dairy cattle (Khatkar et al., 2004).

5.2.3. Protein percentage

Chromosomal regions associated with β for PP were found in CHI3. For this chromosome and close to MS INRA023 (at 59 cM), we detected a chromosomal region affecting β. A QTL for PP located in an area of about 40 cM in BTA3 was reported in several studies (Khatkar et al. 2004).

We did not find significant associations among chromosomal regions of CHI6 and any parameter when looking at PP. Nevertheless, many studies in dairy cattle have detected the presence of a QTL close to MS BM143 in BTA6 that is related to PP, and the marker position agrees in all these studies (Spelman et al., 1996; Zhang et al., 1998; Ron et al., 2001; Viitala et al., 2003). Using longitudinal phenotypic data, Rodriguez-Zas et al. (2002) found significant association between MS BM143 and the scale and the shape parameters at middle and late lactation for PP in dairy cows.

Schnabel et al. (2005) performed a fine-mapping study of BTA6 of dairy cattle and identified the osteopontin (OPN) gene as an ideal functional candidate gene for a QTL very close to BM143. The OPN is a secreted glycoprotein and its expression in the murine mammary gland depends on the stage of postnatal development, which in turn suggests a role for OPN in mammary involution.

5.2.4. Lactation patterns

We use the expression longitudinal mapping model to refer to those statistical models for mapping QTLs of a longitudinal (or functional value) trait. The ability of the model used in the current study to detect associations between markers and lactation stages may contribute to explain the different lactation patterns among individuals.

Although breeders do not usually breed for lactation shape, some traits such as persistence that are described by lactation curve parameters are economically relevant.

6. Conclusions and perspectives

The casein cluster gene is today the most studied chromosomal region related with milk characteristics - quality and quantity -in goats but there are a lot of to come.

The genetic variability and alelle distribution at the CSN1S1 and CSN3 in the Criollo and crosse breed showed in this chapter reveal large differences between breed specialy with those highly selected. Althougth the number of animals, families and the design used in all works presented here are not very large compared with those used in the dairy cattle experiments, they were good enough to show the usefull and potenciallity of both quantitative and molecular data analysis together in differents aspects and with differents methodologies to be used in animal breeding.

Molecular markers are more commonly being used in other ruminant - cattle and sheep - than in goat breeding programmes, due to the animal economic value and the technology costs ratio is more profitable in cattle and sheep than in goat. Nevertheless some laboratories now are offering DNA test mainly for paternity and inherited diseases in goats.

There are a number of research programmes endeavouring to detect additional gene effects most of them to genetic resitance and reproductive traits (i.e. 3SR ; CRC FAO-IAEA) and it is highly likely that these will yield further targets for Marker Asisted Selection(MAS) / Gene Assisted Selection (GAS). The application of MAS/GAS to date has mainly been in the form of introgression programmes or for selection within a few flocks with a high emphasis on the gene of interest.

The strong association found here between some allele at CSN1S1 locus and haplotypes of CSN1S1 and CSN3 and PP suggest that Criollo goat is a breed with good milk quality for cheese production and that they can be use for MAS. The molecular information at the CSN1S1 and CSN3 could be useful for genetic improvement of dairy traits in Criollo goats by contributing to design breeding schemes able to fit selection goals. In addition, this breed is well adapted to harsh environments and it has a great potential for increasing production without loss of local adaptation, being a highly valuable genetic resource for the region.

The methodology used by Roldán et al. (2008) to detect QTL for milk production traits using a longitudinal model identified nine map positions affecting any parameter of the lactation function of dairy goats: one in CHI3, four in CHI6, two in CHI14 and two in CHI20. Some of these results were consistent with QTLs found in dairy cattle while using either aggregate lactation records, or longitudinal-linkage analysis. This is one example in goats of integration between molecular and quantitative information to enhance rates of genetic improvement in small ruminants. As was stated by Notter & Baker (2007) a quantitative-molecular paradigm for livestock improvement is badly needed and is gradually emerging. This paradigm aspires to provide a structure to combine established strategies for prediction of breeding values from performance records on individuals and their relatives with modern molecular techniques to determine parentage, identify major genes or genetic markers associated with desirable phenotypes, and utilize the power of functional genomics to improve understanding of the genetic mechanisms that control expression of complex traits.

When the cattle genome was sequenced (The bovine genome consortium.Science, April 24, 2009), it puts that species on the threshold of a new era in which the challenges of health, production and sustainability can be addressed far more effectively through selective breeding than was previously possible. The whole genome sequencing and the subsequent characterisation of densely spaced genetic markers, such systems may eventually be superceded by 'genome selection-GS' techniques (Meuwissen et al., 2001).

Through use of genome assisted selection improved selection response can be expected for most traits of interest, but especially for traits where the routine collection of accurate phenotypes is difficult, costly or possible only late in an animal's life or only in one sex. Such traits include many health characteristics as well as reproductive efficiency.

Despite a slow start to the development of genomic tools, the goat genome assembly is running. In January 2010 was "officially" constituded the International Goat Genome Consortium (IGGC - http://www.goatgenome.org/doku.php?id=start). Goat whole genome reference sequence began with de novo assembly into high quality contigs and continue with information from EST-based virtual goat genome and BAC clones. Additionally, SNP discovery using next-generation sequencing technology is underway to develop initial genetic polymorphism database and SNP chip(s). A 50-60K goat SNP chip will be available by November 2011 from Illumina company. RH panel, mapping of markers for RH map and HapMap development will be utilized in the following version of the reference genome.

Despite the widespread excitement about the potential for GS to provide new approaches for the improvement of sustainability traits there are a number of reasons to be cautious about this approach in goat. Firstly, there is the challenge of many different breeds and production environments: whilst it is entirely feasible to 'train' GS for Holstein dairy cattle where there are relatively few variations in production environments, very little is currently known about the 'portability' of GS algorithms across different breeds or environments. Early results for beef and dairy cattle indicate that more dense SNP arrays than are currently available are likely to be needed. This brings us to the second concern, viz. cost. Moreover only about 1/3 of the SNP associated with any productive traits in one breed can be used in another breed (Boichard et at., 2010; Lopez-Villalobos, pers com.).

Nevertheless, as technology development will never stop, we will face in a few years to the new challenge, that is the use of the whole genome sequence data for QTL mapping and genomic selection instead of dense SNPs chips. As was stated by Meuwissen (2010), future technologies are predicted to reduce cost by about 100 fold that today and we can expect to have whole genome sequence data available on substancial number of animals at reasonable low cost.

At the same time as these valuable new genomic resources for goat are being developed, livestock breeders are facing real challenges in delivering more balanced breeding objectives that seek to broaden selection goals beyond traditional productivity traits to include sustainability and welfare traits such as disease resistance, robustness, reproductive efficiency and longevity.

Author details

M.E. Caffaro, C. Suárez, D.L. Roldán, and M.A. Poli
Instituto de Genética, CICVyA-INTA, Castelar, Provincia de Buenos Aires, Argentina

Acknowledgement

The authors thank SeCyT-ANPCyT, Argentina, through the National Project BID1201 PICT04226 and BID 1712 PICTO12968; La Serenísima, Argentina, for their cooperation in supplying compositional milk data. Alicia Rabasa, Fernando Holgado, Silvina Saldaño y Jorge Fernandez who took the phenotypic data and milk samples for quality analysis and field staff on CER Leales INTA caretakers of the experimental flock.

7. References

Adnoy, T; Vegarud, F.; Gulbrandsen, T.; Nordblo, R., Colbjorsen, I., Brovold, M., Markovic, B., Roseth, A. & Lien, S. (2003). Effects of the O and F alleles of alpha S1 casein in two farms of northern Norway. International Workshop on Major Genes and QTL in Sheep and Goats. Toulouse (France) 8-11 Dec. CD-ROM communication N° 2-20.

Ali T.E. & Schaeffer L.R. (1987) Accounting for covariances among test day milk yields in dairy cows. Can. J. Anim. Sci., 67, 637–644.

Amils, M. (1996). Diseño y amplificación de técnicas moleculares para el análisis del efecto del polimorfismo de la caseína α_{s1} caprina sobre caracteres productivos. Tesis doctoral, Facultad de Veterinaria, Universidad Autónoma de Barcelona, España, pág 165.

Anderson, L. & George, M. (2004). Domestic animal genomics: deciphering the genetics of complex traits. Nature Genetics, 5: 202-212.

Angulo, C.; Ares, J .L.; Amills, M.; Sanchez, A.; Ilahi, H. & Serradilla, J. M. (2002). Effect of α_{s1}-casein polymorphism on dairy performances of Malagueña goats. 7th World

Congress on Genetics Applied to Livestock Production, August 19-23, Montpellier, France. CD-ROM, communication N 09-17.

Ashwell M.S.; Heyen D.W.; Sonstegard T.S.; Van Tasell C.P.; Da Y.; VanRaden P.M.; Ron M.; Weller J.I. & Lewin H.A. (2004) Detection of quantitative trait loci affecting milk production, health, and reproductive traits in Holstein cattle. J. Dairy Sci., 87, 468–475.

Barbieri, M.E. (1995). Polymorphisme de la caséine αs1. Effets des génotypes sur des performances zootechniques et utilisation en sélection caprine. Thèse Docteur de l'Institut National Agronomique Paris-Grignon, pp. 160.

Barroso, A.; Dunner, S. & Cañon, J. (1999). Polimorfismo genético de las lactoproteinas de los rumiantes domésticos. Separata ITEA. Información técnica económica agraria. Vol. 95A N° 2, 143-179.

Baruch, E.; Weller, J.; Cohen-Zinder, M.; Ron, M. & Seroussi, E. (2006). Efficient Inference of Haplotypes from Genotypes on a large animal pedigree. Genetics. 172: 1757-1765.

Boettcher, P. J.; Caroli, A.; Stella, A.; Chessa, S.; Budelli, E.; Canavesi, F.; Ghiroldi, S. & Pagnacco, G. (2004). Effects of Casein Haplotypes on Milk Production Traits in Italian Holstein and Brown Swiss Cattle. J. Dairy Sci. 87: 4311-4317.

Boichard, D.; Guillaume, F.; Baur, A.; Croiseau, P.; Rossignol, M.N.; Boscher, M.Y; Druet, T.; Genestout, L.; Eggen, A; Journaux, L. Ducrocq, V. and Fritz S. (2010) Genomic Selection In French Dairy Cattle. 9th WCGALP Leipzig, Germany. August 1-6. 0716.

Botstein, D, White, RL, Skolnick, M & Davis, RW (1980) Construction of a genetic linkage map in man using restriction fragment length polymorphisms. American Journal of Human Genetics 32, 314-331.

Bovine Genome Consortium . The Genome Sequence of Taurine Cattle: A window to ruminant biology and evolution. 2009. Science April 24, vol 324 Issue 5926 – 522.

Caffaro, M.E. (2007). Polimorfismo en los genes CSN1S1 y CSN3 en caprinos Criollos y cruzas. Efecto sobre características lecheras. Tesis Maestria en Biotecnología. Universidad de Buenos Aires.

Caroli, A.; Chessa, S.; Bolla, P.; Budelli, E. & Gandini, G. (2004). Genetic structure of milk protein polymorphisms and effects on milk production traits in a local dairy cattle. J. Anim. Breed. Genet.121: 119-127.

Caroli, A.; Chiatti, F.; Chessa, S.; Rignanese, D.; Bolla, P. & Pagnacco, G. 2006. Focusing on the goat casein complex. J. Dairy Sci. 89: 3178-3187.

Caroli, A.; Chiatti, F.; Chessa, S.; Rignanese, D.; Ibeagha-Awemu, E. M. & Erhardt, G. 2007. Characterization of the casein gene complex in West African goats and description of a new αs1-casein polymorphism. J. Dairy Sci. 90: 2989-2996.

Chanat, E.; Martin, P. & Ollivier-Bousquet, M. (1999). αs1-casein is required for the efficient transport of β- and κ-casein from the endoplasmic reticulum to the Golgi apparatus of mammary epithelial cells. J. Cell Sci. 112:3399–3412.

Chessa, S.; Budelli, E.; Gutscher, K.; Caroli, A. & Erhardt G. 2003. Short communication: Simultaneous identification of five kappa-casein (CSN3) alleles in domestic goat by polymerase chain reaction-single strand conformation polymorphism. J. Dairy Sci. 86: 3726-3729.

Milk Casein Alleles, Haplotypes and QTL Effect on Protein and Fat Content and Milk Yield in
Argentinean Criollo and Cross Goats

195

Chiatti, F.; Caroli, A.; Chessa, S.; Bolla, P. & Pagnacco, G. (2005). Relationships between goat
κ-casein (CSN3) polymorphism and milk composition. The role of biotechnology. Turin,
Italy. 5-7 March. Pág:163-164.

Churchill, G.A. & Doerge, R.W. (1994) Empirical threshold values for quantitative trait
mapping. Genetics, 138, 963–971.

Clément, V.; Boichard, O.; Piacere, A.; Barbat ,A. & Manfredi, E. (2002) Genetic evaluations
of French goats for dairy and type traits, Proceedings of the 7th World Congress on
Genetics Applied Livestock Production, Montpellier, France, 19–23 August 2002.
Comm. 1–46.

Coppieters, W.; Riquet, J.; Arranz, J.; Berzi, P.; Cambisano, N.; Grisart, B.; Karim, L.; Marcq,
F.; Moreau L.; Nezer, C.; Simon, P.; Vanmanshoven, P.; Wagenaar, D. & Georges, M.
(1998) A QTL with major effect on milk yield and composition maps to bovine
chromosome 14. Mamm. Genome, 9, 540–544.

Dimassi, O; Neidhart, S.; Carle, R.; Mertz, L.; Migliore, G.; Mané-Bielfeldt, A. & Valle Zárate,
A. (2005). Cheese production potential of milk of Dahlem Cashmere goats from a
rheological point of view. Small Rumin. Res. 57: 31-36.

Falconer, D.S: and Mackey, T.F.C. (1996). Introduction to quantitative genetics, 2nd Ed.
Longman. Chapter 7.

Février, C.; Jaguelin, Y.; Lebreton, Y.; Colleaux , Y. & Pringent, J.P. (2000). Nutritive value for
piglets and ileal amino acid digestibility of goat's milk differing in alpha s1- casein. 7 th.
International Conference on Goats, Tours, France, May 2000, 876- 879.

Fekadu, B.; Soryal, K.; Zeng, S.; Van Hekken, D.; Bah, B. & Villaquiran, M. (2005). Changes
in goat milk composition during lactation and their effect on yield and quality of hard
and semi-hard cheeses. Small Rumin. Res. 59: 55-63.

Gautier, M. (2003). Fine mapping of QTL regions in dairy cattle. These Docteur de l'Institut
Nacional Agronomique Paris-Grignon. Pp.1-284.

Green, P.; Falls, K. & Crooks S. (1990). Cri-map Documentation Version 2.4. Washington
University School of Medicine, St Louis, MO. Available at
http://linkage.rockefeller.edu/soft/crimap/.

Grisart, B.; Coppieters, W.; Farnir, F.; Karim, L.; Ford, C.; Berzi, P.; Cambisano, N.; Mni, M.;
Reid, S.; Simon, P.; Spelman, R.; Georges M. & Snell R. (2002) Positional candidate
cloning of a QTL in dairy cattle: identification of a missense mutation in the bovine
DGAT1 gene with major effect on milk yield and composition. Genome Res., 12, 222–
231.

Grosclaude, F.; Mahé, M.F.; Brignon, G.; Di Stasio, L. & Jeunet, T. (1987). A Mendelian
polymorphism underlying quantitatives variations of goat α_{s1}-casein. Génét. Sél. Evol.
19: 399-412.

Grosclaude, F.; Martin, P.; Ricordeau, G.; Remeuf, F.; Vassel, L. & Bouillon, J. (1994). Du
gene au frommage: le polymorphisme de la caséine alpha S1 caprine, ses effets, son
evolution. INRA Prod. Anim. 7: 3-9.

Gipson, T.A. & M. Grossman. (1990). Lactation curves in dairy goats: A review. Small
Rumin. Res. 3: 383-396.

Gonzalez-Stagnaro, C. 1997. Ovinos de pelo. Ovis N° 48. Luzan.7 Ed Madrid. pág 98.

Hayes, B.; Petit, E.; Bouniol, C. & Popescu, P. (1993). Localisation of the alpha-S2-casein gene (CSN1S2) to the homologous cattle, sheep and goat chromosome 4 by in situ hybridization. Cytogenet Cell Genet. 64: 282-285.

Hearne, CM; Ghosh, S & Todd, JA (1992) Microsatellites for linkage analysis of genetic traits. Trends in Ecology and Evolution 8: 288-94.

Heyen, D.W.; Weller, J.I.; Ron, M.; Band, M.; Beever, J.E.; Feldmesser, E.; Da, Y.; Wiggans, G.R.; VanRaden, P.M. & Lewin H.A. (1999) A genome scan for QTL influencing milk production and health traits in dairy cattle. Physiol. Genomics, 1, 165–175.

Jann, O.; Prinzenbeberg, E.; Luikart, G.; Caroli, A. & Erhardh, G. (2004). High polymorphism in the κ-casein (CSN3) gene from wild and domestic caprine species revealed by DNA sequencing. J. Dairy Res. 71: 188 -195.

Jolliffe, I. (2002). Principal Component Analysis, 2nd Ed. Springer. Chapter1.

Khatkar, M.S.; Thomson, P.C.; Tammen, I. & Raadsma H.W. (2004) Quantitative trait loci mapping in dairy cattle: review and meta-analysis. Genet. Sel. Evol., 36, 163–190.

Kinghorn, B. & J. Van der Werf. (2000). Identify and incorporating genetic markers and major genes in animal breeding programs. Course notes Belo Horizonte (Brazil) 31 May-5 June 2000.

Knott S., Elsen J. M. & Haley C. (1996) Methods for multiple- marker mapping of quantitative trait loci in half-sib populations. Theor. Appli. Genet., 93, 71–80.

Kovac, M. & Groeneveld, E. (2003) VCE-5. User's Guide and Reference Manual. Institute of Animal Science, Neustand-Germany.

Leroux, C.; Le Provost, F.; Petit, E.; Bernard, L.; Chilliard, Y. & Martin, P. (2003). Real-time RT-PCR and cDNA macroarray to study the impact of the genetic polymorphism at the αs1-casein locus on the expression of genes in the mammary gland during lactation. Reprod. Nutr. Dev. 43: 459-469.

Lynch, M. & Walsh, J.B.. (1998). Genetics and Analysis of Quantitative Traits. Sinauer Assocs.; Inc. Sunderland; MA. PP. 980.

Macciota, N.; Vicario, D. & Cappio-Borlino A. (2005) Detection of different shapes of lactation curve for milk yield in dairy cattle by empirical mathematical models. J. Dairy Sci., 88, 1178–1191.

Madisen, L.; Hoar, D.I.; Holroyd, C.D.; Crisp, M. & Hodes, M.E. (1987). DNA banking: the effects of storage of blood and isolated DNA on the integrity of DNA. Am. J. Med. Genetics 27:379-390.

Maddox, J.A., & Cockett, N. E. 2007. An update on sheep and goat linkage maps and other genomic resources. Small Rumin. Res. 70(1):4-20.

Mahé, M.F.; Manfredi, E.; Ricordeau, G.; Piacère, A. & Grosclaude, F. (1993). Effets du polymorphisme de la caséine αs1 caprine sur les performances laitières: analyse intra-descendence des boucs de race alpine. Genet Sel Evol. 26:151-157.

Manfredi, E., (2003). Proceedings of the International Workshop on Major Genes and QTL in Sheep and Goat. Toulouse, 2003.

Martin, P.; Szymanowska, M.; Zwierzchowski, L. & Leroux, C. (2002). The impact of polymorphism on the protein composition of ruminant. Reprod. Nutr. Dev. 42: 433-459.

Meuwissen, T.; Hayes, B.J. and Goddard, M. E. (2001). Prediction of total genetic value using genome-wide dense marker maps. Genetics, Vol. 157, 1819-1829.

Meuwissen, T. (2010). Use of whole genome sequence data for QTL mapping and genomic selection. 9th WCGALP Leipzig, Germany. August 1-6. 0018.

Moioli, B.; D'Andrea, M. & Pilla, F. (2007). Candidate genes affecting sheep and goat milk quality. Small Rumin. Res. 68: 179-192.

Morand-Fehr, P. & Boyazoglu, J. (1999). Present state and future outlook of the small ruminant sector. Small Rumin.t Res. 34: 175-188.

Muller, C.J.C.; Cloet, S.W.P. & Schoeman, S.J. (2002) Estimation of genetic parameters for milk yield and milk composition of South African Saanen goats. Proceedings of the 7th World Congress on Genetics Applied Livestock Production, Montpellier, France, 19–23 August 2002. Comm. 1–52.

Notter, D.R & Baker, R.L. (2007) The outlook for quantitative and molecular genetic applications in improving sheep and goats. Small Rumin. Res. doi:10.1016/j.smallrumres.2007.01.009

Peña Blanco, F; García Martínez, A. & Martos Peinado, J. (2005). Revisión bibliográfica sobre producción de leche, control lechero y curvas de lactación. Prod. Animal y Gestión DT4, vol. 2 pp. 54.

Plante, Y.; Gibson, J.P.; Nadesalingam, J.; Mehrabani-Yeganeh, H.; Lefebvre, S.; Vandervoort, G. & Jansen G.B. (2001) Detection of quantitative trait loci affecting milk production traits on 10 chromosomes in Holstein cattle. J. Dairy Sci., 84, 1516–1524.

Prinzenberg, E.; Gutscher, K.; Chessa, S.; Caroli, A. & Erhardt, G. (2005). Caprine ®-Casein (CSN3) Polymorphism: New Developments in Molecular Knowledge. J. Dairy Sci. 88: 1490-1498.

Rodero, A., Delgado, J.V., Rodero, E. (1992). Primitive Andalusian Livestocks and their Implication in the discovery of America. Arch. Zoot. vol. 41 (154): 383-400.

Rodriguez-Zas, S.L.; Southey, B.R.; Heyen, D.W. & Lewin H.A. (2002) Detection of quantitative trait loci influencing dairy traits using a model for longitudinal data. J. Dairy Sci., 85, 2681–2691.

Roldán, D.L.; Rabasa, A.E.; Saldaño, S.; Holgado, F.; Poli, M.A. & Cantet, R.J.C. (2008). QTL detection for milk production traits in goats using a longitudinal model. J. Anim. Breed. Genet., 125, 187-193.

Ron, M.; Kliger, D.; Feldmesser, E.; Seroussi, E.; Ezra, E. & Weller J.I. (2001) Multiple quantitative trait locus analysis of bovine autosome 6 in the Israeli Holstein population by a daugther design. Genetics, 159, 727–735. SAS Institute Inc, Cary, NC, USA (2005) Statistical Analysis Systems.

Sacchi, P.; Chessa, S.; Budelli, E.; Bolla, P.; Ceriotti, G.; Soglia, D.; Rasero, R.; Cauvin E. & Caroli A. (2005) Casein haplotype structure in five Italian goat breeds. J. Dairy Sci., 88, 1561–1568.

SAS Institute Inc, (2005) Statistical Analysis Systems.SAS, Cary, NC, USA.

Schnabel, R.D.; Kim, J.J.; Ashwell, M.S.; Sonstegard, T.S.; Van Tassell, C.P.; Connor, E.E. & Taylor J.F. (2005) Finemapping milk production quantitative trait loci on BTA6: Analysis of the bovine osteopontin gene. Proc. Natl. Acad. Sci., 102, 6896–6901.

Seaton, G.; Haley, C.; Knott, S.; Kearsey, M. & Visscher P. (2002) QTL Express: mapping quantitative trait loci in simple and complex pedigrees. Bioinformatics, 18, 339–340.

Serradilla, J.M. (2003). The goat alphas1-casein gene: A paradigm of the use of a major gene to improve milk quality? . In Gabiña D. (ed.), Sanna S. (ed.). Breeding programmes for improving the quality and safety of products. New traits, tools, rules and organization?. Zaragoza : CIHEAM-IAMZ, p. 99-106.

Spelman, R.J.; Coppieters, W.; Karim, L.; van Arendonk, J.A.M. & Bovenhuis H. (1996) Quantitative trait loci analysis for five milk production traits on chromosome six in the Dutch Holstein-Friesian population. Genetics, 144, 1799–1808.

Suárez, C. (2003). Polimorfismo en el gen CSN1S1 en caprinos Criollos de los valles Calchaquíes. Su distribución y efecto sobre caracteres de producción de leche. Universidad de Buenos Aires. Tesis Maestría en Biotecnología. 66 - 72 pp.

Soryal, K.; Beyene, F.A.; Zeng, S.; Bah, B. & Tesfai, K. (2005). Effect of goat breed and milk composition on yield, sensory quality, fatty acid concentration of soft cheese during lactation. Small Rumin. Res. 58: 275-281.

Vandervoort G. & Jansen G.B. (2002) Comparison of QTL mapping models with multiple traits and multiple intervals in a halfsib design in dairy cattle. Proceedings of the 7th World Congress on Genetic Applied Livestock Production, Montpellier, France, 19–23 August 2002. CD Comm. 21–49.

Vassal, L.; Delacroix-Buchet, A. & Bouillon, J. (1994). Influences des variants AA, EE, FF de la caséine αs1-caprine sur le rendement fromager et les caractéristiques sensorielles des fromages traditionnels: premières observations. Lait 74: 89-103.

Viitala, S.M.; Schulman, N.F.; de Koning, D.J.; Elo, K.; Virta, A.; Virta, J.; Mäki- Tanila, A. & Vilkki J.H. (2003) Quantitative trait loci affecting milk production traits in Finnish Ayrshire dairy cattle. J. Dairy Sci., 86, 1828–1836.

Weppert, M. & Hayes, J.F. (2004) Direct genetic and maternal genetic influences on first lactation production in four breeds of dairy goats. Small Rumin. Res., 52, 173– 178.

Winter, A.; Kra"mer, W.; Werner, F.A.; Kollers, S.; Kata, S.; Durstewitz, G.; Buitkamp, J.; Womack, J.E.; Thaller, G. & Fries, R. (2002) Association of a lysine-232 / alanine polymorphism in a bovine gene encoding acyl- CoA:diacylglycerol acyltransferase (DGAT1) with variation at a quantitative trait locus for milk fat content. Proc. Natl. Acad. Sci., 99, 9300–9305.

Yahyaoui, M. H.; Angiolillo, A.; Pilla, F.; Sanchez, A. & Folch, J. M. (2003). Characterization and genotyping of the caprine kappa casein variants. (Goat kappa casein polymorphism). J. Dairy Sci. 88: 2715-2720.

Zhang, Q.; Boichard, D.; Hoeschele, I.; Ernst, C.; Eggen, A.; Murkve, B.; Pfister- Genskow, M.; Witte, L.A.; Grignola, F.E.; Uimari, P.; Thaller, G. & Bishop M.D. (1998) Mapping quantitative trait loci for milk production and health of dairy cattle in a large outbred pedigree. Genetics, 149, 1959–1973.

Chapter 9

Sire X Herd and Sire X Herd-Year Interactions on Genetic Evaluation of Buffaloes

Antonia Kécya França Moita, Paulo Sávio Lopes, Robledo de Almeida Torres, Ricardo Frederico Euclydes, Humberto Tonhati and Ary Ferreira de Freitas

Additional information is available at the end of the chapter

1. Introduction

According to FAO, (2012), World's Total Milk Production is 703,996,079 tonnes per year considering Cow milk (585,234,624 tonnes per year), Buffalo milk (92,140,146 tonnes per year), Goat milk (15,510,411 tonnes per year), Sheep milk (9,272,693 tonnes per year), Camel milk (1,840,201 tonnes per year). This gives us an average milk consumtion of 108 kg per person per year (FAO, 2012).

In 2010, the largest milk producers in the world were USA (87,461,300), India (50,300,000), China (36,022,650), Russian Federation (31,895,100), Brazil (31,667,600), Germany (29,628,900), France (23,301,200), New Zealand (17,010,500), UK (13,960,000) and Turkey (12,480,100) tonnes (DAIRYCO, 2012).

The USA was the largest cow's milk producer in the world in 2010 accounting for 14.6% of world production. World cow's milk production in 2010 stood at nearly 600 million tonnes, with the top ten producing countries accounting for 55.7% of production. The USA is the largest cow's milk producer in the world accounting for 14.6% of world production, producing over 87 million tonnes in 2010, an increase of 1.8% when compared to 2009 (DAIRYCO, 2012).

India is the second largest cow's milk producer, accounting for 8.4% of world production and producing over 50 million tonnes in 2010. The UK is the 9th largest producer in the world producing nearly 14 million tonnes in 2010 and accounting for 2.3% of world cow's milk production (DAIRYCO, 2012).

According to the Brazilian Institute of Geography and Statistics - IBGE (2011), the buffalo herd in 2009 was 1,135,191 heads. Since these animals were distributed by the five regions of Brazil, in the following amounts and proportions: North: 714.852 (62,97%); Northeast:

125.603 (11,06%); South: 121.251 (10,68%), Southeast: 105.615 (9,3%); e Midwest: 67.870 (5,98%).

Compared to dairy cattle herds, the information on buffaloes is considered small. Informations cited below refer to livestock grazing, conducted in Brazil. Studies make an assessment of information generated from the control of the major Murrah buffalo cows herds.

Almost all the buffalo farms in Brazil adopts the extensive regime, characterized by deficient in of control livestock, health and nutrition (Tonhati, 1997). These animals are distributed among the five regions of Brazil, consequently the response to selection is different, considering the weather conditions and objectives of the farmers.

In Brazil the largest amount of bufallo is found on the Marajó Island, but it is in the Southeast that we will find the best quality in dairy products and in the south the best meat production. The first Buffalo breeders Association from the world was criated in Brazil. Despite this and despite of efforts from several breeders and researchers, few are the properties that do proper zootechnical bookkeeping of their herd thus hindering the research and the improvement of the species.

However, the buffalo culture is showing great expansion in Brazil, as shown by Ramos (2003), who observed an yearly 10% growth, therefore, there is an expectation to the development of well structured breeding programs. This large growth is due to the capacity of these animals in being rustical, good food converters when explored for milk and meat, long-living and being able to occupy spaces not inhabited by other animal species economically exploited by man. Based on this information, Ramos (2003) proposed a genetic improvement program of the species to be conducted together with Brazilian Association of Buffalo Breeders.

Buffalo has milk yield per lactation ranging from 500 to 4000 kg, depending on breed and management used. The age of the buffaloes at birth ranged from 24 to 266 months. In Brazil, Tonhati et al. (2000) found this feature to an average of 1259.47 ± 523.09 kg of milk per lactation in cattle exploited in the state of Sao Paulo. The lactation length can vary from 260 to 327 days, being greatly influenced by management type (Pereira et al., 2007).

In literature, we observed the average heritability and comparing several studies, we conclude that there is sufficient genetic variation among individuals in the populations so that they can apply to the methods of selection. However, there has also been observed that, due to different populations, seasons and regions, the coefficient values of heritability for milk production are quite varied, indicating a possible effect of the interaction from genotype x environment.

The environment for the selection of animals, as well as the interaction effect is of interest for animal breeding and has been widely discussed in literature. According to Reis & Lôbo (1991) the environment conditions the results of the improvement process. According to Hammond (1947), selection should be made in superior environment so that the animal could show its full potential, on the other hand, Lush (1964) argues that the selection of future breeders should be done in similar environmental conditions to which their progeny will be submitted.

Falconer (1952) suggests that, depending on the environment, the gene cluster responsible for the expression of a particular feature may not be the same. Thus, including the genotype-environment interaction in genetic evaluation models would end up in better results.

In recent years an increase in the production of buffalo milk has been observed, but also an increasing in the number of zoothecnic, health and nutrition controlling. As well as an increased number of daughters of sires in distinct regions of Brazil, providing greater possibility of studies, especially studies involving the genotype environment interaction.

Most procedures for genetic evaluation in dairy cattle, assumes homogeneous genetic and residual variances between herds. Thus, if the variances increases with the increase in average production and are assumed homogeneous, they could induce to bias in genetic evaluation; and so animals could be wrongly classified. For this reason, several studies made in Brazil (Costa et al., 2002a; Costa et al., 2002b; Melo et al., 2005; Freitas, 2003; Cobuci et al., 2006, Araujo et al. 2008) have been performed to identify a better model in milk production using different breeds and species.

Facing this problem, and it's evident importance, the objectives of this study were:

- to evaluate the effect of the inclusion of the sire x herd or sire x herd-year interaction over the milk production in lactating buffalo records, and;
- to determine the most appropriate model in genetic evaluation of animals.

2. Revision about study

Milk production in Brazil in 2010 was of acquired raw or cooled milk (thousand liters), 20,966,731, of which 99.51% of raw milk production was industrialized or cooled by the establishment (Brazilian Institute of Geography and Statistics - IBGE, 2011). In the fourth quarter of 2010, there were 5.557 billion liters of milk formula. Comparing the fourth quarter of 2009 there was an increase in the industrialization of 2.0% and 7.4% from the third quarter of 2010.

The purchase of buffalo milk in Brazil showed an ascending character in the last 13 years, the biggest change in 2005 (12.3%), and lowest in 2002 (0.0%). These values are due to unfavorable weather conditions and increased production costs in some regions of Brazil (Brazilian Institute of Geography and Statistics - IBGE, 2011).

From all of the milk acquired in Brazil 26.0% comes from the state of Minas Gerais and 14.3%, of Rio Grande do Sul The Southeast Region with 40.3% of production, followed by the South with 33.4% Midwest 14.7%, 6.0% North with the production, and the Northeast with 5.6%. When comparing to the previous year's production, the North, Midwest and Southeast regions showed a decline of milk yield, respectively 13.6%, 6.7% and 0.7% (Brazilian Institute of Geography and Statistics - IBGE, 2011).

It has been observed that the highest daily production of buffalo milk was recorded in April, and lowest in January (Macedo et al., 2001). The reproductive seasonability of buffalo is conditioned by the decrease in daytime light (Pereira et al. 2007). It was observed that the highest birth frequency occurr in the period from February to April.

Ramos (2003) points out that the genetic progresses have also been gained for the milk production in several dairy herds in Brazil. Marques (1991) noted that crossings have also been adopted as a way to promote improved performance. This author found heritability estimatives for various productive and reproductive traits, among which may be cited values range from 0.249 to interpartos, 0.39 for service period, 0.304 for milk and 0.412 for weaning weight.

Boldman & Freeman (1990) using milk producing records of cows in the U.S., from 1976 to 1984, in an animal model that ignored the relationship between animals, to estimate genetic and residual variances at different levels of production, ignoring the covariance between levels found an increase of genetic and residual variance and of permanent environment with the level of production. They also noted that the estimates of heritability and repeatability were lower at low level but similar in levels of medium and high production.

2.1. Variances heterogeneity

Lofgren et al. (1985) estimated heritability for milk yield in three levels of standard deviation for Holstein and Jersey cows. The increase in the variance of sire and herd standard deviation was proportionally greater than the growth of the residual variance, producing higher heritability in herds with a high standard deviation. Heritabilities increased with the standard deviation of herd, approximately 17% (0.18 to 0.21) for Holstein, and 46% (0.25 to 0.37) for Jersey animals.

The variability among herds, or the variability between levels of stratification adopted in the data, especially in dairy cattle, has been disregarded, causing the concentration of animals in these selected herds, with greater expression of phenotypic variability (Hill et al., 1983).

The alternative proposals that may solve the problem of heterogeneous variances are: grouping the data based on some criteria and subsequent analysis of multiple traits, as did Hill et al. (1983), has in principle the proposal by Falconer (1952), considering the expression of genotype in different environments as different traits.

According to the idea of Falconer (1952), if the genetic correlation is high, the performance in both environments represent approximately the same characteristica, determined by the same genes group, and if there are no special circumstances affecting the heritability or the selection intensity, there will be little difference in the environment in which the selection will be performed. If the genetic correlation is low, the characteristics will be considered different, and high or low performance will require different sets of genes. So, it would be advantageous to conduct the selection in the environment where the animals should live. As long as heritability or selection intensity in other environment, be considerably higher.

Van Vleck (1986) noted that the fraction of animals selected from an environment determines selection intensity as a factor in direct response and correlated in various environments. A high heritability leads to a more accurate assessment of the low heritability of the same number of daughters, or equal accuracy with few daughters. The effects of different residual standard deviations when heritability is the same in different environments can also be calculated. According to the author, the genetic correlation between the expressions of a genotype in different environments can also be considered if the correlation is too different from the unity.

Garrick & Van Vleck (1987) exposed that the consequences, for evaluation, of incorrectly assumed homogeneity are demonstrated by progeny testing and by an artificial breeding program that test cows and bulls of heterogeneous populations. Selection assuming homogeneity can be very efficient as heritability and then the accuracy of the selection is greater in the most feasible environment. Therefore, there is appreciable reduction in the results when the heritability is greater in less variable environments.

These authors observed in their progeny test results, that the effect of incorrect variance components in the equation of mixed models is greater when the most variable environment has lower heritability. When heritability increases with residual variance, there is little loss in selection efficiency of selection assuming homogeneity. Furthermore, there is a negligible effect to take a correlation between the same genetic performance in different environments when in fact little interaction is present.

They also consider that it is worth investigating the reduction in efficiency of the selection assuming the heterogeneity for applications in individual breeding. The study with bulls also showed that the bulls sampling is represented in the variety of different livestock environments.

The accuracy in the estimation of genetic parameters depends on the influence of a large set of factors. Freitas (2000) gave more emphasis on the estimation method used, but also stated that the variance heterogeneity is a problem which, if not treated properly, may result in biased prediction of genetic value. According to this author, the quality of data collected in the field also contributes to decrease the accuracy of genetic evaluations. This problem is more serious in small data sets, because in this case, the estimation of some variance components and covariance is fully committed. In a genetic evaluation program, the economic relevance of the characteristics to be evaluated is as important as the quality of solutions.

In a study of milk production in buffaloes it was observed that herds with bigger variability interact with more animals, and if the variability between herds is not considered, the sire selection of the performance of their daughters, may be due not only to its potential, but also to the environment in which their progeny express the phenotype (Cardoso, 2005).

The similarity of the treatment used in females can reduce the environmental variability, if they receive the same amount of food or may increase if the cows in a herd start to receive more differentiated diets as adults. These factors may be responsible for changes in variance components from one level to another, in one year or one year to another (de Veer & van Vleck, 1987).

According to Winkelman & Shaeffer (1988), when the heterogeneity is caused by environmental factors, the genetic evaluation of the animal should be higher due to its environment than their genetic composition. Among the environmental factors that cause variance heterogeneity are herd-period, region, year of birth and management (Ibáñez et al., 1996).

Researches have shown that, in a selection of the best animals in a high quality management environment, there is a tendency to select genotypes with high response value. In contrast, in the selection of the better individuals in an environment of low quality, there is a tendency to select, genotypes with low responsiveness to environmental improvement. Based on this information, it emphasizes the tendency of the responses of animals to associate negatively with phenotypic means in environments of low quality and positively to the averages in high quality environments. In environments of intermediate quality, there is a clear trend of association between phenotype and the average response of animals (Reis & Lôbo, 1991).

Hammami et al. (2009) to overcome the lack of detailed information obtained from routine milk recording data, genotype×environment interaction measure can be based on experiments. These authors confirmed that genotype×environment interaction was essentially when high differences between production environments and/or genotypes (genetically distant genotypes) were observed. Environmental effects were aggregated in most studies and identification of the components of the environment was largely unresolved, with only a few studies based on more definite-descriptors of environment.

Hill et al., (1983) herds were stratified by milk production , according to the average, the variance and variation coefficient in high and low levels, to estimate milk, fat and protein heritability, and fat and protein percentage. Heritabilities to the productions caracteristics were higher in higher production levels with greater heritability differences between the variance levels above mentioned. Results for milk components were similar, except in herds with low variation coefficient, where the heritability estimates were slightly higher.

Moita et al. (2010) observed that when selecting animals without considering the variance heterogeneity tends to favor those belonging to high variance herds, to the detriment of lower phenotypic variability. They also noted that the stratification of the buffalo herds on high and low standard deviation corrected for variance heterogeneity.

In evaluating the effects of variance heterogeneity the use of different criteria is recommended to compare the results, considering that such effects may not be evident depending on the methodology used. Thus, it is indicated that the effect of variance heterogeneitys is considered to reduce the risk, because they have equal or superior results to those obtained by the evaluation that assumes homogeneity of variance (Carvalheiro, et al. 2002).

Araújo, et al. (2008b) concluded that there is variance heterogeneity among herds that exploit the milk production in Murrah buffaloes and also that the nature of this variance heterogeneity is a result of environmental factors. So the sires are being selected because of the variable environment in which their progeny are created, than to its own genetic merits.

In a study evaluating variance heterogeneity, Balieiro et al. (2004) found out that the genetic variance components were higher, in the stratified data, than the ones obtained in the univariate analysis, and residual variance components were lower, resulting in higher heritability estimates for the analysis of multiple traits. Such joint analysis allowed higher redemptions of portions of additive genetic variances, which would be directed to the residual component in the analysis of unique feature.

Araújo, et al. (2008b) observed in their study evaluating herds of Murrah buffaloes that constant variances among herds in genetic evaluation of animals, disregarding the level of production or the estimated variance among herds, according to results obtained in the overall analysis there may be a misclassification of animal genetic merit. With the increasing of variances, the production of daughters of sires, raised in herds with greater variability and in consequence less productive, as in the case of this study, would influence more the essessment of the breeders than the daughters raised in herds with less variability.

Torres'work (1998) characterized the pseudo-interaction described by Dickerson (1962), which occurs when the genetic correlations are high, therefore, the ranking of the animals would not be altered, but the magnitude of the variability of the components, as well as the heritability evaluation, would vary from the phenotypic standard-deviation to selection programs.

Normally, genetic variance heterogeneity, additive or residual, within the herd is due to matings and to preferencial treatments given to specific reproducers (Norman,1974). One alternative to restrict the effect of the preferential treatment between herds is the inclusion of sire-herd interaction in the statistic model, in the form of environmental-correlation. However, it recalls that the heterogeneity of the variance can be responsible for part of the variability component attributed to interaction (Norman,1974).

When there is variance heterogeneity and the same is ignored, the progeny of the daughters of a given sire will be weighted in proportion to the standard deviations of the herds in which they were raised. The result is that the progeny of the daughters in herds with higher variabilities will influence the evaluation of the breeders more than progeny of the daughters originated from herds with lesser variability (Vinson, 1987). Also, there is the risk of erroneous sorting from animals through their genetic values and, therefore, the genetic progress can be affected (Araújo, et al. 2008b).

In the evaluation of females, to ignore the effect of variance heterogeneity would tend to favor females that had their production in herds with high variance, the opposite occurs with females in herds with low variance (Araújo, et al. 2008b). Of practical importance, would be the likely trend existing in the evaluations of high-producing cows, chosen as parents of future breeders in artificial insemination. In the long run, we can say that any trend in the evaluations of the females would accumulate along the time, because daughters and mothers tend to express productions in the same herd (Torres, 1999).

Costa et al., (2000) used records for milk and fat production to study the variance heterogeneity between sites, which were classified into two groups (low and high) based on phenotypic standard deviation of milk production per herd-year. They noted that the variance components of sire and residual for milk and fat in environments of low standard deviations were lower than in environments with high phenotypic standard deviation.

The variance heterogeneity has a greater effect on the genetic evaluation of females, because these are evaluated within herds, and their predicted genetic values would be greatly affected by the variance within herd, which tends to be uniform over time (Torres, 1999).

Assuming homogeneity variances has little effect in the evaluations of breeders, if these are used in herds of different production levels, and heritability increases with the increasing of residual variances (Garrick & Van Vleck, 1987; Vinson, 1987; Winkelman & Schaeffer, 1988). However, when heritabilities are smaller in the environment in which the residual variances are greater reductions in the efficiency of selection of bulls may happen for falsely considering that the variances are homogeneous.

Carvalheiro, et al. (2002) observed in their study evaluating the variance heterogeneities that calves and cows have their breeding genetic values most affected by the variance heterogeneity of the bulls.

In a simulation study, Garrick & Van Vleck (1987) observed a reduction in genetic gain of up to 3.4% when the effect of variance heterogeneity was ignored in progeny testing schemes resulting in the selection of environments with low heritability, as they are more variable. When the heritability increased with the increase of variability, the reduction was only 0.1%.

Variance heterogeneity between herds results in the reduction of the selection and implies in unequal genetic progress between environments classified by the standard deviation of the herd. So to ignore the variance heterogeneity has consequences in the selection and in the genetic gain, reducing the effectiveness of a breeding program (Van Vleck, 1987; Vinson, 1987).

According to Visscher et al. (1991) The initial adjustment for variance within herd-year-season phenotypic variance was more effective in reducing the variance heterogeneity, and seems to be the most practical way to variance heterogeneity for genetic evaluation per animal model.

According to Vinson (1987), the proportion of selected animals from more variated environments increases with the intensity of imposed selection. Hill (1984) in a normal data distribution, showed that the standard-deviation of more variable populations is 50% higher than the least variable, and the proportion of selected animals is 0,01, indicating that 95% from the selected individuals are provenient of more variable populations.

The genetic correlations for milk and fat production between the two groups of environments in a study of variance heterogeneity between places, were close to 1.0. The authors concluded that genetic evaluations in Brazil should consider the variance heterogeneitys to increase the accuracy of evaluations and the selection efficiency for the milk and fat production in Holstein (Costa et al., 2000).

Carvalheiro, et al. (2002) observed that effects of heterogeneity of residual variance on genetic evaluations are associated with the selection pressure and heterogeneity levels. If corrections are not made, herds that practice intense selection and that present accentuated levels of residual variance heterogeneity can have their animals incorrectly classified and, consequently, less responsive to selection.

The effect of heterogeneity, if any, in the genetic evaluation should be best determined by examining the position of animals on the magnitude of the relative difference in the genetic values in levels of variances in herds (Dong & Mao, 1990).

Van der Werf et al. (1994) observed that the correction for variance heterogeneity within herds did not remove all biases of the average in parents, but the improvement in bias and accuracy of genetic value can be expected. They also observed an increase of phenotypical standard-deviation of milk production in herds over the years, probably due to increased production. They found that the variation coefficient within herd-year was 31%, indicating that the method of genetic evaluation could be improved with corrections for heterogeneity variance. Correction methods of heterogeneity probably would correct all the biases in EBV (Expected breeding value), especially if these biases were due to preferential treatment.

2.2. Genotype by environment interaction

The differences between livestock in the variability of production, usually results from differences in climatic factors, regional or local, and types of management, including such factors as intensity of power, supply voltage according to the production, success of diseases control programs and quality procedures for the creation of young breeders (Vinson, 1987).

According to Van Vleck (1987), if the genetic and residual variances and covariances were known in each environment, represented by a set of herds, then selection based on the results obtained from analysis of multiple traits in mixed models, would produce an assessment quality that could be used to select optimally bulls or cows to produce in herds or in specific environments.

According to Stanton et al. (1991), there are two situations in which differences in responses to selection would be verified in different environments, characterizing the genotype x environment interaction. The first situation occurs when the genetic correlation between the phenotypic expression in two environments would be substantially less than 1.0, indicating that different genetic bases would be acting in different environments.

The genetic correlations derive informations about the genotype x environment interaction and can be used to quantify the loss of information when using production records from a different environment from where the animals selected will be used.

Another case cited by Stanton et al. (1991) that characterizes the genotype x environment interaction would result from the variance heterogeneities, as in the case of evaluations of simple features, in which sires have the same classification in each environment, but differences in response to selection of the daughters and the genetic values from the bulls would be lower in an environment with less variability.

Houri Neto (1996) evaluated genotype x environment interaction between Brazil and USA. through 332,617 lactations obtained from the State of New York (USA) and 115,547 productions observed in Minas Gerais, Espírito Santo, São Paulo, Santa Catarina and Rio Grande do Sul, in the period 1979 to 1991. Genetic correlations of milk productions obtained between the two countries ranged from 0.46 to 0.68. This result indicates that sires evaluated in the USA did not show the same performance in Brazil.

Paula, et al. (2009) studying the effect of genotype by environment interaction for production of Holstein noted different bull genetic values when environment was modified, what characterizes genotype x environment interaction and proves that the genotype x environment interaction alters the classification from the animals and can lead to the inappropriate choice of bulls in certain regions, damaging the genetic improvement of livestock.

Cienfuegos-Rivas et al., (1999), when comparing the performance of daughters of Hollstein bulls in Mexico and in the northeast from the USA found different and lesser answer from the daughters of bulls selected in USA and raised in mexican environment, suggesting that the better answers are predicted to Mexico, when using information from daughters raised in american environment of low standard deviation of herd-year milk production.

These authors concluded that the genotype x environment interaction, evidenced both by the variance heterogeneity, and the genetic correlation coefficients, not only prevents genetic gain, but contributes severely to the reduced and different net economic benefit from the biotechnology of artificial insemination in the countries of Latin America, using American semen.

Cardoso (2005) in a study evaluating the genotype x environment interaction for milk production in buffaloes observed that there is variance heterogeneity among herds that exploit the production of buffalo milk, predominantly Murrah, and still that the nature of variance heterogeneity is resulting from environmental factors. He also noted that the admission of constant variances among herds in genetic evaluation of animals leads to a misclassification of the genetic merit of the best animals.

2.3. Study of variances

Ratings of genetic (co) variances between groups of herds indicated that the correlations between the three major types of variances were above 0.80. These correlations were high, indicating little benefit in the calculations for evidence of bulls if a non singular matrix of genetic variances and covariances was used instead of a singular matrix (Winkelman & Schaeffer 1988).

Higher values for genetic and residual variances, as the average production or the standard deviation within environments increases, have been presented in several papers (Boldman

& Freeman, 1990; & Dong Mao, 1990, Torres, 1998; Araújo , 2000, Araújo et al., 2002). Most of these studies show high heritability estimates, as the variability of production within the environment increases (Hill et al., 1983, Lofgren et al., 1985; de Veer & Van Vleck, 1987; & Dong Mao, 1990; Meuwissen & van der Werf, 1993; Araujo, 2000).

Dong & Mao (1990) observed that the increase in the percentage of residual variance in the group of herds with values considered low to medium production were equal to or greater than the percentage increase in the group of herds with medium to high values of production. However, the rate of increase of the residual variance in the group of herds with values considered low to high was not as great as that observed in the sire variance.

Dorneles, et. al. (2009) observed an increase in estimates of heritability partially associated with the reduction of permanent environmental variance, which promoted the reduction in total variance and, consequently, increasing the proportional contribution of additive genetic variance. This growing trend is similar to those reported by Melo (2003) and Cobuci et al. (2005), for Holstein, using the Wilmink function to model the genetic variance and permanent environment. Decreasing trend was observed, in Brazil, by Costa et al. (2005), for Gyr, by Freitas (2003), for Gyr-Holstein, and Dionello et al. (2006), for the Jersey breed.

2.4. Sire by herd interaction

Mayer (1987) observed that the variance components due to the effect of bull-environment interaction on the effect of environment, have been found with values as high as the variance between sires for milk, fat and protein production. A variety of factors may have contributed to greater similarity between the daughters of the same sire in different herds or identified in subclasses in herd-year-season. One of them is a difference in variation between herds, which can be reduced by transforming the data.

Mohammad et al. (1981) observed that the variance components of sire-herd interaction accounted for 10% of the phenotypic variance in milk yield and these presented a negative effect. Araújo (2000) found that the use of herd-sire interaction was not effective as a way to set the variance heterogeneities.

The increasing in the number of daughters for breeding using artificial insemination, include several herd-year-season and several stable matings, and could also reduce the interaction influence next to errors caused by other factors and thus be used in the selection of bulls (Kelleher et al., 1966).

The study of the production of daughters of a certain sire that does not consider the variance heterogeneities will be adjusted by the ratio of standard deviation of the herds to where they were created. The production of the daughters of breeders raised in more variable herds will influence more in the evaluation of sires than in the production of daughters derived from herds less variable (Vinson, 1987).

Winkelman & Schaeffer (1988), with the aim to estimate variance components of sire and residual to Canadian herds, noted that the accuracy of the estimated variance components within herds, could influence the effectiveness of assessment for variance heterogeneity, since small sample sizes could lead to large sampling errors on the estimates.

This variance behavior led to higher heritability estimates during the initial phase of lactation, with a maximum value of 0.30 at the third test-day. Probably, this pattern is due to the fact that only more persistent cows remain producing until the tenth month of lactation which would decrease genetic variability (Tonhati et al. 2008).

Stratifying records intrarebanho-year variance proved to be the most effective way to estimate the variance components and heritability (Dong & Mao, 1990). When the data were divided by production levels, the residual variance components of sire in high production were less than twice than those of low production (Dong & Mao, 1990).

The knowledge of residual genetical variances covariances in each herd or on the environment represented by a set of livestock, represents for selection based on results obtained by multiple characteristics analysis in mixed models, the production of a quality evaluation that could be used to properly select bulls and cows,to produce in herds or in specific environments (Van Vleck,1987).

Bueno et al. (2005) working with Brown Swiss cows found that the inclusion of the effects from the interaction of sire x herd or sire x herd-year in genetic evaluation models changed very little the estimates of the components of addictive and residual genetic (co) variance and consequently, heritabilities.

2.5. Methods

Information obtained from the measurements of production of buffalo milk in Brazil were used to estimate variance components and foresaid genetic parameters and breeding values. For this, the month of delivery of buffaloes were grouped in two seasons, season 1, from April to September, which corresponds to the months of lower rainfall, and season 2, corresponding to the months from October to March, in which there is more rainfall precipitation, except for a herd located in the Northeast, where the opposite was considered.

Subsequently, the effects of herd and year were grouped into subclasses, and required at least four observations per herd-year subclass. So that it was possible to investigate the effect of interaction of sire x herd or sire x herd-year on milk production it was required that each sire had at least two daughters distributed in more than one herd.

After the necessary deletions were made, the data file consisted of 1774 lactations, from 754 Murrah buffaloes, daughters of 39 sires, which, calved in the period 1987 to 2005, and were distributed in 13 herds. The duration of lactation was maintained from 64 to 305 days and age at birth ranged from 24 to 185 months. The connectedness of the data was performed to assess breeding-herd according to the daughters.

Estimates of variance components and genetic parameters and breeding values were obtained using MTDFREML (Multiple Trait Derivative Free Restricted Maximum Likelihood), described by Boldman et al. (1995), using an animal model.

The pedigree file that caused the numerators parenthood matrix (NRM) coefficients, used in all analysis, contained 864 different animals, 1776 non-zero elements in the NRM and no endogamous animal, consequently, the mean coefficient of inbreeding was equal zero.

The existence of genotype x environment interaction has been tested by the estimation of variance components, comparing models including or not the effect of genotype x environment interaction (Banos & Shook, 1990, Bueno, et al. 2005). Another way to detect the presence of genotype x environment interaction would be the comparison between the classifications of animals, according to breeding values.

Some studies have shown that if sire x herd interaction is included in the model of genetic evaluation, the influence of observations from a few herds will be more limited in the evaluation of breeding and may not significantly affect the evaluation of animals with progeny in many herds. However, ignoring it would increase the estimates of addictive genetic variance, which would underestimate the breeding values of sires and its accuracy.

In the present study the analysis were conducted using six models of simple traits, in which, were considered as fixed effects season of calving and herd-calving year and cow age as a covariate (linear and quadratic effects). The random effects of the six models are described below: 1 - additive model (additive effect and error), 2 - repeatability model (model 1 plus permanent environmental effect), 3 - additive model with the sire x herd interaction (model 1 more sire x herd interaction) 4 - additive model with interaction of sire x herd-year (including a model interaction of sire x herd-year), 5 - repeatability model with the sire x herd interaction (model 2 including the interaction of sire x herd), 6 - repeatability model with the interaction of sire x herd-year (model 2 including the interaction of sire x herd-year).

The importance of including the sire x herd interaction and sire x herd-year and permanent environment effect model in genetic evaluation of animals was performed using the likelihood test of models sequentially reduced (Rao, 1973). The test statistic of the likelihood estimate (LR) was compared with the value obtained through the distribution of chi-square ($\chi 2$), with 1 degree of freedom. The estimate was obtained by the expression

$$LR_{ij} = -2\log_e\left(L_j / L_i\right),$$
$$LR_{ij} = 2\log_e\left(L_i\right) - 2\log_e\left(L_j\right),$$
$$LR_{ij} = 2\left[\log_e\left(L_i\right) - \log_e\left(L_j\right)\right],$$

Where

LRij= test statistic is the likelihood ratio for models sequentially reduced;
Li= is the maximum likelihood for the complete model i, and
Lj= is the maximum likelihood for the reduced model j.

The null hypothesis test implies that the functions of likelihood of full and reduced models did not differ among themselves, ie.,

$$H_0 : -2\log_e\left(L_i\right) = -2\log_e\left(L_j\right),$$

The decision rule used was:

If LRij > χ^2_{tab}, the test was significant and the full model provided the highest likelihood function over the reduced model;

If LRij < χ^2_{tab}, the test was not significant and the full model did not provide the greatest value of likelihood function over the reduced model.

The breeding values of animals for milk production have been organized into files, in order to assess possible changes in the magnitude of the predictions of breeding values and ranking of sires based on such predictions, when the sire x herd interaction and sire x herd-year was included in the model by means of the Pearson and Spearman correlations.

3. Result and discussion

The highest yield per lactation in buffaloes, observed in this study was observed around the 108º month of age. The observed average milk production up to 305 days, referring to 1774 lactations from 754 buffaloes in the period 1987 to 2005, was 1736.66 kg, with a standard deviation of 705.85 and a variation coefficient of 40.64%. Smaller values were found by Fraga et al. (2006), with an average production of 820 kg per lactation, in Cuba, for Tonhati et al. (2000), 1259.47 kg in Brazil; by Patel & Tripathi (1998) that in Surti buffaloes, found an average of 1442.6 kg.

However, higher values were found by Baghdasar & Juma (1998), 5419.95 kg, Iraq, for Rosati and Van Vleck (1998), 2286.8 kilograms, Italy and Moioli et al. (2006), 2184 kg in Italy. However, Khan (1998) study the Nili-Ravi buffaloes in Pakistan found values ranging from 1835 to 2543 Kg.

It is essential to monitor cumulative genetic progresses in selection programs not only to quantify the occurrence of genetic changes but mainly to evaluate benefits and to perform adjustments when necessary. Thus, it is necessary to know the genetic trend of the population studied (Euclides Filho et al. 1997).

The effects of variance heterogeneity on response to the selection depends on the magnitude of differences in genetic variance for milk production and their relations with the phenotypic variances. In the study in models of simple features it was found that the inclusion of the permanent environment (model 2 versus model 1, P = 0.052), the interaction reproductive years herd-(4 versus model a model, P = 0.047) in additive model were significant. The inclusion of sire x herd interaction in the repeatability model (model 5 versus. model 2, P = 0.219) was not significant, whereas the inclusion of the interaction of sire x herd-year (model 6 versus. model 2, P = 0.025) was significant (Table 1).

The study of these interactions lead us to consider how much the environment has influence in the effect of the features, which would be the best way to perform the study. The presence of genotype x environment interaction (G*A) is characterized by a different response of the genotypes to environmental variations, which can cause changes in the ordonance of the performance of the genotypes in different environments (Falconer & Mackay, 1996).

Silva (2002) working with buffaloes, also found no significant effect of inclusion of sire x herd interaction in the model of repeatability. However, Araújo (2000), working with Holstein cattle, and Sirol et al. (2005), with cattle breed Brown Swiss, found a significant effect in the inclusion of sire x herd interaction in the model of repeatability.

Models	-2log$_e$ L	LR	Significance level
(1)Additive	22,658.5037	-	-
(2)A + P	22,654.7100	3.793797	0.052099609
(3)A+RR	22,655.3691	3.134660	0.081110901
(4)A+RRA	22,654.5854	3.918380	0.046616186
(5)A+P+RR	22,654.7099	3.04E-05	0.219078768
(6)A+P+RRA	22,650.3313	4.378671	0.026356906

[1] A + P= additive + permanent environment, A+RxR= additive + interaction sire x herd; A+RxRA = additive + sire x herd-year interaction, A+P+RxR= additive + permanent environment + sire x herd interaction; A+P+RxRA=additive + permanent environment + sire x herd-year interaction.

Table 1. Values of -2 times the natural logarithm of the likelihood function (-2 loge L) and the likelihood ratio test (LRij) to sequentially reduced models, obtained for milk production in buffaloes, in the analysis of simple traits

Breeders produced in these herds are then distributed to be used in various environments. The knowledge of the importance of this interaction allows, in some cases, to set the environmental conditions in which the animals will be selected (Correia et al. 2007). Based on the fact that genotype x environment interaction can affect the genetic populations of beef cattle by inappropriate use of breeding is very important to consider this interaction in genetic evaluation. It is important to know in which of the additive or repeatability models its inclusion would have a significant effect, that in this study was observed in the additive model.

Correa, et. al. (2007) observed that the genotype environment interaction is significant when the differences in genetic and environmental levels are significant. The existence of this type of interaction implies in the optimal combination between genotypes and environments in order to maximize the production. This interaction can be influenced in situations where the selection herds are located in the best regions, using management practices well above average.

Estimates of additive genetic variance components for milk production were higher in models 1, 3 and 4 when compared to models that consider the effect of permanent environment 2, 5 and 6, respectively. Thus, it appears that the inclusion of permanent environmental effect caused a reduction in additive genetic variance.

Silva (2002) working with some of these data found higher values for additive genetic variance 66601.81, 66187.01 and 64989.44, respectively, for additive models + permanent environment, additive + permanent environment + sire x herd interaction, and additive + permanent environment + sire x herd-year interaction. He also noted that the inclusion of sire x herd caused a reduction in additive genetic variance. While Sirol et al. (2005), was working with data on milk production in Brown Swiss breed, found that estimates of residual (co) variance and additive genetic (co) variance did not change when the models were adjusted for the effects of sire x herd interaction.

Models	Variance Components				
	σ_a^2	$\sigma_{c_1}^2$	$\sigma_{c_2}^2$	$\sigma_{c_3}^2$	σ_e^2
(1)Additive	135,186.081	-	-	-	165,086.573
(2)A + P	57,414.559	63,919.5	-	-	164,142.767
(3)A+RR	132,176.705	-	7,870.55		163,389.444
(4)A+RRA	135,545.905	-	-	9,988.00	156,675.128
(5)A+P+RR	57,502.2693	63,911.2	0.038309	-	164,144.258
(6)A+P+RRA	53,060.472	67,674.9	-	10,651.6	155,285.904

[1] A + P= additive + permanent environment, A+RxR= additive + interaction sire x herd; A+RxRA = additive + sire x herd-year interaction, A+P+RxR= additive + permanent environment + sire x herd interaction; A+P+RxRA=additive + permanent environment + sire x herd-year interaction.

Table 2. Estimates of additive genetic variance components (σ_a^2), permanent environmental ($\sigma_{c_1}^2$), the sire x herd interaction ($\sigma_{c_2}^2$), interaction of sire x herd-year ($\sigma_{c_3}^2$) and residual (σ_e^2) for milk production in buffaloes in an analysis of simple trait

The reduction in the estimated variance component of the permanent environmental effect was 0.013%, when the term of sire x herd interaction was included in the repeatability model (model 5 versus. 2). However, the term includes the sire x herd-year interaction increased by 5.87% (model 6 versus. 2), this result shows the importance of including the sire x herd-year interaction and confirms the results of the likelihood ratio test applied in this study (Table 2).

Silva (2002) in working models 2, 5 and 6, in buffaloes, found lower values for the variance component for permanent environmental effect, 37942.52, 37905.22 and 37983.33, respectively. The inclusion of sire x herd interaction caused a decrease of 0.98% in the component, whereas, by including the sire x herd-year interaction, an increase of 0.107%.

Estimates of residual variance components were smaller in models that included the sire x herd-year interaction (about 5% in model 1 versus. 4, and 6 versus 2). While the inclusion of sire x herd interaction caused a reduction of 1.03% in the estimate of residual variance component, the additive model and in the repeatability model the change was only 0,000009%. These results indicate that the inclusion of interaction of sire x herd-year is more important than the sire x herd interaction in models of buffaloes genetic evaluation.

The heritability estimate obtained for the characteristic of milk production decreased from 0.45 to 0.20 when we included the effect of permanent environment of additive model. So when the effect of permanent environment is not included, the expected genetic gains are inflated. By adding the sire x herd interaction or sire x herd-year interaction in model, the heritability estimates did not show large variations between the models studied (Table 3). Similar results were obtained by Araújo (2000), when working with Holstein cows and Silva (2002), when working with buffaloes.

Tonhati et al. (2004) found low values in the heritability estimative in four correcting types for buffalo production, showing that the genetic gain through selection were low. Genetic correlations between productions were high,therefore, to select through production periods permitted that breeders were selected among the remaining ones.

Dionello et al, (2006) in his research using milk production in Jersey cows noted that the values obtained by heritability estimates suggested the repeatabilty model as adjustment alternative for the milk production on controlling day, however, the repeatability estimate does not support the hyphotesis that the production during lactation should be considered with the same characteristics.

The heritability estimates were 0.22 for total milk yield and 0.19 for 305 days. For test-day yields, the heritability estimates ranged from 0.12 to 0.30, with the highest values being observed up to the third test month, followed by a decline until the end of lactation. The present results show that test-day milk yield, mainly during the first six months of lactation, could be adopted as a selection criterion to increase total milk yield (Tonhati et al. 2008).

Models	Milk Production				
	h^2	C^1	C^2	C^3	E
(1)Additive	0.45 (0.031)	-	-	-	0.55 (0.031)
(2)A + P	0.20 (0.091)	0.22 (0.085)	-	-	0.57 (0.030)
(3)A+RR	0.44 (0.033)	-	0.026 (0.019)	-	0.54 (0.031)
(4)A+RRA	0.45 (0.030)	-	-	0.033 (0.018)	0.52 (0.033)
(5)A+P+RR	0.20 (0.094)	0.22 (0.086)	0.13E-06	-	0.57 (0.030)
(6)A+P+RRA	0.19 (0.090)	0.24 (0.084)	-	0.037 (0.019)	0.54 (0.033)

[1] A + P= additive + permanent environment, A+RxR= additive + interaction sire x herd; A+RxRA = additive + sire x herd-year interaction, A+P+RxR= additive + permanent environment + sire x herd interaction; A+P+RxRA=additive + permanent environment + sire x herd-year interaction.

Table 3. Heritability estimated (h^2) and phenotypic variance proportion about permanent environment effect (C^1), sire x herd interaction effect (C^2), sire x herd-year interaction effect (C^3) and environment proportion for total variance (e), for buffaloes milk production, in an analysis of simple trait.

The heritability estimates for milk production in buffaloes, found by Khan (1998), in Pakistan (0.18), and Raheja (1998), India (0.19) were close to those found in Models 2, 5 and 6 of this work. Higher estimates in Brazil, were found by Tonhati & Vasconcelos (1998), 0.25; Tonhati et al. (2000), 0.28, and Silva (2002), 0.31. Values below the estimated in this study were obtained Rosati & Van Vleck (1998), 0.14, and Kalsi Dhilon & (1984), 0.13.

The proportion of phenotypic variance due to permanent environmental effects (C1) was 0.22 for model 2 and 5, and 0.24 for model 6. While Silva (2002) found values slightly below

(0.18) for the same models in buffalo. Rosati & Van Vleck (1998) found similar values (0.24) to the present study.

The proportion that represents of sire x herd interaction decreased with the inclusion of permanent environmental effect in the model, 0.027 (model 3) and 0.000013 (model 5), fact not observed with the inclusion of sire x herd-year interaction, 0.033 (section 4) and 0.037 (model 6) (Table 3). Silva (2002) found lower values for the models that consider the effects of permanent environment and of sire x herd interaction or sire x herd-year interaction, respectively 0.0000026 (model 5) and 0.02 (model 6), buffalo. Sirol et al. (2005), was working with cattle breed Brown Swiss also found that the proportion of total variation explained by sire x herd were near zero for milk production.

The range between the predicted breeding values for animals was very different between models 1 and 2. However, this difference is smaller when comparing models 1, 3 and 4 or 2, 5 and 6.

Models	Breeding values				
	Average	SD	Amplitude	Minimum	Maximum
(1)Additive	-6.70903	254.2341	2,009.420	-813.4204	1196
(2)A + P	-6.21746	135.7058	995.0106	-439.9113	555.0993
(3)A+RR	-4.71076	243.4701	1,953.337	-777.3368	1,176
(4)A+RRA	-6.26055	253.4086	2,007.501	-804.5006	1,203
(5)A+P+RR	-6.21974	135.8125	995.7977	-440.2247	555.5730
(6)A+P+RRA	-5.69773	126.1884	918.6691	-408.5506	510.1185

[1] A + P= additive + permanent environment, A+RxR= additive + interaction sire x herd; A+RxRA = additive + sire x herd-year interaction, A+P+RxR= additive + permanent environment + sire x herd interaction; A+P+RxRA=additive + permanent environment + sire x herd-year interaction.

Table 4. Average breeding values for all animals, standard deviations, maximum, minimum and maximum amplitudes in buffalo milk production in an analysis of simple trait

The Pearson correlation coefficient was equal to unity between the two models (additive over environment) and 5 (more additive plus permanent environmental sire x herd interaction) and close to 1 between models 5 and 6 (more additive plus permanent environmental interaction sire x herd-year). The same happened to the Spearman correlation. This implies that there is no difference between models 2 and 5. Despite the correlation between the 5 and 6 models to be high, statistically significant differences when using of sire x herd-year interaction (Table 1). Thus, despite the likelihood ratio test was significant (P <0.03), no significant change in the expected magnitude of the predicted values when the model is or is not adjusted for the effect of sire x herd interaction or sire x herd-year interaction (Table 5).

Falcão, et al. (2006) found low values of genetic correlation and demonstrate the effect of genotype x environment interaction, showing that the best sires may not be the same in different environments. Therefore, the choice of the breeder, the genotype × environment interaction should be considered.

Models	1	2	3	4	5	6
1	1	0.95076	0.99621	0.99873	0.95088	0.94719
2	0.94735	1	0.93433	0.94568	1.00000	0.99847
3	0.99527	0.92927	1	0.99725	0.93447	0.93263
4	0.99826	0.94184	0.99646	1	0.94581	0.94453
5	0.94748	1	0.92942	0.94197	1	0.99847
6	0.94301	0.99778	0.92731	0.94068	0.99777	1

Table 5. Pearson's correlation coefficient above the diagonal and Spearman below the diagonal, between breeding values for milk production, in analysis of simple trait

Similar results were obtained by Silva et al. (2002). Sirol et al. (2005) also found correlations from Pearson and Spearman for animals from the Swiss-Brown race, obtained by different models next to 1. The same was observed by Mohammad et al. (1982), that found Pearson and Spearman correlation above 0,99 for models that included and that ignored the interaction sire x herd effect and concluded that to ignore the interaction effect would not cause major alterations in the classification in Holstein cattle.

(1)Additive		(2)A + P		(3)A+RR		(4)A+RRA		(5)A+P+RR		(6)A+P+RRA	
AI	BV	AI	BV	AI	BV	AI	BV	AI	BV	AI	BV
145	1,196.199175	417	555.099273	145	1,176.458158	145	1,202.885413	417	555.573032	417	510.118508
417	1,096.918037	145	534.741842	417	1,021.933134	417	1,083.087238	145	535.263843	145	500.446971
648	932.169385	830	443.041801	535	925.531012	648	946.064351	830	443.419160	830	406.857072
535	916.531689	648	417.798173	648	904.522785	535	940.492253	648	418.269010	648	389.193901
158	866.897851	535	407.532504	158	880.667644	158	874.750424	535	407.951156	535	388.678508
830	829.170969	633	399.643734	830	783.759571	830	815.939882	633	400.032722	158	373.762665
763	804.776426	158	396.902144	372	757.467738	763	797.078757	158	397.264084	633	368.644302
633	789.345535	780	387.983107	763	756.916825	633	783.553783	780	388.305834	780	361.965814
707	729.145247	88	377.104287	633	735.288004	707	755.150389	778	377.345364	88	347.851767
372	704.552849	778	377.029700	707	714.328625	372	716.738100	88	377.246737	778	343.801016
780	692.384619	763	355.236708	599	687.152329	780	694.440688	763	355.629186	533	324.039457
533	681.656841	533	352.189245	780	674.324676	599	684.550793	533	352.536469	708	320.667190
778	670.211976	708	342.453984	533	655.734153	156	675.159165	708	342.801505	763	318.772181
156	666.651135	498	340.036913	156	635.988657	533	674.976642	498	340.275739	707	313.277979
708	654.593767	781	330.849318	778	635.410788	708	659.764026	781	331.076036	498	312.627546
638	650.632615	829	330.147344	708	635.173974	778	651.799147	829	330.387198	781	307.745756
599	647.109638	494	324.896616	638	615.855574	135	629.955574	494	325.188312	494	301.316618
135	615.840639	707	324.539277	135	600.695858	638	624.301576	707	324.866834	829	300.288924
494	594.167074	92	315.587774	412	583.553355	412	592.398731	92	315.884408	92	290.646106
276	589.146008	413	303.638671	276	573.741561	494	590.895416	413	303.826051	156	285.113078

[1] A + P= additive + permanent environment, A+RxR= additive + interaction sire x herd; A+RxRA = additive + sire x herd-year interaction, A+P+RxR= additive + permanent environment + sire x herd interaction; A+P+RxRA=additive + permanent environment + sire x herd-year interaction.

Table 6. Breeding values for 20 the best animals, animal identification (AI); breeding values(BV), in models 1, 2, 3, 4, 5 e 6

There were no great changes in the ranking of animals based on predicted breeding values when using different models. In comparing the top 20 animals one sees that 65% were common in all models (Table 6). This result was confirmed by the Spearman correlation estimates was equal to unity between models 2 and 5 and close to unity between models 2 and 6. Similar behavior was observed for the Pearson correlation.

Comparing the templates 2 and 5, it is observed that the top 20 animals were common, and that 95% were common between the models 1 and 4, 2 and 6, 3 and 4, and 5 and 6. The lowest percentage of animals was observed in common between the models 2 and 3, 3 and 5, 3 and 6, and 4 and 5. What shows us that even considering the six models studied, 95% of the best animals were common to most models which shows that the environmental variance influenced animal performance when added to the model the effect of environment.

4. Conclusion

This study concluded that the inclusion of sire x herd interaction and sire x herd-year interaction, had no significant effect. Although the heritability estimates have been low, estimates of phenotypic and genetic correlations between all the characteristics show the possibility of developing a selection scheme to improve the characteristics of milk production in buffaloes. More studies are needed to select the best animals for they can be used in various environments.

Author details

Antonia Kécya França Moita
Department of Animal Science, Universidade Federal da Bahia, Brazil

Paulo Sávio Lopes, Robledo de Almeida Torres and Ricardo Frederico Euclydes
Department of Animal Science, Universidade Federal de Viçosa, Brazil

Humberto Tonhati
Department of Animal Science, Universidade Estadual Paulista, Brazil

Ary Ferreira de Freitas
Embrapa Gado de Leite, Brazil

Acknowledgement

To the buffalo breeders, for the perseverance in dedication of animal breeding; and for the Association of Buffalo Breeders.

5. References

Araújo, C. V.; Cardoso, A. M. C.; Ramos, A. A.; Araújo, S. I.; Marques, J. R. F.; Tomazini, A. P. I.; Chaves, L. C. (2008a) Heterogeneidade de variâncias e parâmetros genéticos para produção de leite em bubalinos da raça Murrah, mediante inferência Bayesiana. *Revista Brasileira de Saúde Produção Animal*, v.9, n.3, p. 416-425, jul/set.

Araújo,C.V.; Cardoso, A.M.C.; Ramos, A. A.; Araújo,S. I. Marques,J. R. F.; Inoe, A. P.; Chaves, L. C. (2008b) Genetic parameters and variance heterogeneity to milk yield in Murrah breed using Bayesian inference. *Archivos Latinoamericanos de Produccion Animal*, Vol. 16, No. 4, October-December, pp. 234-240.

Araújo, C. V.; Torres, R. A.; Rennó, F. P.; Pereira, J. C.; Torres Filho, R. A.; Araújo, S. I.; Pires, A. V.; Rodrigues, C. A. F. (2002) Heterogeneidade de variância na avaliação genética de reprodutores da raça pardo-Suíça no Brasil. *Revista Brasileira de Zootecnia*, v. 31, n. 3, p. 1343-1349.

Araújo, C. V. de. (2000) *Efeito da interação reprodutor x rebanho sobre a produção de leite na raça holandesa*. Viçosa, MG: UFV, 2000. 80 p. Dissertação (Mestrado em Zootecnia) – Universidade Federal de Viçosa.

Associação Brasileira Dos Criadores De Búfalos (ABCB) (2011). Avalilable from www.bufalo.com.br.

Baghdasar, G.A.; Juma, K.H. (1998) Some aspects of life performance of Iraqi buffalo (Bubalus bubalis) cows. In: World Congress on Genetic Applied to Livestock Production, 6, Armidale, Austrália. *Proceedings...* Armidale, v.24, p. 474-478.

Balieiro, J.C.C., Lopes, P.S., Eler, J.P., Ferraz, J.B.S., Euclydes, R.F.,Cecon, P.R. (2004) Efeito da heterogeneidade da variância na avaliação genética de bovinos da raça Nelore: análises de características múltiplas para peso à desmama. In: Reunião Anual Da Sociedade Brasileira De Zootecnia, 41, 2004, Campo Grande, SBZ, p. 1-4 (CD ROM).

Banos, G.; Shook, G.E. (1990) Genotype by environment interaction and genetic correlations among parities for somatic cell count and milk yield. *Journal of Dairy Science*, v.73, n.9, p.2563-2573.

Boldman, K. G.; Kriese, L. A.; Van Vleck, L. D.; Van Tassell, C. P.; Kachman, S. D. (1995) *A manual for use of MTDFREML: a set of programs to obtain estimates of variances and covariances (DRAFT)*. Lincoln: Department of Agriculture / Agriculture Research Service.

Boldman, K. G. & Freeman, A. E. (1990) Adjustment for variance heterogeneity by herd production level in dairy cow and sire evaluation. *Journal of Dairy Science*, v. 73, n. 2, p. 503-512,.

Bueno, R. S., Torres, R. A., Rennó, F.P., Pereira, J. C., Araújo, C. V., Lopes, P. S., Euclydes, R. F. (2005) Efeito da Interação Reprodutor x Rebanho sobre os Valores Genéticos de Reprodutores para Produção de Leite e Gordura na Raça Pardo-Suíça. *Revista Brasileira Zootecnia*, v.34, n.4, p.1156-1164.

Cardoso, A. M. C. (2005) Interação genótipo x ambiente para a produção de leite na espécie bubalina utilizando inferência bayesiana por meio de amostradores de GIBBS. Dissertação: Universidade Federal do Pará.

Carvalheiro, R, Fries, L. A. Schenkel, F. S. Albuquerque, L. G. (2002) Efeitos da Heterogeneidade de Variância Residual entre Grupos de Contemporâneos naAvaliação Genética de Bovinos de Corte. *Revista Brasileira de Zootenia.*, v.31, n.4, p.1680-1688.

Cobuci, J.A.; Costa, C.N.; Teixeira, Freitas, A.F. (2006) Utilização dos polinômios de Legendre e da função de Wilmink em avaliações genéticas para persistência na lactação de animais da raça holandesa. *Arquivo Brasileiro de Medicina Veterinária e Zootecnia*, Belo Horizonte.

Cobuci, J.A.; Euclydes, R.F.; Lopes, P.S., Costa, C. N.; Torres, R. A.; Pereira, C. S. (2005) Estimation of genetic parameters for test-day milk yield in Holstein cows using a random regression models. Genetic Molecular Biology, v.28, p.75-83.

Corrêa, M. B. B.; Dionello, N. J. L; Cardoso, F. F. (2007) Efeito da interação genótipo-ambiente na avaliação genética de bovinos de corte. *Revista Brasileira de Agrociência*, Pelotas, v.13, n.2, p.153-159, abr-jun.

Costa, C.N.; MELO, C.N.R.; Machado, C.H.C.; Freitas,A. F.; Packer, I. U.; Cobuci, J. A. (2005) Parâmetros genéticos para a produção de leite de controles individuais de vacas da raça Gir estimados com modelos de repetibilidade e regressão aleatória. *Revista Brasileira de Zo tecnia*, v.34, p.1520-1531.

Costa, C.N.; Melo, C.M.R.; Machado, C.H.C. Freitas, A.F.; Martinez, M.L.; Packer, I.U. (2002a) Avaliação das funções polinomiais para o ajuste da produção de leite no dia do controle de primeiras lactações de vacas Gir com modelo de regressão aleatória. In: Reunião Da Sociedade Brasileira De Zootecnia, 39., 2002, Recife. *Anais...* Recife. (CD-ROM)

Costa, C.N.; Melo, C.M.R.; Martinez, M.L.; Machado, C. H. C.; Packer, I.U. (2002b) Estimation of genetic parameters for test day milk records of first lactation Gir cows in Brazil using random regression (compact disc). In: World Congress Of Genetics Applied Livestock Production, 7, 2002, Montpellier. *Proccedings...* Montpellier: INRA, Communication n.17-07.

Costa, C.N.; Blake, R.W.; Pollak, E.J. (2000) Genetic analysis of Holstein cattle populations in Brazil and United States. *Journal Dairy Science*, v.83, n.12, p.2963-2974.

Dairyco. (2012) Avalilable from http://www.dairyco.org.uk/datum/milk-supply/milk-production/world-milk-production.aspx.

De Veer, J. C. & Van Vleck, L. D. (1987) Genetic parameters for first lactation milk yields at three levels of herd production. *Journal of Dairy Science*, v. 70, n. 7, p. 1434-1441.

Dionello. N.J.L; Silva, C. A. S.; Costa, C. N.; Cobuci, J.A. (2006) Estimação de parâmetros genéticos utilizando-se a produção de leite no dia do controle em primeiras lactações de vacas da raça Jersey. *Revista Brasileira de Zootecnia*. vol.35 no.4 suppl.0 Viçosa July/Aug.

Dong, M.C.; Mao, I.L. (1990) Heterogeneity of (co)variance and heritability in different levels of intraherd milk production variance and of herd average. *Journal of Dairy Science*. v. 73, p. 843-851.

Dorneles, C.K.P.; Cobuci J.A; Rorati, P.R.N.; Weber, T.; Lopes, J.S.; Oliveira, H.N. (2009) Estimação de parâmetros genéticos para produção de leite de vacas da raça Holandesa via regressão aleatória. *Arquivo Brasileiro de Medicina Veterinária e Zootecnia*. vol.61 no.2 Belo Horizonte Apr.

Euclides Filho, K.; Silva, L.O.C.; Alves, R.G.O. (1997a.) Tendências genéticas na raça Indubrasil. In: Reunião Anual Da Sociedade Brasileira De Zootecnia, 34., 1997, Juiz de Fora. *Anais...* Juiz de Fora: Sociedade Brasileira de Zootecnia, p.171-172.

Falcão, A. J.S.; Martins, E. N.; Costa, C. N.; E. S. Sakaguti, Mazucheli, J. (2006) Heterocedasticidade entre estados para produção de leite em vacas da raça Holandesa, usando métodos bayesianos via amostrador de Gibbs. *Revista Brasileira de Zootecnia*, v.35, n.2, p.405-414.

Falconer, D. S. (1952) The problem of environment and selection. *American Nature*, v. 86, n. 830, p. 293-298.

Falconer, D. S.; Mackay, T. F. C. (1996) Introduction to quantitative genetics. Harlow: Longman group Ltd. 464p.

FAO. FAOSTAT. Data Base 2004. Avalilable from http://faostat3.fao.org/home/index.html#SEARCH_DATA =0.

Fraga, L.M.; De León, R.P.; Gutiérrez, M.; Fundora, O.; Mora, M.; González, M.E. (2006) Sire dam evaluation in buffaloes considering total milk production ability. *In:* 8th World Congress on Genetics Applied to Livestock Production, 2006, *Anais...* Belo Horizonte, MG, Brasil. (CD ROM)

Freitas, M.S. (2003) Utilização de modelos de regressão aleatória na avaliação genética de animais da raça Girolando. Viçosa, MG: Universidade Federal de Viçosa, 2003. 78p. *Dissertação* (Mestrado em Zootecnia) - Universidade Federal de Viçosa.

Freitas, A. R. de. (2000) Avaliação de procedimentos na estimação de parâmetros genéticos em bovinos de corte. *Revista Brasileira de Zootecnia*, v. 29, n. 1, p. 94-102.

Garrick, D. J., Van Vleck, L. D. (1987) Aspects of selection for performance in several environments with heterogeneous variances. *Journal of Animal Science*, v. 65, n.2, p.409-421.

Hammami, H.; Rekik, B.; Gengler, N. (2009) Genotype by environment interaction in dairy cattle. Biotechnol. Agron. Soc. Environ. 13(1), 155-164

Hammond, J. (1947) Animal breeding in relation to nutrition and environmental conditions. *Biological Review*, v. 22, n. 2, p. 195-213.

Hill, W. G. (1984) On selection among groups with heterogeneous variance. Animal Production, v. 39, n. 3, p. 473-477.

Hill, W. G., Edwards, M. R., Ahmed, M. K. A., Thompson, R. (1983) Heritability of milk yield and composition at different levels and variability of production. *Animal Production*, v.36, n.1, p. 59-68.

Houri Neto, M. (1996) Interação genótipo-ambiente e avaliação genética de reprodutores da raça Holandesa, usados no Brasil e nos Estados Unidos da América. Belo Horizonte, MG: UFMG, 1996. 204p. *Tese* (Doutorado em Melhoramento Animal) – Universidade Federal de Minas Gerais.

Ibañez, M. A., Carabaño, M. J., Foulley, J. L., Alenda, R. (1996) Heterogeneity of herd-period phenotypic variances in the Spanish Holstein-Friesian cattle: sources of heterogeneity and genetic evaluation. *Livestock Production Science*, v. 45, n. 01, p. 137-147.

Indicadores IBGE: Estatística da Produção Pecuária. Instituto Brasileiro de Geografia e Estatística – IBGE. (2011) Avalilable from http://www.ibge.gov.br/home/estatistica/indicadores/agropecuaria/producaoagropecua ria /abate-leitecouro-ovos_201004 _publ _completa.pdf.

Kalsi, J.S.; Dhillon, J.S. (1984) *Indian Journal Dairy Science*, v. 37, p. 269-271.

Kelleher, D. J., Freeman, A. E., Lush, J. L. (1966) Importance of bull x herd-year-season interaction in milk production. *Journal of Dairy Science*, v.50, n.10, p. 1703-1707.

Khan, M. S. (1998) Animal model evaluation of Nili-Ravi buffaloes. In: World Congress on Genetic Applied to Livestock Production, 6, Armidale, Austrália. *Proceedings...* Armidale, v.24p. 467-470.

Lofgren, D. J.; Vinson, W. E.; Pearson, R. E. (1985) Heritability of milk yield at different herd means and variance for production. Journal of Dairy Science, v. 68, n. 10, p. 2737-2739.

Lush, J.L. (1964) Melhoramento genético dos animais domésticos. Rio de Janeiro, SEDEGRA. 570p.

Macedo, M.P., Wechsler, F.S., Ramos, A.A., Amaral, J.B., Souza, J.C., Resende, F,D., Oliveira, J.V. (2001) Composição Físico-Química e Produção do Leite de Búfalas da Raça Mediterrâneo no Oeste do Estado de São Paulo. *Revista Brasileira de Zootecnia*, v. 30, n.3, p.1084-1088, (Suplemento 1).

Marques, J.R.F. (1991) Avaliação genético-quantitativa de alguns grupamentos raciais de bubalinos (Bubalus bubalis, L.). Botucatu: FMVZ, 1991. 134p. *Tese* (Doutorado em Genética) - Instituto de Biociências - Universidade Estadual Paulista.

Melo, C.M.R.; Packer, I.U.; Costa, C.N. Machado, P.F. (2005) Parâmetros genéticos para produção de leite no dia do controle e de primeira lactação de vacas da raça holandesa. *Revista Brasileira de Zootecnia*, v.34, n.3, p.796-806.

Melo, C.M.R. (2003) Componentes de variância e valores genéticos para as produções de leite no dia do controle e da lactação na raça Holandesa com diferentes modelos estatísticos. 2003. 97f. *Tese* (Doutorado) – Escola Superior de Agricultura Luiz de Queiroz, Universidade de São Paulo, Piracicaba.

Meyer, K. (1987) Estimates of variance due to sire x herd interactions and environmental covariances between paternal half-sibs for first lactation dairy production. *Livestock Production Animal Science*, v.17, n.1, p.95-115.

Meuwissen, T. H. E. & Van Der Werf, J. H. J. (1993) Impact of heterogeneous within herd variances on dairy cattle breeding schemes: a simulation study. *Livestock Production Science*, v. 33, p. 31-41.

Mohammad, W.A.; Lee, A.J.; Grossman, M. (1981) Interactions of sires with feeding and management factors in Illinois Holsteins cows. *Journal of Dairy Science*. v. 65, p. 625-631.

Moioli, B.; Coletta, A.; Fioretti, M.; Khan, M.S. (2006) Genetic improvement of dairy buffalo: constraints and perspectives. In: 8th World Congress on Genetics Applied to Livestock Production, 2006, *Anais...* Belo Horizonte, MG, Brasil. (CD ROM)

Moita, A. K. F., Lopes, P.S., Torres, R. A., Euclydes, R.F., Tonhati, H. Freitas, A. F. (2010) Heterogeneidade de variâncias na avaliação genética de búfalas no Brasil. *Revista Brasileira de Zootecnia*, v. 39, n.7, p. 1443-1449.

Nauta, W. J.; Veerkamp, R.F.; Brascamp, E.W., Bovenhuis, H. (2006) Genotype by Environment Interaction for Milk Production Traits Between Organic and Conventional Dairy Cattle Production in The Netherlands. *Journal of Dairy Science*. Volume 89, Issue 7, Pages 2729-2737.

Norman, H. D. (1974) Factors that should be considered in a national sire summary model. *Journal of Dairy Science*, v.57, n.7, p. 955-962.

Patel. A.K.; Tripathi, V.N. (1998) Genetic studies on lifetime performance in Surti buffalo. In: World Congress on Genetic Applied to Livestock Production, 6, Armidale, Austrália. *Proceedings...* Armidale, v.24p. 471-473.

Paula, M. C.; Martins, E. N.; Silva, L. O. C.; Oliveira, C. A. L.; Valotto, A. A.; Ribas, N. P. (2009) Interação genótipo × ambiente para produção de leite de bovinos da raça Holandesa entre bacias leiteiras no estado do Paraná. *Revista Brasileira Zootecnia,* v.35, n.2, p.405-414.

Pereira, R.G.A, Townsend, C. R., Costa, N.L., Magalhães, J. A. (2007) Eficiência reprodutiva de búfalos. Porto Velho, RO: Embrapa Rondônia. 15 p. – *(Documentos/* Embrapa Rondônia, ISSN 0103-9865; 123).

Promebul. (2011) Projeto de Melhoramento Genético dos Bubalinos. Promebul-Cabul. Centro de Avaliação dos Bubalinos (CABUL) FMVZ/UNESP/Botucatu. Avalilable from http://www.sernet.com.br/canais/agropec .asp.

Raheja, K.L. (1998) Multi-variate restricted maximum likelihood estimates of genetic and phenotypic parameters of lifetime performance traits for Murrah buffalo. In: World Congress on Genetic Applied to Livestock Production, 6, Armidale, Austrália. *Proceedings...* Armidale, v.24p. 463-466.

Ramos AA. (2011) Projeto de Melhoramento Genético dos Bubalinos-PROMEBUL-D.P.E.A./FMVZ/UNESP. 2/3/2003. Avalilable from http://www.ruralnews.com.br/.

Rao, C. R. (1973) *Linear statistical inference and its applications.* 2. ed., New York: John Wiley e Sons. 552 p.

Reis, J. de C., Lôbo, R. B. (1991) *Interação genótipo-ambiente nos animais domésticos.* Ribeirão Preto: J.C.R./R.B.L. 194p.

Rosati, A.; Van Vleck, L. D. (1998) Estimation of genetic parameters for milk, fat, protein and mozzarella cheese production in the italian river buffalo population. In: World Congress on Genetic Applied to Livestock Production, 6, Armidale, Austrália. *Proceedings...* Armidale, v.24. p.459-462.

Silva, H.C. (2002) *Efeito da interação reprodutor x rebanho sobre a produção de leite em búfalos.* Dissertação de mestrado. Universidade Federal de Viçosa. 51p.

Sirol, M.L.F.G.; Euclydes, R.F.; Torres, R.A.; Lopes, P.S.; Pereira, C.S.; Araújo, C.V.; Rennó, F.P. (2005) Efeito da interação reprodutor x rebanho sobre as produções de leite e de gordura na raça Pardo-Suiça. *Revista Brasileira de Zootecnia,* v.34, n.5, p.1573-1580.

Stanton, T. L., Blake, R. W., Quaas, R. L., Van Vleck, L. D., Carabaño, M. J. (1991) Genotype by environment interaction for Holstein milk yield in Colombia, México and Porto Rico. *Journal of Dairy Science,* v.74, n.5, p.1700-1714.

Tonhati, H.; Muñoz, M.F.C.; Oliveira J.A.; , Faro, L.; Lima, A.L.F.; Albuquerque, L.G. (2008) Test-day milk yield as a selection criterion for dairy buffaloes (Bubalus bubalis Artiodactyla, Bovidae). *Genetics and Molecular Biology,* 31, 3, 674-679.

Tonhati, H.; Muñoz, M.F.C.; Duarte, J.M.C.; Reichert, R.H.; Oliveira J.A.; Lima, A.L.F. (2004) Estimates of correction factors for lactation length and genetic parameters for milk yield in buffaloes. *Aruivo Brasileira de Medicina Veterinária e Zootecnia.* vol.56 no.2 Belo Horizonte.

Tonhati, H.; Muñoz, M.F.C.; Oliveira, J.A.; Duarte, J.M.C.; Furtado, T.P.; TSEIMAZIDES, S.P. (2000) Parâmetros Genéticos para a Produção de Leite, Gordura e Proteína em Bubalinos1. *Revista Brasileira de Zootecnia*, v. 29, n.6, p. 2051-2056, (Suplemento 1).

Tonhati, H.; Vasconcellos, B.F. (1998) Genetic aspects of productive and reproductive traits in Murrah buffaloes herd in São Paulo, Brasil. In: World Congress on Genetic Applied to Livestock Production, 6, Armidale, Austrália. *Proceedings...* Armidale, v.24, p.485-488.

Tonhati, H. (1997) Melhoramento genético de bubalinos para carne e leite. In: Oliveira, G.J.C, Almeida, A.M.L., Souza Filho, U.A.S. *O búfalo no Brasil.* Cruz das almas: UFBA – Escola de Agronomia. p. 101-113.

Torres, R. A., Bergmann, J. A. G, Costa, C. N., Pereira, C. S., Valente, J., Penna, V. M., Torres Filho, R. A. (1999) Ajustamento de variâncias para a produção de leite entre rebanhos da raça holandesa no Brasil. *Revista Brasileira de Zootecnia*, v. 28, n. 2, p. 295-303.

Van Der Werf, J.H.J., Meuwinssen, T.H.E., JONG, G. DE. (1994) Effects of correlation for variance heterogeneity on bias and accuracy of breeding value estimation for Dutch dairy cattle. *Journal of Dairy Science*, v.77, p. 3174-3184.

Van Vleck, L. D. (1987) Selection when traits have different genetic and phenotypic variances in different environments. *Journal of Dairy Science*, v. 70, n. 1, p. 337-344.

Vanvleck, L.D. (1986) Selection when traits have different genetic and phenotypic variances in different environments. *Journal of Dairy Science*, v. 70. p. 337-344.

Vinson, W. E. (1987) Potential bias in genetic evaluations from differences in variation within herds. *Journal of Dairy Science*, v.70, n.9, p.2450-2455.

Visscher, P. M. (1991) On the estimation of variances within herd-mean production groups. *Journal of Dairy Science,* v. 74, n. 6, p. 1987-1992.

Winkelman, A., Schaeffer, L. R. (1988) Effect of variance heterogeneity on dairy sire evaluation. *Journal of Dairy Science.* v.71, p. 3033.

Animal Management, Nutrition and Husbandry

Dissemination of Scientific Data for Sustainable, Organic Milk Production Systems

Sezen Ocak and Sinan Ogun

Additional information is available at the end of the chapter

1. Introduction

Most ecologically sustainable or organic agricultural projects concerning "Milk Production" have not been monitored comprehensively or been subjected to stringent evaluation methods to allow the outcomes to be regarded as scientifically acceptable. There is a strong need to not only promote such dairy projects but to collect, scrutinize and share the information in an appropriate scientific manner to encourage more widespread acceptance.

Researchers dealing with organic production have been more interested in solving practical problems than publishing papers, as such it makes it difficult to do comparative analysis with conventional productions systems and draw general conclusions regarding the nutritional value of dairy produce or health and well-being of livestock.

Despite the lack of reliable data on the benefits of organic produce from around the globe, the organic dairy sector is reported to be growing at between 15 to 23 percent a year in the countries assessed in this study, namely; Europe, USA, Australia and New Zealand. It is expected that by 2015, 25-30 percent of consumers of dairy products in the above nations will ensure that their regular purchases will be organic (Lohr, 1998).

In addition to the human health benefits, organic dairy farming aims to create more holistic, agro-ecological systems. Therefore the aim is not only to produce nutritious and chemical free produce but to ensure a sustainable management of natural resources where these products are produced. Animal health and welfare are also important elements of such a system (Lund & Rocklinsberg, 2001). This is a factor that is clearly recognized by the International Federation of Organic Agricultural Movements (IFOAM), the organization setting the basic standards for what can be labeled as organic.

The aim of this study is to review some of the relevant research carried out in this field and to assess suitable methods to disseminate such scientific data in a standardized, reliable and an easily accessible manner to farmers needing this information. The information should be reliable enough to inform the conventional farmer of the benefits and shortfalls if they are considering organic production, or at least provide resource material for them to pursue further research into the topic.

Marketing and consumption of organic produce also being an important factor in the decisions of a potential producer; the study also evaluated numerous factors that may be the cause of the global shift toward the increased consumption of organic produce.

Besides the generally accepted health benefits of organic products, the contemporary, educated and well informed consumer is also trying to make a statement about the environmental benefits of organic production. They are asking from the producer/supplier to further inform them about food miles, efficient energy use, water pollution, soil conservation, carbon credits, natural resource management and alike. The study discovered that further effort was required on how information about consumer preference could be relayed back to the farmer.

Besides general scientific literature on organic dairy farming carried out by few very committed organizations and some institutions the authors also looked at a number of farms in the USA and Australia as case study farms to see if they were open to sharing relevant information about their outcomes. These farms were selected based on the importance they gave to the principles of sustainable organic production, such as; diversified farming measures and the use of crop rotations, resilient production systems based on diverse cultural system, minimal use of external agrochemical inputs, minimal use of pharmaceuticals for animal health issue by encouraging preventative measures rather than curative ones, emphasis on the use of local resources, recycling organic wastes, reduced environmental impact, reduced use of mechanization, utilization of small areas for cultivation and encouraging the use and preservation of traditional skills.

The producers who apply these principles have a very limited database of information on which to rely on, in addition the information would not stand up to the rigors of intensive scientific scrutiny. The study has provided numerous resources in its acknowledgment and reference list, which supports the principles of a holistic and dynamic production systems, but would strongly encourage further research in the field.

2. Farming trends

In the present market place with heightened scrutiny on how food is produced, the conventional dairy farm unit is going through a major shift from mid-size farms to very small and very large ones - this is particularly the case in USA (Benbrook, 2009). In the very large size farms the cows do not have access to sufficient pasture to contribute significantly to their daily feed intakes, therefore grain-based rations and high-quality alfalfa (lucerne) hay form the backbone of the cows' diet. However in contrast to this trend the number and importance of small to moderate-scale organic dairy farms is increasing. The most successful operations grow all or most of their feed on or near the home farm. In addition, pasture and grazing contributes significantly to daily feed intakes in those parts of the year when the weather supports active forage plant growth. On conventional, feedlot-based industrial dairies, corn and other corn-based feeds typically account for around two-thirds of a cow's diet, whereas on grazing-based organic farms, pasture and forage-based feeds typically account for at least two-thirds of daily feed intake, and corn in all its forms less than one-quarter.

On large conventional dairies, artificial insemination is used on mostly purebred Holstein cows. Each milking dairy animal is expected to produces about 10 to 11,000 liters of milk during a 305-day lactation period (Benbrook, 2009). A range of drugs are routinely administered to these animals to help them fight infections, efficiently digest their energy-dense, low-fiber feed, and to help synchronize artificial insemination breeding attempts.

On most small and moderate scale organic dairies, production levels are lower, averaging closer to 8,000 liters per year. Breeds of cattle other than Holsteins, as well as crossbreed cattle are common and artificial insemination is a tool used on many farms, but has not replaced bulls and traditional breeding programs.

However the authors have struggled to find sufficient research results in the agricultural economic domain to support this concept of a more long-term productive and a cost efficient production system in that of organic dairies. The few case study farms that were looked at certainly had the figures but were reluctant to share the results assumedly with the fear of losing their premium market edge. Due to the lack of actual accounting figures, its difficult to ascertain whether a shift in the supply – demand curve of organic milk in the market place would affect the profitability of organic farming in the long run. Presently the price premium is based solely on the availability of organic milk. It would appear however that if more people consumed organic milk and as such the health benefits became more evident than this could result in increase in consumption of organic milk products, meaning higher demand maintaining the high value in the market place.

3. Organic vs. conventional vs. biodiversity

Organic farming demonstrates clear advantages for biodiversity over conventional farming. Depending on altitude, organic farms have between 46 and 72 percent more semi-natural habitats and host 30 percent more species and 50 percent more individuals than non-organic farms (Source: FiBL). The lower farming intensities and higher proportion of semi-natural areas enable site-typical plant and animal species to exist on organic farms and allow farmers to benefit from an intact and therefore sustainably functioning ecosystem.

Agricultural policies are increasingly promoting ecologically-oriented farming methods that preserve biodiversity and conserve natural resources (FAO, 2002). Intensive farming by introduction of exotic species, land clearing, vegetation fragmentation, habitat change and soil erosion has been one of the main causes of biodiversity decline. (Bengtsson et al., 2005; Hole et al., 2005). Specific contributing factors to this decline with conventional farming have been; indiscriminate use of pesticides and synthetic fertilizers, land consolidation, drainage as well as the use of heavy machinery.

Numerous comparative studies showing the impact of conventional and organic farming systems verify the positive effect organic farming has on flora and fauna on field and also farm level (Fuller et al., 2005, Hole et al., 2005). A comprehensive analysis of 66 scientific studies shows that organically farmed areas have on average 30 percent more species and 50 percent more individuals than non-organic areas (Bengtsson et al., 2005). The positive effect of organic farming is most significant in cleared landscapes, but is also seen in structurally rich regions (Gabriel et al., 2006; Gabriel et al., 2010).

In particular birds, predatory insects, spiders, soil dwelling organisms and field flora benefit most from organic management as can be seen in **Figure 1** below. Pests and indifferent organisms on the other hand occur in similar numbers in the various farming systems. The differences in species diversity are especially noticeable with arable and horticulture crops in valleys – the differences seen in grassland are less pronounced. Comparison studies in mountainous regions are scarcely existing.

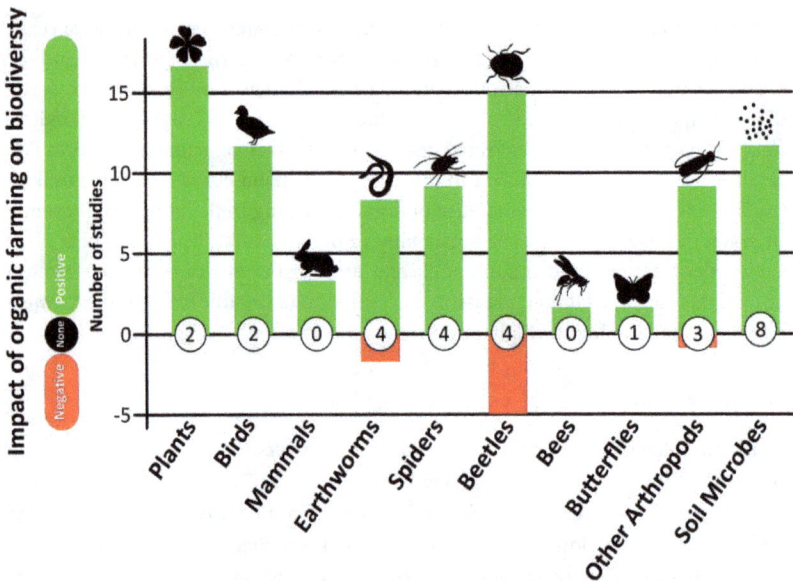

Figure 1. Impact of Organic Farming on Biodiversity - Number of studies that show organic farming having a positive (green bar), negative (red bar) or no effect (number in white circle) on biodiversity of various animal and plant groups in comparison to non-organic farm management. Summary of 95 scientific publications. **(Source: FiBL)**

To preserve rare and endangered species, adapted species protection programs are frequently necessary. The typical ecological compensation programs for farmland are not sufficient. Organic farming in combination with valuable semi-natural areas can therefore significantly contribute to improving species numbers (Pfiffner et al., 2003). Sky larks (Photo 1), a typical species that have been suppressed through intensification of farming, as well as the now rare lapwings, partridges, and whinchats, achieve higher population densities on organically managed farms (NABU, 2004; Neumann et al., 2007). Rare plant species on agricultural land (Gabriel et al., 2006; Gabriel et al., 2007) and ground beetles (Pfiffner et al., 2003) are also proven to be in higher diversity and density on organic farms.

Photo 1. Ground-nesting birds can only survive in less intensively used areas (Skylark).

Habitats with numerous species are shown to better adapt or are more resilient to environmental changes. For instance, species rich mountain meadows erode less and allow for more stable yields during dry periods.

Critical ecological processes are influenced by the higher biodiversity and larger population densities of various species seen on organic farms. Organic farming shows significant improvements for functions such as:

- Pollination (Gabriel et al., 2007; Holzschuh et al., 2007, 2008; Moradin et al., 2005)
- Reduction in soil erosion on arable land (Siegrist et al., 1998)
- Decomposition of dung in pastures (Hutton et al., 2003)
- Natural pest reduction in soil (Klingen et al., 2002) and crops (Crowder et al., 2010; Zehnder et al., 2007)

Flower-visiting insects such as honeybees, wild bees, and bumblebees benefit from the higher coverage and diversity of secondary flora in organic grain fields. Biodiversity is 3 times higher and the number of bees 7 times higher than in conventional areas (Holzschuh et al., 2007). With organic farming areas increasing, populations of wild bees, honeybees, and bumblebees are also markedly climbing in the surrounding farmland and semi-natural areas (Holzschuh et al., 2008). Organic agriculture thus improves the pollination of flowering plants in the surrounding environment (Gabriel et al., 2007).

The higher diversity of flora and fauna also encourages beneficial organisms that naturally reduce pests (Zehnder et al., 2007). Organic farming leads to a significantly more balanced number of beneficial insects that reduce pests and yield losses in potato crops (Crowder et al., 2010). Organic pastures allow richer fauna to exist in dung than conventional pastures as they are not contaminated by chemical veterinary drugs (Holzschuh et al., 2007). Dung fauna considerably adds to the degradation and recycling of dung and in turn makes for better-feed quality.

A more diverse flora and fauna in organic soil result in a revitalized, more active soil life (Mäder et al., 2002). Research from Norway shows a stronger reduction in soil pests in organic soils than in conventional soils due to richer fungal fauna (Klingen et al., 2002).

Various farm practices and landscaping measures are implemented in organic farming that have a proven positive influence on biodiversity as shown in Figure 1 above. The following measures typically carried out on organic farms that most notably promote biodiversity are:

• Non use of herbicides
• Non use of chemically-synthesized pesticides
• Less and purer organic fertilizers
• Fewer cattle per square meter
• More diversified crop rotation with higher clover grass percentage
• Conservation tillage
• Higher percentage of semi-natural areas
• Higher percentage of arable and ecological areas
• More diversified farm structure

These factors enhance not only biodiversity, but strengthen natural cycles and improve environmental performance that in turn increase the sustainability of organic farms (FAO, 2002; Pimentel et al., 2005). To optimally promote biodiversity, cross-farm and landscaping measures need to be instituted – ideally on extensively managed habitats within landscapes (e.g., bioregions) (Gabriel et al., 2010). It would seem from the amount of scientific literature albeit limited, that organic farming with its strong principles of natural resource management significantly protects biodiversity by maintaining species numbers as compared to conventional farming.

4. Consumer trends

Retailing of organic milk has changed since the early 1990s, when most organic food was sold in specialty shops. Since then, organic food products have become available in a wide range of venues, with trends in retailing organic milk following those of conventionally produced food, including a growing reliance on private-label products (Dimitri, et al., 2007).

Two other factors have affected retailing of organic milk and other organic products: *Stringent Certification Codes* (Marketing of organic products is facilitated through the use of global organic standards, which establish rules for the use of the label "organic" and the accompanying logo) and *Price Premiums* (Organic milk, like most organic products, receives a price premium over conventional products).

4.1. Price premium

Consumers have been willing to pay premium prices in the market for certified organic dairy products, with the understanding that the food has been raised in a sustainable, environmentally sound manner and that they are helping support and keep family farmers on the land. As stated earlier, consumers also assume that humane animal husbandry practices are employed by organic farmers, and they may believe that organic food is more nutritious. Those able to make the difficult three-year transition (*2) to organics have been rewarded by top commodity prices at the farm-gate something that stands in stark contrast to the intense price squeeze that has driven many of their conventional neighbours from the business. The success story of organic products in present day agriculture could be the catalyst for doing more reliable scientific research in this field. Unless off course there may be fear in the industry that more comprehensive scientific research may reveal that the product is not as distinctly more beneficial to warrant the current price premiums to that of the conventionally farmed milk. As a result this could then affect the marketability of the product. The present day mystique around organic produce can be attributed to the lack of conclusive scientific evidence proving its benefits to human health.

4.2. Marketing

The boost in organic milk sales is part of a wider growing interest in organic products, which resulted in an average annual growth rate of retail sales of organic food of nearly 18 percent between 1998 and 2005 (Dimitri et al., 2007). Rising consumer interest in organic milk has been accompanied by a newfound widespread availability of the product, and organic milk is now available in nearly all food retail venues, including conventional supermarkets.

In 2006 media reports in USA, the largest organic milk-consuming nation in the world indicated that supermarkets experienced significant shortages of organic milk during 2005 and 2006 (Oliver, 2006; Weinraub & Nicholls, 2005), suggesting that consumer demand is unmet at current market prices.

To date, most characterizations of consumers who purchase organic products result from industry studies and offer conflicting views. The studies have focused on consumers of organic foods in general, not just consumers of organic milk.

These market analyses use consumer surveys to gather information and have focused on trends in consumer purchases of organic foods (For instance in the USA - Whole Foods

Market, 2005) and demographic characteristics of organic consumers (Hartman Group, 2004, 2002, 2000). The Whole Foods 2005 survey indicates that 65 percent of consumers have tried organic foods, 27 percent bought more organic food in 2005 than in 2004, and 10 percent consume organic food several times a week.

Supply responses necessarily lag behind increases in consumer demand because it takes 3 years to convert farmland to meet organic standards so that they can provide organic feed. The cows have to be managed organically and fed organic feed for 1 year.

The authors have studied the most recent Hartman report 2006 which indicated that the majority of organic milk consumers are of ethnic origin with an annual income of less then $50,000.

In contrast to the Hartman 2006 and 2004 results, earlier studies characterize organic consumers as White, affluent, well-educated, and concerned about health and product quality (Lohr, 2001; Richter et al., 2000; ITC, 1999; Thompson, 1998). These studies also cluster the average age of organic consumers in two age groups: 18-29 years and 45-49 years (Thompson, 1998; Lohr & Semali, 2000). One element that has remained generally accepted through the years is that parents of young children or infants are more likely than those without children to purchase organic food.

Figures 2 to 5 show the distribution of organic households compared to conventional households by income (Figure 2), education (Figure 3), household size (Figure 4) and share of households (Figure 5). The percentages were calculated by dividing the number of organic (or conventional) households in each region by the total number of organic (or conventional) households in the sample. This information is useful in that it provides insight into the characteristics that differentiate the typical organic household from the typical conventional household, plus it allows for a comparison of these results with those published by industry groups.

Household income and education of the head of the household seem to be associated with the likelihood that a household will buy organic or conventional milk. The data indicates that the share of organic households across income categories rises as income increases, and the high-income group is the only category where the proportion of households purchasing organic milk exceeds the proportion purchasing conventional milk (Figure 2). Most (80 percent) organic milk consumers have at least attended some college, and those who have graduated from college or completed some post-graduate education make up 51 percent of organic milk consumers. The share of organic households with the highest two levels of education (graduated from college or completed some post-graduate studies) is greater than the share of conventional households with the same level of education (Figure 3). What accounts for the association between income and education and purchasing organic milk? Household income and education are correlated, so income could be the factor driving the association with purchasing organic milk. Alternatively, education could be the driving factor, in that greater education may enhance one's understanding of the relationship between organic production techniques and environmental impacts.

Reasonable explanations are lacking as to the association (or lack thereof) between some demographic factors and the distribution of organic milk households. For example, one might expect that larger households would buy much less organic milk than smaller households, particularly since smaller households have greater disposable income, household size appears to have little relationship with the propensity to purchase organic milk (Figure 4).

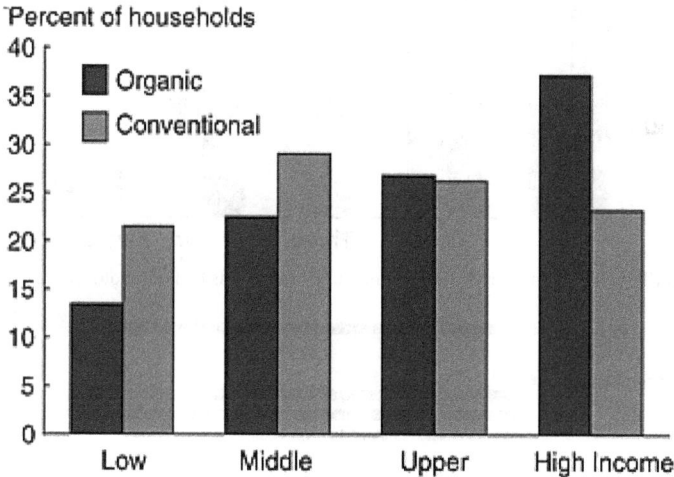

Figure 2. Distribution of organic and conventional milk households by income, **2004

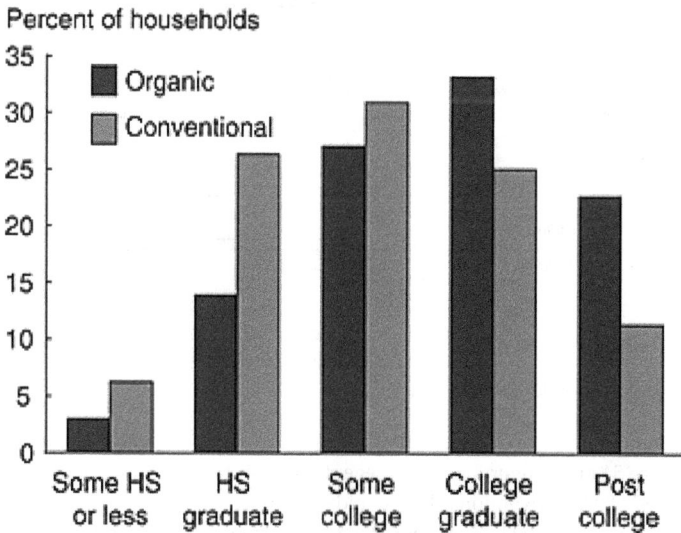

Figure 3. Distribution of organic and conventional milk households by education, 2004

Percent of households

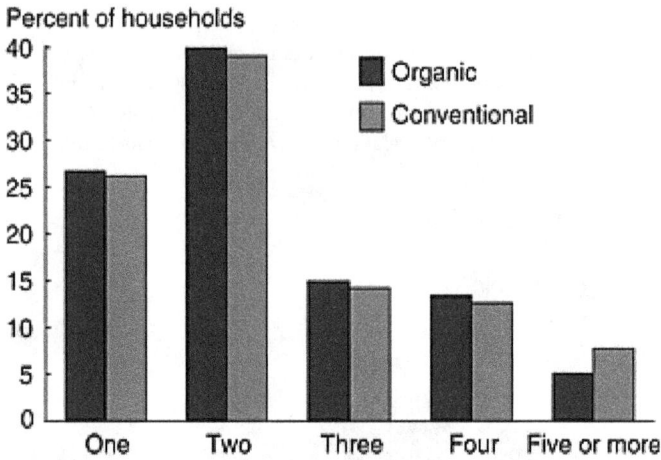

Figure 4. Distribution of organic and conventional milk households by household size, 2004

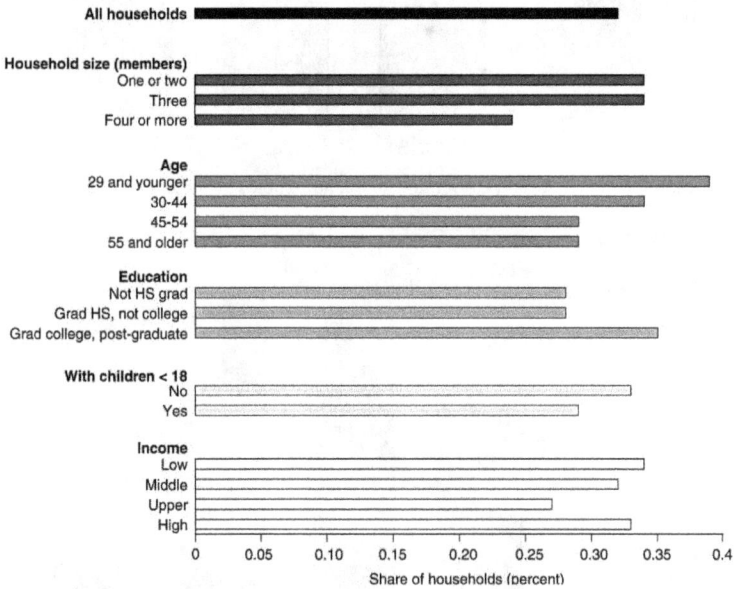

Figure 5. Share of Households (percent - %)

In sum, the demographic data indicate that organic households are most likely to be headed by someone age 54 or younger, have a college degree, and have annual household incomes of at least US$70,000. Conventional households are more likely to have annual household income less than US$70,000, have not graduated from college, and be headed by a household head age 55 or older. Household size has little bearing on whether a household purchases only conventional milk, and the presence of children under age 18 has no bearing on the likelihood of a household to purchase organic or conventional milk. The importance of factors such as high income and a college degree together suggest that organic households have higher discretionary income than conventional households and, thus, are able to afford and are willing to purchase higher priced organic milk.

5. Conclusion

Certain concepts became quite evident during data collection and literature assessment;

- The motivation for implementing organic agricultural systems may be diverse but persuading farmers to change and maintain organic systems required some financial incentive;
- Successful initiatives may have diverse origins, but significant impact required the harnessing of resources and commitment of numerous stakeholders, both private and public sector on a complementary mission;
- the verifying role of organic certification services is both a burden and also a means of delivering truly sustainable agriculture;
- projects based on organic agriculture are more subtle than chemical agriculture and therefore, situation specific, successful organic agriculture is 'knowledge intensive' requiring more design and management from the outset, as opposed to the 'just in time' approach of chemical agriculture.

It is the view of the authors that training, extension and demonstration are perhaps even more critical here than with conventional projects, benefits from organic agriculture may not be immediate. Small farmers will require considerable support or incentive over the initial years if the system is to gain momentum and be maintained, some agro-ecological situations, such as agro-forestry, will convert more easily to organic systems than others.

Farmers appear to resist conversion to organic agriculture when:

- they have been heavily exposed to the chemical message;
- they currently operate high input, high output systems;
- previous extension services have been effective;
- production is relatively mechanized;
- labour costs are high or labour is not available;
- the system is thrust upon them;

Farmers appear more receptive to conversion to organic agriculture when:

- they have not been exposed to the chemical message;
- their farming system is traditional or nil input;
- previous extension services have not been effective;
- production is relatively labour intensive;
- labour costs are low or labour is readily available;
- the concept is developed by them or with them;

In conclusion; organic projects involving milk production systems should be as rigorously identified, designed, implemented, monitored and evaluated as any other agricultural development project with strong stakeholder participation. The organic context does not make the project immune to the potential problems with project implementation from misidentification of issues, political influences and weak institutional support. Extra emphasis should be placed on human resource and institutional development, recognizing that organic dairy farming is knowledge intensive rather than input intensive. Dissemination of information is one of the main drawbacks for appreciating the benefits of organic milk production systems.

Author details

Sezen Ocak and Sinan Ogun
Zirve University / Middle East Sustainable Livestock,
Biotechnology and Agro-Ecology Research and Development Centre, Turkey

Acknowledgement

The authors would like to extend their appreciation to Northeast Organic Dairy Producers Alliance, Cornucopia Institute, Dairy Economic Research Service - USDA, Organic Farming Research Foundation, The Organic Centre, Research Institute of Organic Agriculture (FiBL) and would also like to ackowledge the following authors as their publication, although not directly quoted, were never the less used as a guide in assessing some of the concepts discussed in this study and Emin Basar Ozdemir for his graphic abilities.

6. References

Amaral-Phillips, D.M. & McAllister, J. Planning the Yearly Forage and Commodity Needs for a Dairy Herd, Cooperative Extension Service, Univ. of Kentucky, ASC-160; http://www.uky.edu/Ag/AnimalSciences/ dairy/extension/nut00106.pdf

Barry, M. (2004). The New World Order: Organic Consumer Lifestyle Segmentation. [N] sight. Volume VI, Number 2.

Baxter, B. (2006). Who's buying organic? Demographics 2006, hartbeat taking the pulse of the marketplace, May 17. Accessed at
http://www.hartman-group.com/products/HB/2006_05_17.html; August 3,

Benbrook, C. (2009). Shades of Green : Quantifying the Benefits of Organic Dairy Producti The Organic Centre.

Bengtsson, J.; Ahnström, J.; & Weibull, A.C., (2005). The effects of organic agriculture on biodiversity and abundance: a metaanalysis. *Journal of App. Ecology* Vol.42, pp.261-269.

Budgar, L. (2006). Convenience and Health Drive Natural Food Sales. *The Natural Foods Merchandiser*, June.

Chalupa, et al. (2004)."Dairy nutrition models: their forms and applications, http://www.docstoc.com/ docs/4530729/Dairy-nutrition-models

Crowder, D.W.; Northfield, T.D.; Strand, M. & Snyder, W.E. (2010). Organic agriculture promotes evenness and natural pest control. *Nature*, Vol.46, pp.109-112.

Demeritt, L. (2004). Organic Pathways. [N]Sight. Hartman Group, Inc. Bellevue, WA.

Dimitri, C. & Venezia, K.M. (2007). Retail and Consumer Aspects of the Organic Milk Market . A Report from the Economic Research Service, USDA.

Eastridge, M.L. Bucholtz, H.F.; Slater, A.L. & Hall C.S. (1998). Nutrient Requirements for Dairy Cattle of the National Research Council Versus Some Commonly Used Ration Software, *J. Dairy Science*, Vol.81, pp: 3059-3062.

Economic Research Service, USDA. (2005). Dairy Agricultural Resource Management Survey, tables generated for The Organic Center by Purdue University.

El-Zarkouny, S.Z.; Cartmill, J.A.; Hensley, B.A. & Stevenson, J.S. (2004). Pregnancy in dairy cows after synchronized ovulation regimens with or without presynchronization and progesterone, *J. Dairy Science*, Vol. 87, pp. 1024-1037.

FAO, 2002. Organic agriculture, environment and food security. *Environ. Nat. Res.* No.4, FAO Rome.

Fricke, P.M. Ovsynch, Pre-synch, the Kitchen-Synch: What's Up with Synchronization Protocols, Dept. of Dairy Science, University of Wisconsin-Madison, http://www.wisc.edu/dysci/uwex/brochures/brochures/ fricke.pdf

Fuller, R.J.; Norton, L.R.; Feber, R.E.; Johnson, P.J.; Chamberlain, D.E.; Joys, A.C.; Mathews, F.; Stuart, R.C.; Townsend, M.C.; Manley, W.J.; Wolfe, M.S.; Macdonald, D.W. & Firbank, L.G. (2005). Benefits of organic farming to biodiversity vary among taxa. *Biology Letters* Vol.1, pp. 431-434.

Gogoi, P. (2006). Wal-Mart's Organic Offensive. Business Week. March 29.

Gibson, R. H.; Pearce, S.; Morris, R.J.; Symondson, W.O.C. & Memmott, J. (2007). Plant diversity and land use under organic and conventional agriculture: a whole-farm approach. *Journal of Applied Ecology*, Vol.44, pp. 792-803.

Hoard's Dairyman. (2007). What we've learned from grazing, 57th Annual Hoard's Dairyman Round Table.

Howie, M. (2004). Research Roots Out Myths Behind Buying Organic Foods. *Feedstuffs*. March 29.

Kellogg, D.W. (2001) .Survey of Management Practices Used for the Highest Producing DHI Herds in the United States, *J. Dairy Science*, Vol. 84, pp.120-127.

Leibtag, E. (2005). Where You Shop Matters: Store Formats Drive Variation in Retail Prices. Amber Waves. Vol. 3. Issue 5. November. www.ers.usda.gov/amberwaves/november05/features/whereyoushop.htm

Mäder, P.; Fließbach, A.; Dubois, D.; Gunst, L.; Fried, P. & Niggli, U. (2002). Soil fertility and biodiversity in organic farming. *Science*, Vol. 296, pp. 1694-1697.

Miller, J. J. & Blayney, Don P. (2006). Dairy Backgrounder. Outlook Report No. LDP-M-145-01. U.S. Department of Agriculture, Economic Research Service. July. www.ers.usda.gov/publications/ldp/2006/ 07jul/ldpm14501/

New York Times. (2006). When Wal-Mart Goes Organic. Editorial, May 14.

Nutrition Business Journal (NBJ). (2006). U.S. Organic Food Sales ($Mil) 1997-2010e-Chart 22. Penton Media, Inc.

Nutrition Business Journal (NBJ). (2004). NBJ's Organic Foods Report 2004, Penton Media, Inc.

Nielsen, A.C. (2005). The Power of Private Label 2005: A Review of Growth Trends Around the World. September. Accessed April 2006 at: http://www2.acnielsen.com/reports/documents/2005_privatelabel.pdf

Olynk, N.J. & Wolf, C.A. (2008). Economic Analysis of Reproductive Management Strategies on US Commercial Dairy Farms, *J. Dairy Science*, Vol. 91, pp.4082-4091.

Onazaka, Y.; Bunch, D.; & Douglas, L.(2006). What Exactly Are They Paying For? Explaining the Price Premium for Organic Produce. Agricultural and Resource Economics Update. Vol.9, No.6. University of California Giannini Foundation. July/August.

Organic Trade Association (OTA). (2006). Organic Trade Association's 2006 Manufacturer's Survey. Produced by the Nutrition Business Journal. Greenfield, MA.

Santos, J.E.P.; Juchem, S.O.; Cerri, R.L.A.; Galvão, K.N.; Chebel, R.C.; Thatcher, W.W.; Dei, C.S. & Bilby, C.R. (2004). Effect of bST and reproductive management on reproductive performance of Holstein dairy cows, *J. Dairy Science*, Vol.87, pp. 868-881

U.S. Department of Agriculture. (2002a). Organic Food Standards and Their Labels: The Facts. National Organic Program, April 2002, updated January 2007. Accessed February 7, 2007, at http://www.ams.usda.gov/ nop/consumers/brochure.html

U.S. Department of Agriculture. (2002b). Background Information. National Organic Program, October 2002, updated January 2007. Accessed April 30, 2007, at http://www.ams.usda.gov/nop/FactSheets/Backgrounder.html

Windig, J.J.; Calus, M.P.L. & Veerkamp, R.F. (2005). Influence of herd environment on health and fertility and their relationship with milk production, *J. Dairy Science*, Vol Vol.88, pp. 335-347.

Wal-Mart. (2006). Organics for Everyone. Accessed August 2006 at http://walmart.triaddigital.com/OrganicsContent.aspx?c=Organics

Gabriel, D.; Roschewitz, I.; Tscharntke, T. & Thies, C. (2006). Beta diversity at different spatial scales: plant communities in organic and conventional agriculture. *Ecological Applications* Vol.16, pp. 2011-2021

Gabriel, D.; Sait, S. M.; Hodgson, J. A.; Schmutz, U.; Kunin, W. E. & Benton, T.G. (2010). Scale matters: the impact of organic farming on biodiversity at different spatial scales. *Ecology Letters*, Vol.13, No.7, pp. 858-869

Gabriel, D.; Tscharntke, T. (2007). Insect pollinated plants benefit from organic farming. *Agriculture, Ecosystems and Environment* Vol.118, pp. 43-48

Lewrene, K. G. & Thompson, G. (2000). The Demand for Organic and Conventional Milk. *Presented at the Western Agricultural Economics Association meeting*, Vancouver, British Columbia

Hartman Group, (2000). *The Organic Consumer Profile*, Bellevue, WA.

Hartman Group, (2002). *Hartman Organic Research Review: A Compilation of National Organic Research Conducted by the Hartman Group*, Bellevue, Washington

Hartman Group, (2004). *Organic Food and Beverage Trends*, Bellevue, Washington

Hole, D.G.; Perkins, A.J.; Wilson, J.D.; Alexander, I.H.; Grice, P.V. & Evans, A.D. (2005). Does organic farming benefit biodiversity? *Biological Conservation*, Vol.122, pp. 113-130.

Holzschuh, A.; Stefan-Dewenter, I.; Kleijn, D. & Tscharntke, T. (2007). Diversity of flower-visiting bees in cereal fields: effects of farming system, landscape composition and regional context. *Journal of Applied Ecology* Vol.44, pp. 41-49

Holzschuh, A.; Stefan-Dewenter, I. & Tscharntke, T. (2008). Agricultural landscapes with organic crops support higher pollinator diversity. *Oikos*, Vol.117, pp. 354-361

Hutton, S.A. & Giller, P.S. (2003). The effects of the intensification of agriculture on northern temperate dung beetle communities. *Journal of Applied Ecology*, Vol.40, pp. 994-1007

International Trade Centre (ITC), (1999). Organic Food and Beverages: World Supply and Major European Markets. *ITC/UNCTAD/WTO*, Geneva

Klingen, I.; Eilenberg, J. & Meadow, R. (2002). Effects of farming system, field margins and bait insect on the occurrence of insect pathogenic fungi in soils. *Agriculture, Ecosystems and Environment* Vol.91, pp. 191-198

Lohr, L. (2001). Factors Affecting International Demand and Trade in Organic Food Products. In Changing Structure of Global Food Consumption and Trade, A. Regmi (ed.). *Agriculture and Trade Report No. WRS01-1. U.S. Department of Agriculture, Economic Research Service*, pp. 67-79, www.ers.usda.gov/publications/wrs011/

Lohr, L. & A. Semali. (2000). Retailer Decision Making in Organic Produce Marketing. In W.J. Florkowski, S.E. Prussia, and R.L. Shewfelt (eds.). *Integrated View of Fruit and Vegetable Quality*. Technomic Pub. Co., Inc., pp. 201-208, Lancaster, PA

Mäder, P., Fließbach, A., Dubois, D., Gunst, L., Fried, P. & Niggli, U. (2002). Soil fertility and biodiversity in organic farming. Science 296: 1694-1697.

Moradin, L.A. & Winston, M.L. (2005). Wild bee abundance and seed production in conventional, organic, and genetically modified canola. *Ecological Applications* Vol.15, pp. 871-881

NABU, (2004). Vögel der Agrarlandschaft – Bestand, Gefährdung, Schutz. *Naturschutzbund Deutschland*, p.44, Berlin

Neumann, H.; Loges, R. & Taube, F. (2007). Fördert der ökologische Landbau die Vielfalt und Häufigkeit von Brutvögeln auf Ackerflächen? *Berichte über Landwirtschaft*, Vol.85, pp. 272-299

Oliver, Hilary. (2006). Organic Dairy Demand Exceeds Supply. *Natural Foods Merchandiser*, (August2006)

Pfiffner, L. & Luka, H. (2003). Effects of low-input farming systems on carabids and epigeal spiders – a paired farm approach. *Basic and Applied Ecology*, Vol.4, pp. 117-127

Pimentel, D.; Hepperly, P.; Hanson, J.; Douds, D. & Seidel. R. (2005). Environmental, energetic, and economic comparisons of organic and conventional farming systems. *Bioscience*, Vol.55, No.7, pp. 573-582

Richter, T.; Schmid, O.; Freyer, B.; Halpin, D. & R. Vetter. (2000). Organic Consumer in Supermarkets - New Consumer Group With Different Buying Behavior and Demands!, *In Proceedings 13th IFOAM Scientific Conference, T. Alfödi, W. Lockeretz, U. Niggli (eds.). vdf Hochschulverlag AG and der ETH Zürich*, pp. 542-545

Schader, C.; Pfiffner, L.; Schlatter, C. & Stolze, M. (2008). Umsetzung von Ökomassnahmen auf Bio- und ÖLN-Betrieben. *Agrarforschung*, Vol.15, pp. 506-511

Siegrist, S.; Schaub, D.; Pfiffner, L. & Mäder, P. (1998). Does organic agriculture reduce soil erodibility? The results of a longterm field study on loess in Switzerland. *Agriculture, Ecosystems and Environment*, Vol.69, pp. 253-265

Thompson, G.D.; (1998). Consumer Demand for Organic Foods: What We Know and What We Need to Know. *American Journal of Agricultural Economics*, Vol.80, pp. 1113-1118

Weinraub, J. & W. Nicholls. (2005). Organic Milk Supply Falls Short. *Washington Post*, (June2005) Vol.1, pp. F01

Whole Foods Market, (2005). Nearly Two-Thirds of Americans Have Tried Organic Foods and Beverages. Accessed (August 2006) at
http://www.wholefoodsmarket.com/company/pr_11-18-05.html

Zehnder, G.; Gurr, G.M.; Kühne, S.; Wade, M.R.; Wratten, S.D. & Wyss, E. (2007). Arthropod pest management in organic crops. *Annual Review of Entomology* No.52, pp. 57-80

Chapter 11

Determination of the Herd Size

Rocky R.J. Akarro

Additional information is available at the end of the chapter

1. Introduction

Scientists can work out breeds to be kept for greatest milk yield. Such breeds can differ from area to area and breed to breed. Akarro (1995; 2009) developed a simulation model to identify breeds to be bred for greatest milk yield in selected well managed farms in Tanzania. Having found out the breed to keep for the greatest milk yield, the problem that follows is to work out the herd size or the stocking rate for a particular farm. Also a stocking rate should render a profitability to the farm, otherwise there is no need of carrying out the enterprise of keeping cows. Indeed a simulation model of the form presented in Fig 1 can be used, but the problem is that most of the animal activities can not be quantified neither can they be approximated by the well known probability distributions. Furthermore, even for the approximation of distribution of animal activities, there is a problem of quantifying certain parameters on the quality and quantity of forage on offer and feeds in general. In view of this, a linear programming model can be developed as a proxy to determine the herd size or the stocking rate so that an enlightenment on ways to organize the farm for a profitable farm operation can be achieved. This is done for one farm only which is believed to have the necessary data input for the development of the linear model. The farm selected is Uyole. This method can be adapted to similar farms in the world which operate on the basis of 'zero' grazing.

Zero grazing is hereby defined as

1.1. The Uyole Agricultural Centre (UAC)

The Uyole Agricultural Centre (UAC) is of particular importance in this respect. The predominant dairy breed at Uyole is the Friesian / Zebu cross. Natural pastures around Uyole have been observed to produce around 2500 kg. dry matter (DM/ha/year). Considering a 400 kg cow requiring 8.5 kg. DM/day, then this cow needs 3100 kg DM/year. One hectare cannot therefore, maintain such a cow (Kifaro & Akarro, 1987). It was therefore decided that in agricultural high potential areas of Rungwe district (this is the area surrounding Uyole) more productive pasture species were required. Sensing this, the Uyole Agricultural Centre established a pasture and forage research programme. It commenced its work in 1970 with the aim to improve the phytomass and quality of pastures. Initial work involved fertilization of natural pastures, introductions and testing of grass/legume mixtures, special purpose pastures and short term crops like oats, lupine, maize, and fodder sugar beets.

2. The Linear Programming (L.P.) model

Simulation of the cow activities and feeding regimes as shown in Fig. 1 would probably be a more appropriate method for establishing the stocking rate.

However, given the intricacies of implementing this simulation especially with reference to management policies of a particular farm, a linear programming (L.P.) method is suggested as a plausible alternative. Here L.P. is defined as a mathematical structure, involving particular mathematical assumptions that can be solved using a standard solution technique, called the *simplex method*. It is the purpose of this section to formulate a mathematical model that would enlighten us on the type of pastures to be grown, when and what supplementation level is required and the number of dairy animals to keep for the farm to be profitable. It was assumed that a known section of the farm was to be developed entirely to dairy enterprise and the problem was to find how to organize this farm so that its annual net profit would be maximized. An L.P. is suggested. The herd size was kept constant throughout the year. In formulating the L.P. there arose a need of identifying major constraints to dairy cow needs as discussed below.

2.1. Nutritional (energy, protein, minerals and vitamins) requirements of a cow

2.1.1. Energy requirements

Organic nutrients obtained from different sources of feed available to an animal are used for a variety of purposes, including maintenance of body functions, the construction of body tissues, the synthesis of milk, and the conversion to mechanical energy used for walking and other work. All these diverse functions require the transfer of considerable quantities of energy, so that in most situations when the energy requirement of the animals' different needs are met, it may be assumed that animal's non-energy requirements (protein-minerals and vitamins) are also met. Hence, the nutritive value of different feeds can be expressed by their energy content or by their ability to supply energy with high coefficient of conversion into usable energy for the different body functions. The gross energy contents of different forages are very similar at about 18 MJ/kg (Hunt, 1966). A portion of this energy is lost as faeces while the remaining digestible energy (DE) proportional to the digestibility (d) of the consumed feed is converted into metabolized energy (ME) after additional losses of about 19% of DE as urine and methane (Armstrong, 1964; MAFF, 1975).

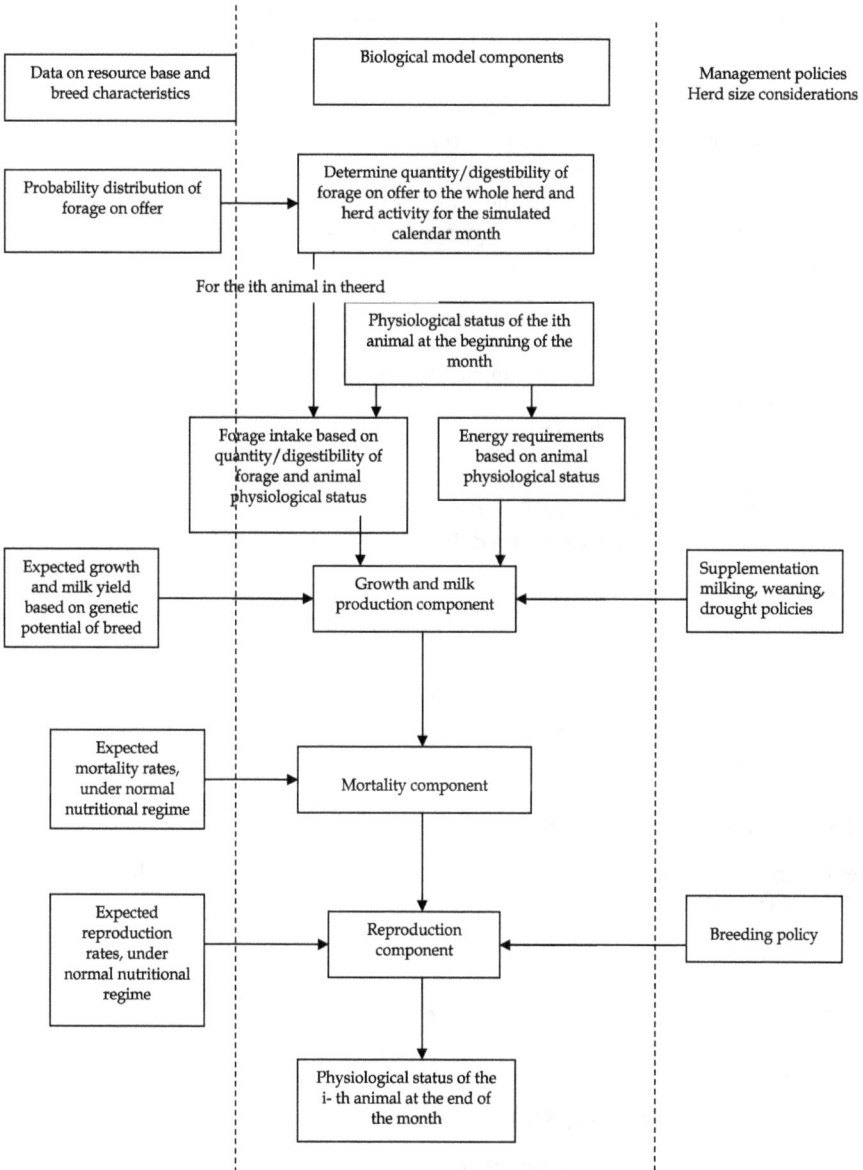

Figure 1. Components of a Dairy Cow Operational System : Source: Modified from Konandreas and Anderson (1982)

2.1.2. Energy requirements for maintenance

Maintenance can be defined as the state of the animal in which there is neither a net gain nor loss of nutrients (Kay, 1976). Maintenance requirements are estimates of the amount of nutrients required to achieve such an equilibrium. One component of the energy requirements for maintenance is referred to as basal metabolism and is proportional to the body size of the animal. The second component of the energy requirements for maintenance is related to the level of the animals activity and can be expressed approximately by live weight and the daily distance walked. Thus following Blaxter (1969) and Webster (1978), total net energy requirements for maintenance can be obtained from the relationship:

$$E_m = 0.376 \, W^{.73} + 0.0021WD \tag{1}$$

Where E_m= net energy requirements for maintenance (MJ/day).
        W= live weight (kg)
        D= distance walked (km/day).

The efficiency with which metabolizable energy is used for maintenance is a function of metabolizability of the consumed forage (see for example Blaxter, 1974; Van Es, 1976; Ministry of Agriculture Fisheries and Food (MAFF); 1975; Pigden et. al; 1979).

According to MAFF when distance walked is negligible, equation (2.1) reduces to

$$E_m = 8.3 + 0.091 \, W. \left(MAFF, 1975\right) \tag{2}$$

Where E_m= net energy requirements for maintenance.
        = metabolizable energy for maintenance (MJ/day).
        W= body weight

2.1.3. Energy requirements for lactation

Net energy requirements for lactation are approximately proportional to the quantity of milk produced (ILCA, 1978), and this is given by

$$E_L = e_L M \tag{3}$$

Where E_L= net energy requirements for lactation (MJ/day).
        e_L= energy content of milk.
        M= milk yield (kg/day).

The energy content of milk is approximately given by the relationship (MAFF, 1975):

$$e_L = 0.03886 \, BF + 0.0205 \, SNF - 0.0236 \tag{4}$$

where   BF= Butterfat content (g/kg).
       SNF= solids not fat content (g/kg).

2.1.4. Protein requirements

Protein is an essential nutrient for animals. This nutrient however cannot be synthesized in sufficient quantities by animals to meet their requirements. Fortunately it is synthesized by plants and stored in plant cells. Through this means, a source of protein is provided for use by ruminants. An animal's requirement for protein is based on the protein stored in its body; its products such as milk, eggs, or wool; the products of conception, and the metabolic losses in faeces, endogenous losses in urine and by other losses (hair, skin, hoofs, etc.). To maintain an animal in protein equilibrium, these losses must be off set. The sum of these becomes the protein requirement for that animal. Protein requirements can be determined through nitrogen balance studies. In these studies, healthy adult animals are fed an adequate amount of energy and other nutrients in diets that contain different levels of protein. The minimum protein intake that will support nitrogen equilibrium is the maintenance requirement. The protein requirement for lactation is easily calculated by determining the amount of protein present in the milk and multiplying this by 1.25 (Kearl, 1982). Dairy animals seem to adapt very well to a wide range of protein intakes without any ill effects. The protein contained in milk, however, represents a direct loss of protein by the body and obviously this must be replaced.

2.1.5. Protein requirements for maintenance

The Digestible Protein (DP) maintenance requirements have been quite well defined. Orskov (1976) stated that the rate of protein deposition by young ruminants is appropriately expressed as the nitrogen retained per unit of energy digested, and that the retention of protein per unit of energy digested increases with the level of feeding and decreases as the animal matures. Balch (1976) suggested that at any given intake of protein, the response of the animal may vary greatly depending on the intake of energy. Poppe & Gabel (1977), after reviewing the literature concerning protein requirements for cattle, cited a DP requirement of 3g per kilogram of live weight W raised to power .75 for maintenance based on a digestible organic matter (DOM) fermentation rate of 60%.

Nehring (1970) suggested a value of 2.57g of DP per kilogram of live weight raised to the power .75 as the maintenance requirement of cattle weighing 400 to 800 kg. Sen et. al; (1978) whose data are used as the feeding standard in India, recommended 2.84g DP per kilogram of $W^{.75}$ for zebu cross bred cattle and buffaloes.

Additional information is needed to substantiate these results, but on the basis of a wide range of values found in the literature and those suggested as standards to be used in several countries an average value of

$$2.86g \text{ DP per kilogram of } W^{0.75} \tag{5}$$

Where W is the live weight in kg has been used in estimating the DP maintenance requirement. This is the value used in Kearl (1982) which is also used in the formulation of feed values in food stuffs.

2.1.6. Protein requirement for lactation

Many studies have been done to determine the amount of Digestible Protein (DP) required to produce one kilogram (kg) of milk. Generally, the recommended amounts of DP per kilogram of milk have been correlated with the fat content of the milk. Nehring (1970) proposed a DP requirement of 50 to 80 g of DP for milk containing butterfat content from 3 to 6 percent. Ranjhan et. al; (1977) suggested 4.17 g of DP per kilogram of milk. Patle & Mudgal (1976) agreed with Ranjham et.al; (1977).. The National Research Council (1971) recommends a DP requirement of 42 to 60 g per kilogram of milk containing 2.5 to 6% fat. The Ministry of Agriculture Fisheries and Food (MAFF, 1979) noted a DP requirement of 48 to 63 g of DP per kilogram of milk containing 3.6 to 4.9% butterfat.

The MAFF (1979) values are the ones used in our estimates because they are regarded as standards in the formulation of feed values.

2.1.7. Estimation of nutritive values

On the average, a Friesian cow or a Friesian cross weighs 450 kg and produces milk whose composition is 3.6 percent butterfat (BF) and 8.6 percent solids not fat (SNF) at UAC (Myoya, 1980). Using results (2.2), (2.3), (2.4), (2.5) and (2.6), energy and protein requirements per dairy cow can be calculated on monthly basis for the available milk yields as shown in Table 2.1. The monthly yield figures were obtained from Uyole Agricultural Centre (UAC).

Month	Monthly Yield in Kg	Net Energy Lactation MJ/Month	Metabolizable Energy Required MJ/Month	Protein Required (DCP Kg per Month)
November	206	1018	2496	18.26
December	270	1334	2812	21.33
January	316	1561	3039	23.54
February	319	1576	3054	23.68
March	376	1858	3336	26.42
April	343	1695	3173	24.83
May	462	2283	3761	30.55
June	389	1922	3400	27.04
July	355	1754	3232	25.41
August	315	1556	3034	23.49
September	346	1709	3187	24.98
October	407	2011	3489	27.91

Table 1. Monthly Nutrition Requirements per Dairy Cow

Using equation (2.2), the Metabolizable Energy (ME) requirements for maintenance is 1478 Metabolizable Energy in Megajoules (MEMJ) per month. Using equation (2.4) the energy content e_L of milk is 4.94 megajoules (MJ) per kilogram. Using equation (2.3). net energy for lactation E_L is obtained. This is column 3 of Table 2.1. Using result (2.5), the Digestible Crude Protein (DCP) for maintenance is 279g per day. The value given by MAFF (1979) is 275g per day for a dairy cow of the same weight. DCP for maintenance required in a month is therefore 8.37 kg.

Using expression (2.6) the DCP allowances for milk production per kg for a Friesian cow with Butter fat percentage (BF%) of 3.6% is 48g (MAFF, 1979).

Column 5 in Table 2.1 is obtained by multiplying 48 by milk yield in kilograms plus DCP for maintenance which is 8.37.

2.1.8. Energy and protein supply

The main energy and protein source of dairy cows is obtained from the bulky food eaten by the cow. The bulky foods can either be grown on a farm or be purchased. At Uyole, land is scarce and the nutritive value of natural pastures and DM yield is low. Subsequent research involved evaluation of improved pasture and legume species. These included Rhodes grass (*Chloris gayana*), Napier grass (*Pennisetum purpureum*) *Desmodium spp.*; Nandi setaria, Lucerne, oats, Lupins etc. fertilizer application, cutting frequencies and grass/legume mixture.

Invariably, fertilizer application improved the quality and quantity of the production of feeds/ha but also the cost of production was increased due to input costs.

2.1.9. Fertilizer efficiency

An increase in nitrogen application leads to increase in Dry matter yield per hectare as can be seen from Table 2.2. However it is desired not to apply infinite amounts of fertilizer but to apply the amounts that will give the maximum yield (kg DM) per unit of fertilizer applied. Such amount will be termed 'efficient' fertilizer applications. Such quantities will be used in our model for the input costs in pasture/crop production. Efficient fertilizer applications were worked out as yield increase per amount of fertilizer applied. Data for yields and fertilizer applications were obtained from UAC. These are presented in Table 2.2

Nitrogen kg/ha/year	Rhodes grass yield kg DM/ha	Fertilizer efficiency kg Dm/kg N	Nandi setaria yield kg DM/ha	Efficiency kg DM /kg N
0	3700		2370	
60	6020	39	5000	44
120	8720	42	7820	45
240	14120	43	14820	52
380	17860	39	18160	44
480	21630	37	21460	40

Table 2. Dry Matter Yield (kg/ha) of Rhodes Grass and Nandi setaria under Six levels of Nitrogen (Mean of Three Years): Source: Myoya (1980).

The higher yield in Rhodes grass without nitrogen and with rates up to 120kg N/ha suggest that Rhodes grass requires less nitrogen than Nandi setaria. At higher nitrogen rates, the difference disappears.

2.2. Forage supply at UAC

Due to climatic variations at Uyole, certain types of crops are available only in particular months or periods. During the wet season one expects surplus fodder which is not the case in the dry season. Table 2.3 shows the forage/feed supply sequence of some of the animal food stuffs grown at Uyole during the year.

Crop	Season						
	January-February	March-April	May	June	July-August	September-November	December
X_1 - Natural pasture	V	V	V	X	X	X	X
X_2 - Rhodes grass pasture	V	V	V	V	X	X	V
X_3 - Rhodes/Desmodium pasture	V	V	V	V	X	X	V
X_4 - Napier grass green feed	X	V	V	V	V	V	V
X_5 - Lupins green feed	X	V	V	V	V	X	V
X_6 - Napier /Desmodium green feed	X	V	V	V	V	V	X
X_7 - Oats green feed	X	V	V	V	X	X	V
X_8 - Rhodes grass hay	X	X	X	X	X	V	X
X_9 - Maize silage	X	V	V	V	V	V	V
X_{10} - Rhodes grass silage	X	X	X	X	V	V	V
X_{11} - Rhodes/Desmodium silage	X	X	X	X	V	V	V
X_{12} - Napier grass silage	X	X	X	X	V	V	X
X_{13} - Oats silage	X	X	X	X	V	X	X
X_{14} - Lupins silage	X	X	X	X	V	V	V
X_{15} - Napier/Desmodium silage	X	X	X	X	V	V	V

Key: V Available
           X Not available

Table 3. Availability of Crops

Hay and silage could also be grown and these can be fed at any time during the year although they are usually fed during the dry season.

Thus, the year could be divided into seven periods, in each of which a different combination of crops or grazing output was available as shown in Table 2.3.

2.3. Estimation of Metabolizable Energy (ME) and Digestible Crude Protein (DCP) in feed stuffs

At UAC only a few feed stuffs have had their ME and DCP estimates done. In this case the figures available from the literature are assumed to be similar for the same feed stuffs where ME and DCP estimates are lacking. These are used because even in the formulation of quantity of food required, the estimates found in the literature, especially MAFF (1979), are used (Kurwijila, 1991). Table 2.4 gives Metabolizable Energy in Megajoules (MEMJ), Crude Protein Percentages and Crude Protein Digestibility Coefficient percentages estimates based on one kilogram of Dry matter (Mbwile, et. al; 1981; Gohl, 1981; Bredon, 1963; MAFF, 1979).

Feed	CP %	CPDC %	MEMJ/Kg DM
Natural Pastures	9.9	69.5	9.2
Rhodes Grass	7.5	62.0	8.7
Rhodes/Desmodium	11.4	52.1	7.9
Green Feed (forage)			
Lupins	15.5	73.4	10.3
Oats	13.4	76.0	10.5
Napier Grass ***	15.3	77	10.4
Napier/Desmodium	26.5	85	12.1
Silages			
Rhodes Grass	6.0	44.7	7.7
Rhodes/Desmodium	7.2	35.3	7.2
Oats	8.0	57.9	8.4
Maize	5.7	48.6	8.9
Napier Grass	16.0	64	8.8
Napier/Desmodium	16.0	64	8.8
Hay			
Rhodes Grass**	8.5	46	8.4
Rhodes Desmodium*	10.1	57	9.0

* Approximated the same as grass moderate digestibility silage.
** Approximated the same as grass with high digestibility.
*** Approximated the same as pasture grass, set stocking, close grazing.
Note: CP% means Crude Protein percentage, CPDC% means Crude Protein Digestibility Coefficient, ME MJ/kg DM means metabolizable energy in mega joules per kg of dry matter.

Table 4. Metabolizable Energy, Protein Content and Digestibility of Some of the Common Feeds at UAC

2.4. Nutrient value for the purchased concentrates

Supplementary feeding by purchased concentrates is usually done to the milking cows in order to increase their milk output. This is done throughout the year. These concentrates are in the form of energy feeds and protein feeds. Their nutrient values are given in Table 2.5.

Type	CP %	CPDC %	MEMJ/Kg DM
Energy Feeds			
Maize meal	10.6	86	14.2
Maize bran	9.6	65	12.5
Rice Polishing	14.9	87	15.5
Protein Feeds			
Cotton seed case (undecorticated)	23.1	77	8.5
Cotton seed case (decorticated)	41.7	72	10.8
Sunflower cake (undecorticated)	20.6	90	9.5
Sunflower cake (decorticated)	31.0	75	11.9
Lupin grain	33.7	81	14

Table 5. Metabolizable Energy, Protein Content and Digestibility of the Concentrates

The yields of different crops and grasses for various fertilizer application levels are shown in Table 2.6 and 2.7. The total ME and DCP on the basis of ha can be estimated (see Table 2.8). The total ME and DCP for the purchased concentrates is estimated on the basis of tonnage (see Table 2.9)

Crop	Fertilizer Applied kg N/ha	DM Yield in kg/ha	Nitrogen Efficiency kg DM/kg N
Natural Pasture	0	3000	
	80	5500	31*
	160	7500	28
	320	12000	28
Rhodes Pasture	60	1410	24*
	120	2083	17
Rhodes Hay	60	3115	52*
	120	3515	29
Rhodes Silage	60	3455	58*
	120	5855	49
Napier Grass Silage	80	4720	59*
	160	7670	48
	320	11370	36
Napier Grass Greenfield	80	4490	56*
	160	7280	46
	320	15220	48

* The most efficient fertilizer application yield per kg. of dry matter.

Table 6. Approximate Nitrogen Efficiency for some of the Crops where Different levels of Fertilizer are applied at UAC

Dry matter yield and fertilizer applied figures were obtained from the UAC.

Type	Use	DM Yield in kg/ha	Fertilizer Applied kgN/ha/ (or P/ha)
Natural Pastures	Pasture	5500	80
Rhodes Grass	Silage	3455	60
	Hay	3115	60
	Pasture	1410	60
Rhodes/Desmodium	Pasture	2100	0
	Silage	2100	0
Napier Grass	Silage	4720	80
	Green Feed	4490	80
Napier/Desmodium	Silage	4000	0
	Green feed	4000	0
Maize	Silage	10000	100N 20P[1]
Oats	Green Forage	2500	50N 20P[1]
	Silage	2500	
Lupins	Silage	6000	40P[1]
	Green Forage	500	

[1] Phosphate was included.

Table 7. Approximate Yield of Different Crops and Grasses (Feeds) for the Most Efficient Fertilizer Levels

The figures for dry matter yield and fertilizer applied were obtained from Uyole Agricultural Centre (UAC).

Metabolizable energy and DCP in Table 2.8 were obtained by multiplying dry matter yield in Table 2.7 by MEMJ/kg DM CP% and CPDC% in Table 2.4 respectively..

2.5. Fertilizer use and pasture production costs

The primary inputs involved in crop production are fertilizer, labour and cost of seeds in certain types of crops.

By using nitrogen, the carrying capacity of the land is increased which directly affects the cost of production. The profitability of applying nitrogen depends on the relationship between the cost of inputs and the value of realized output in the form of livestock and livestock products.

Based on records from production at the UAC, the following costs (Table 2.10) are incurred in pasture production – seed, cultivation and planting costs distributed over five years (the leys/grown pasture assumed life time) and fertilizer application and harvesting costs for three yearly harvest.

Type	Fertilizer Level N kg	Metabolizable Energy values in MEMJ/ha	DCP kg/ha
Pasture			
Natural Pastures	80	50600	382.3
Rhodes Grass	60	12267	65.6
Rhodes/Desmodium	-	16590	124.7
Green Feed (Forage)			
Lupins	20P	57500	568.9
Oats	25N 10P	26250	254.6
Napier Grass	80N	46696	529
Napier/Desmodium	-	48400	901
Hay			
Rhodes Grass	60N	28035	174
Silages			
Rhodes Grass	60N	26604	92.7
Rhodes/Desmodium	-	15120	53.4
Maize	100N 20P	89000	277
Oats	25N 10P	21000	115.8
Lupins*	20P	85200	1650
Napier Grass	40N	41536	483
Napier/Desmodium		35200	409.6

* Assumed same as lupin grain.

Table 8. Estimated Total Metabolizable Energy (ME) and Digestible Crude Protein (DCP) of some of the Commonly Grown Feeds per hectare at UAC

Type	ME/ton	DCP/ton
Energy Feeds		
Maize Meal	14200	91.2
Maize Bran	12500	62.4
Rice Polishing	15000	129.6
Protein feeds		
Cotton Seed Cake (undecorticated)	8500	178
Cotton Seed Cake (decorticated)	10800	296.6
Sunflower Cake (undecortimated)	9500	185.4
Sunflower Cake (decorticated)	11900	232.5
Lupin Grain	14200	273

Note: Digestible Crude Protein (DCP) is a measure of the useful protein potential of the feed and has been calculated from the crude protein content and the crude protein digestibility of the feed. The nutritive value of each feed has been expressed in terms of its Metabolizable energy and the Digestible Crude Protein (DCP).
ME and DCP in Table 2.9 were obtained by multiplying 1000 by MEMJ/kg Dm, CP% and CPDC% in Table 2.5 respectively..

Table 9. Estimated Total Metabolizable Energy (ME) and Digestible Crude Protein (DCP) of the Purchased Concentrates per ton at UAC

Input Operation	Units per ha	Cost/unit	Total cost per ha T shs	Cost per Year T shs
Rhodes Grass Seed	5 kg	40	200	40
Desmodium Seed	5 kg	80	400	80
Napier Grass Seed (Assumed for Nandi setaria)	7 kg	40	280	56
Cultivation	1.8 hrs	165	297	223
Fertilizer Application	0.8 hrs	165	132	528
Harvesting	2.0 hrs	225	450	1350
Interest on Working Capital				100

Table 10. Pasture Production Input Cost at UAC. Source: Myoya ., 1980, p. 41.

The production costs are based on the following:

1. That the harvesting costs per hectare are those obtained at UAC farm where the rates of 100-150 kg. N/ha per year are used.
2. That with yield increase due to the increase in nitrogen application, more dry matter per hectare are handled and as such the harvesting costs for 120 kg. N/ha will be taken as 100%. For 0 kg. N/ha as 50%, for 60 kg. N/ha as 75%, for 80 kg. N/ha as 80%, for 160 kg. N/ha as 120%, for 240 kg. N/ha as 150%, for 320 kg. N/ha as 180% and for 480 kg N/ha as 200%.
3. The production costs for silages and hay are assumed to be 50% higher than those of the corresponding grass or crop. Taking into account the harvesting costs of 450 T shs. For 120 kg. N/ha, phosphorous fertilizer cost 8.5 T shs. and nitrogen cost 6.5 T shs. The following were the prices for common feeds at UAC (Table 2.11). The prices for supplementary feeds which are usually bought, are given on tonnage basis.
4. The official currency of Tanzania is Tanzanian shillings hereby abbreviated as T shs.

Note that at the time of this research, 100 T shs was approximately equivalent to 1 U.S. $. Thus the estimated cost of production of various feeds is depicted in Table 2.11 below.

Feed	Fertilizer Applied	Cost (T shs)
Pasture		
Natural Pasture (per ha)	80N	1048
Rhodes Grass (per ha)	80N	2353
Rhodes/Desmodium (per ha)	-	2032
Green Feed		
Napier Grass (per ha)	80N	2566
Napier/Desmodium (per ha)	-	1653
Lupins (per ha)	20P	3544
Oats (per ha)	25N 10P	3661
Hay		
Rhodes Grass (per ha)		3530

Silage		
Maize (per ha)		5820
Lupins (per ha)		5316
Oats (per ha)		5492
Napier Grass (per ha)		3849
Napier/Desmodium (per ha)		2480
Rhodes Grass (per ha)		3530
Rhodes/Desmodium		3048
Purchased Foods Cost in Tshs (per ton)		
Energy Feeds		
Maize Meal		10000
Maize bran		4000
Rice Polishing		12000
Protein Feeds		
Cotton Seed Cake (undecorticated)		6000
Cotton seed Cake (decorticated)		6000
Cotton Seed Cake (undecorticated)		6000
Sunflower Cade (undecorticated)		6000
Lupin Grain		8000

Table 11. Costs of Feeds per Hectare or per ton Depending on the Nature of Feed (grown or purchased) for most Efficient Fertilizer Applications.

3.

3.1. Setting up the Linear Programming (L.P.) model

All together 23 different feeds were available at UAC under the land utilisation programme. The 23 different feeds include concentrates and minerals which are fed according to milk production. We denote the acreage of the different crop types for the most 'efficient' fertilizer application by X_j in hectares for the grown crops and by Y_j in tons for the purchased feeds.

3.1.1. The objective function

The objective of the model is to determine the herd size that would maximize the net profit at UAC.

3.2. The objective function coefficient

3.2.1. Milk output and its revenue

According to the annual livestock report of 1984-85 of UAC, Gross income was 3.5 million Tshs. (90% was from dairy, excluding butter processing and cream). Gross income from dairy was 3.15 million T shs. The variable costs of production were 1.575 million T shs. Gross profit was therefore 1.575 million T shs. According to the same report the average number of cows was 100. Therefore the profit per cow was 15,750 T shs per annum.

3.2.2. Milk production input costs

In order to find the optimum herd size, it is important that the inputs i.e. crops involved in dairy production are included in the programme. As already seen earlier, various costs of production could be worked out.

The objective function is therefore to Maximize the Net Profit. Denote the different acreage for the grown crops types by X where j = 1, 2, …. 15 are grown crops in hectares, Y_j for the purchased concentrates in tons where j ≥ 17 and by Z_j for the herd size when j = 16 where,

For grown crops
X_1 hectares of Natural grass pasture.
X_2 hectares of Rhodes pasture.
X_3 hectares of Rhodes/Desmodium pasture.
X_4 hectares of Napier grass green feed.
X_5 hectares of Lupins green feed.
X_6 hectares of Napier/Desmodium green feed.
X_7 hectares of Oats green feed.
X_8 hectares of Rhodes grass hay.
X_9 hectares of Maize silage.
X_{10} hectares of Rhodes grass silage.
X_{11} hectares of Rhodes/desmodium silage.
X_{12} hectares of Napier grass silage.
X_{13} hectares of Oasts silage.
X_{14} hectares of Lupins silage.
X_{15} Hectares of Napier/Desmodium silage.
For Purchased Feeds
Y_{17} tons Maize meal.
Y_{18} tons of Maize bran.
Y_{19} tons of Rice polishing.
Y_{20} tons of Lupin grain.
Y_{21} tons of Cotton seed cake (undecorticated).
Y_{22} tons of Cotton seed cake (decorticated).
Y_{23} tons of Sunflower cake (undecorticated).
Y_{24} tons of Sunflower cake (decorticated).
Z_{16} the Herd size or the Stocking rate.

Table 12.

Using the cost values in Table 2.12, the objective function will be to

Maximize

$-15750 \quad Z_{16} - 1048X_1 - 2353X_2 - 2032X_3$
$- 2566X_4 - 3544X_5 - 1653X_6 - 3661X_7 - 3530X_8 - 5820X_9 - 3530X_{10}$
$- 3048X_{11} - 3840X_{12} - 5492X_{13} - 5316X_{14} - 2480X_{15} - 10000Y_{17}$
$- 4000Y_{18} - 12000Y_{19} - 8000Y_{20} - 6000Y_{21} - 6000Y_{22} - 6000Y_{23} - 6000Y_{24}$

At UAC, the objective is formulated on the basis of one type of breed only since there is only one breed at UAC for dairy production. In situations where multiple breeds are involved, a multiple objective function can be formulated in line with modified costs and profits accordingly.

The Constraints

3.2.3. Land constraint

Let the total acreage available be A. The acreage constraint ensures that the amount of land available for the growth of various crops is not exceeded.

$$\sum_{j=1}^{15} X_j \leq A$$

where A is the total acreage in hectares
X_j is the acreage for different crop types j in hectares.

In the case of UAC, A is 790 hectares.

3.2.4. Maintenance energy requirement constraint

As a cow needs a minimum quantity of bulky food in its diet, it was decided that at least sufficient energy to supply maintenance requirements should come from food of this type, and should be grown on the farm.

Suppose crop j supplies a_j kg of energy (MJME) per ha. If one cow requires E_m MJME for maintenance, then

$$\sum_{j=1}^{15} a_j X_j \geq E_m Z_{16}$$

3.2.5. Total energy requirement constraint

Suppose crop j supplies a_j megajoules of metabolizable energy per hectare and suppose the purchased concentrates do supply b_j megajoules of metabolizable energy per ton. If one cow requires E_l metabolizable energy for maintenance and lactation then.

$$\sum_{j=1}^{15} a_j X_j + \sum_{j=17}^{24} b_j Y_j \geq E_l Z_{16}$$

3.2.6. Maintenance protein requirement constraint

Suppose crop j supplies p_j kg. of digestible crude protein per hectare. If one cow requires q kg. of Digestible Crude Protein for maintenance then.

$$\sum_{j=1}^{15} p_j \, X_j \; \geq q \, Z_{16}$$

3.2.7. Total protein requirement constraint

Suppose crop j supplies p_j kg. of digestible crude protein per hectare and suppose the purchased concentrated do supply r_j kg. of digestible crude protein per ton. If one cow requires t kg. of digestible Crude Protein for maintenance and lactation

Then $\displaystyle\sum_{j=1}^{15} p_j \, X_j + \sum_{j=17}^{24} r_j \, Y_j \; \geq t \, Z_{16}$

3.2.8. Space constraint

Let the space needed for a cow on the average be s m 2. If the available space has a total area of h square metres then this particular farm can accommodate a maximum of M = h/s animals. Thus

$$Z_{16} \leq M$$

At UAC s = 6 m 2 area needed by one cow. H = 24,000 m 2 is area of the available shelter at UAC. Then M = h/s = 4000. The number of animals that can be 'accommodated'. Therefore

$$Z_{16} \leq 4000$$

Three L.P. models were run with different assumptions for each model. In the first model, the model was run for grown crops only. In the second model the imposed restriction was that maize should not be grown for the purpose of feeding animals (with an intuitive idea that maize should be for humans only). This was removed from the programme in the usual way by making its cost of production exorbitantly high. The problem was unchanged except for the coefficient C9 which was changed from 5820 to 99999. In the third model, the model was run for grown crops and purchased concentrates. The third model was feasible and gave the maximum profit. Thus the third model was adopted for our study. The solutions to the third model is presented in section 3. Together with the solution, post optimality analysis i.e. how sensitive is the optimal solution - and the appropriate interpretation are given for the this model.

We shall use the cost values in Table 2.12 and the net profit of 15,750 Tshs. per cow as calculated in section 3.2.1 for the objective function coefficients. The feed values in Table 2.9

and 2.10 will be used as the coefficients of the left hand side of crop and purchased constraints respectively while feed requirement values in Table 2.1 will be used as the coefficient of the Z on right hand sides of the constraints for the available feed supply periods as shown in Table 2.4. Our linear programming problem involving all the feeds (grown foods and purchased concentrates) is presented as follows:

Maximize

15750 Z_{16} - 1048X_1 - 2353X_2 - 2032X_3 - 2566X_4 - 3544X_5 - 1653X_6 - 3661X_7
- 3530 X_8 - 5820X_9 - 3530X_{10} - 3048X_{11} - 3840X_{12} - 5492X_{13} - 5316X_{14}
- 2480X_{15} - 10000Y_{17} - 4000Y_{18} - 12000Y_{19} - 8000Y_{20} - 6000Y_{21} - 6000Y_{22}
- 6000Y_{23} - 6000Y_{24}

Subject to

$$\sum_{j=1}^{15} X_j \leq 790 \quad \text{land constraint} \tag{6}$$

$$Z_{16} \leq 4000 \text{ Fencing space constraint} \tag{7}$$

Total energy requirement constraint in January and February

$$50600X_1 + 12267X_2 + 16590X_3 + 14200Y_{17} + 12500Y_{18} = 15000Y_{19} + $$
$$+ 14200Y_{20} + 8500Y_{21} + 10800Y_{22} + 9500Y_{23} + 11900Y_{24} \geq 6073\ Z_{16} \tag{8}$$

Maintenance energy requirement constraint in January and February

$$50600X_1 + 12267X_2 + 16590X_3 \geq 2956\ Z_{16} \tag{9}$$

Total protein requirement constraint in January and February

$$382.3\,X_1 + 65.6X_2 + 124.7X_3 + 91.2Y_{17} + 62.8Y_{18}\,129.6\ Y_{19} + $$
$$+273Y_{20} + 178Y_{21} + 296.6Y_{22} + 185.4Y_{23} + 232.5Y_{24} \geq 47.22\ Z_{16} \tag{10}$$

Maintenance protein requirement constraint in January and February

$$382.3\,X_1 + 65.6X_2 + 124.7X_3 \geq 16.74\ Z_{16} \tag{11}$$

Total energy requirement constraint in December

$$50600X_1 + 12267X_2 + 16590X_3 + 46696X_4 + 48400X_6 + 28035X_8 + 89000X_9 + $$
$$+ 26604X_{10} + 15120X_{11} + 41536X_{12} + 35200X_{15} + 14200Y_{17}\,12500Y_{18} + $$
$$+15000Y_{19} + 14200Y_{20} + 85000Y_{21} = 10800Y_{22} + 9500Y23 + 11900Y24 \geq $$
$$\geq 2812\ Z_{16} \tag{12}$$

Maintenance energy requirement constraint in December

$$50600X_1 + 12267X_2 + 16590X_3 + 46696X_4 + 48400X_6 + 28035X_8 +$$
$$+ 89000X_9 + 26604X_{10} + 15120X_{11} + 41536X_{12} + 35200X_{15} \geq 1478 Z_{16} \quad (13)$$

Total protein requirement constraint in December

$$382.3X_1 + 65.6X_2 + 124.7X_3 + 529X_4 + 901X_6 + 174X_8 + 277X_8 +$$
$$+ 92.7X_{10} + 53.4X_{11} + 483X_{12} + 409.6X_{15} + 91.2Y_{17} + 62.4Y_{18} + 129.6Y_{19} +$$
$$+ 273Y_{20} + 178Y_{21} + 296.6Y_{22} + 185.4Y_{23} + 232.5Y_{24} \geq 21.33 Z_{16} \quad (14)$$

Maintenance protein requirement in December

$$382.3X_1 + 65.6X_2 + 124.7X_3 + 529X_4 + 901X_6 +$$
$$+174X_8 + 277X_8 + 92.7X_{10} + 53.4X_{11} + 483X_{12} + 409.6X_{15} \geq 8.37 Z_{16} \quad (15)$$

Total energy requirement constraint in March and April

$$50600X_1 + 12267X_2 + 16590X_3 + 46696X_4 + 57500X_5 + 48400X_6 + 26250X_{11} +$$
$$+1400Y_{17} + 12500Y_{18} + 15000Y_{19} + 14200Y_{20} + 8500Y_{21} + 10800Y_{22} + 95500Y_{23} + \quad (16)$$
$$+11900Y_{24} \geq 6509 Z_{16}$$

Maintenance energy requirement constraint in March and April

$$50600X_1 + 12267X_2 + 16590X_3 + 46696X_4 + 57500X_5 + 48400X_6 + 26250X_7 \geq 2956 z_{16} \quad (17)$$

Total protein requirement constraint in March and April

$$382.3X_1 + 65.6X_2 124.7X_3 + 529X_4 + 569X_6 + 901X_6 + 256X_7 +$$
$$+ 91.2Y_{17} + 62.4Y_{18} + 129.6Y_{19} 273Y_{20} + 178Y_{21} + 296.6Y_{22} + 185.4Y_{23} + \quad (18)$$
$$+ 232.5Y_{24} \geq 51.25 Z_{16}$$

Maintenance protein requirement in March and April

$$382.3X_1 + 65.6X_2 124.7X_3 + 529X_4 + 569X_6 + 901X_6 + 256X_7 \geq 16.74 Z_{16} \quad (19)$$

Total energy requirement constraint in May

$$50600X_1 + 12267X_2 + 16590X_3 + 46696X_4 + 57500X_5 +$$
$$+ 48400X_6 + 26250X_7 + 89000X_9 + 14200Y_{17} + 12500Y_{18} + \quad (20)$$
$$+ 15000Y_{19} + 14200y_{20} + 85500y_{21} + 10800Y_{22} + 9500Y_{23} + 11900Y_{24} \geq 3761 Z_{16}$$

Maintenance energy requirement constraint in May

$$50600X_1 + 12267X_2 + 16590X_3 + 46696X_4 +$$
$$+ 57500X_5 + 48400X_6 + 26250X_7 + 89000X_9 \geq 1478 Z_{16} \quad (21)$$

Total protein requirement constraint in May

$$382.3X_1 + 65.6.X_2 + 124.7X_3 + 529X_4 + 569X_4 + 901X_6 + 256X_7 +$$
$$+ 91.2Y_{17} + 62.4Y_{18} + 129.6Y_{19} + 273Y_{20} + 178Y_{21} + 296.6Y_{22} + 185.4Y_{23} + \quad (22)$$
$$+ 232.5Y_{24} \geq 30.55\ Z_{16}$$

Maintenance protein requirement constraint in May

$$382.3X_1 + 65.6.X_2 + 124.7X_3 + 529X_4 + 569X_4 + 901X_6 +$$
$$+ 256X_7 + 277X_9 \geq 8.37\ Z_{16} \quad (23)$$

Total energy requirement constraint in June

$$46696X_4 + 57500X_5 + 48400X_6 + 26250X_7 +$$
$$+ 89000X_9 + 14200Y_{17} + 12500Y_{18} + 15000Y_{19} +$$
$$+ 14200Y_{20} + 8500Y_{21} + 10800Y_{22} + 9500Y_{23} + \quad (24)$$
$$+ 11900Y_{24} \geq 3400\ Z_{16}$$

Maintenance energy requirement constraint in June

$$46696X_4 + 57500X_5 + 48400X_6 + 26250X_7 + 89000X_9 \geq 1478\ Z_{16} \quad (25)$$

Total protein requirement constraint in June

$$529X_4 + 569X_5 + 901X_6 + 256X_7 + 277X_9 9.2Y_{17} +$$
$$+ 62.4Y_{18} + 129.6Y_{19} + 273Y_{20} + 178Y_{21} + 296.6Y_{22} + \quad (26)$$
$$+ 185.4Y_{23} + 232.5Y_{24} \geq 27.04\ Z_{16}$$

Maintenance protein requirement constraint in June

$$529X_4 + 569X_5 + 901X_6 + 277X_9 \geq 8.37\ Z_{16} \quad (27)$$

Total energy requirement constraint in July and August

$$46696X_4 + 57500X_5 + 48400X_6 + 89000X_9 + 26604X_{10} + 15120X_{11} +$$
$$+ 41536X_{12} + 21000X_{13} + 14200Y_{17} + 12500Y_{18} + 15000Y_{19} + 14200Y_{20} +$$
$$+ 14200Y_{20} + 8500Y_{21} + 10800Y_{22} + 9500Y_{23} + 11900Y_{24} + 85200X_{14} + \quad (28)$$
$$+ 35200X_1 \geq 6266\ Z_{16}$$

Maintenance energy requirement constraint in July and August

$$46696X_4 + 57500X_5 + 48400X_6 + 89000X_9 + 26604X_{10} +$$
$$+ 15120X_{11} + 41536X_{12} + 21000X_{13} + 85200X_{14} + 35200X_{15} \geq \quad (29)$$
$$\geq 2956\ Z_{16}$$

Total protein requirement constraint in July and August

$$529X_4 + 569X_5 + 901X_6 + 277X_9 + 92.7X_{10} + 91.2Y_{17} +$$
$$+ 62.4Y_{18} + 129.6Y_{19} + 273Y_{20} + 178Y_{21} + 296.6Y_{22} +$$
$$+ 185.4Y_{23} + 232.5Y_{24} + 53.4X_{11} + 483X_{12} + 115.8X_{13} + 1650X_{14} +$$
$$+ 409.6X_{15} \geq 48.9\, Z_{16}$$
(30)

Maintenance protein requirement constraint in July and August

$$529X_4 + 569X_5 + 901X_X + 277X_9 + 92.7X_{10} +$$
$$+ 53.4X_{11} + 483X_{12} + 115.8X_{13} + 1650X_{14} +$$
$$+ 409.6X_{15} \geq 16.74\, Z_{16}$$
(31)

Total energy requirement constraint in September, October and November

$$46696X_4 + 48400X_6 + OX7 + 28035X_9 + 26604X_{10} +$$
$$+ 15120X_{11}41536X_{12} + 35200X_{15} + 14200Y_{17} + 12500Y_{18} +$$
$$+ 15000Y_{19} + 14200Y_{20} + 8500Y_{21} + 10800Y_{22} + 9500Y_{23} +$$
$$+ 11900Y_{24} \geq 9172\, Z_{16}$$
(32)

Maintenance energy requirement constraint in September, October and November

$$46696X_4 + 48400X_6 + 28035X_8 + 89000X9 +$$
$$+ 26604X_{10} + 15120X_{11} + 41536X_{12} + 35200X_{15} \geq 4434\, Z_{16}$$
(33)

Total protein requirement constraint in September, October and November

$$529X_4 + 901X_6 + 174X_8 + 277X_9 + 92.7X_{17} +$$
$$+ 91.2Y_{17} + 62.4Y_{18} + 129.6Y_{19} + 273Y_{20} + 178Y_{21} +$$
$$+ 296.6Y_{22} + 185.4Y_{23} + 232.5Y_{24}53.4X_{11} + 484X_{12} +$$
$$+ 409.6X_{15} \geq 71.15\, Z_{16}$$
(34)

Maintenance protein requirement constraint in September, October and November

$$529X_4 + 901X_6 + 174X_8 + 277X_9 + 92.7X_{10} +$$
$$+ 53.4X_{11} + 483X_{12} + 409.6X_{15} \geq 25.11\, Z_{16}$$
(35)

ALL X's, Y's and $Z \geq 0$.

3.3. The stocking rate or the herd size model

The L.P. problem for UAC was run using the OR software by Dennis and Dennis (1991).

One could run any number of models with different assumptions for each model. In our case, three LP models were considered with different assumptions for each model. The assumptions considered were running the LP with all grown crops included, concentrates excluded, running the LP with maize and concentrates excluded and running the LP with all grown crops and concentrates included. The purpose of doing this was to find out what combination of foods would give the maximum profit. This and the previous linear programming problems were run on an IBM PC using the OR software by Dennis and Dennis (1991). Here OR (Operations Research) is defined as the systematic application of quantitative methods, techniques, and tools to the analysis of problems involving the operation of the system. The aim is the evaluation of probable consequences of decision choices, usually under conditions requiring the allocation of scarce resources –funds, manpower, time, or raw materials (Daellenbach & George, 1978). The computer output tables are presented in Table 3.1. The simplex tableau for the grown crops and purchased concentrates is discussed.

Inclusion of concentrates results into a big profit of T shs. 58,752,345.70. The option of giving concentrates to cows has a significant impact on profit maximization at UAC as shown in Table 3.1 We recommend this model.

The following results were obtained after 15 iterations of the Simplex Algorithm.

Variable	Quantity	Variable	Quantity	Variable	Quantity
				S_{17}	300499.771
X_1	361.234	S_8	49054442.076	S_{18}	389219.771
X_6	184.378	S_9	328539.771	S_{19}	23088000
X_9	244.388	S_{10}	389219.771	S_{20}	30776000
Y_{22}	85.545	S_{11}	7179873.951	S_{21}	176440
Y_{18}	407.174	S_{12}	21391873.951	S_{22}	251120
Z_{16}	4000	S_{13}	150004.182	S_{23}	11624000
S_4	12468000	S_{14}	288044.182	S_{24}	24864000
S_6	121920	S_{15}	39922442.076	S_{25}	89000
S_7	43718442.076	S_{16}	49054442.076	S_{26}	217640
				S_{28}	18952000
				S_{30}	184160

Table 13. The Final Simplex Tableau displaying all the Feeds (Grown and Purchased Concentrates).

Optimal Profit = T.shs 58,752,345.717. S here refers to slacks.

Results show that land to be allocated for natural pastures is 361.234 hectares, Napier/Desmodium 184.378 hectares and maize production 244.388 hectares. Total land used for their production is therefore 790 hectares i.e. the whole land available is utilized. Concentrate supplementation is 85.545 tons of cotton seed cake (decorticated) and 407.174 tons of maize bran. Fencing land for the cows is fully utilized. Herd size is 4000 cows. Since whatever available land and fencing has been utilized under this programme and profit has been maximized, it was deemed reasonable to adapt this model.

4.

4.1. Sensitivity analysis

4.1.1. Abundant and scarce resources

Associated with every LP; there is a corresponding optimization problem called the Dual Problem. The original problem is called the primal problem. The purpose of the dual in our case is to identify scarce and abundant resources and as such give recommendations if any. Dual values represent quite precisely the per unit increase in the objective function which would follow from an increase in the availability of the corresponding factors or resources.

It should be obvious, first of all, that an increased availability of a factor which is not fully used will only leave more of it unused and add nothing to the objective function and such a constraint has zero dual value – it is a free good. (Note that a good is free, not because it is not used, but because there is more available than is required. Air and water are the classical cases of free goods which would be very far from free if their availability were restricted).

To summarise, we can assert that if Y_k represents the per unit increase in revenue from an increase in the availability of the k th factors, then a change in availability of Δ_k will lead to a change in revenue of $Y_k\Delta_k$.

It is obvious that an increased availability of a factor which is fully utilized can add considerably to the value of the objective function. Land constraint is fully utilized and its dual value is positive (8003.55). Similarly fencing space for cows (constraint 2) is fully utilized, its dual value is positive (13107.383). So an increase in the land for crops and an increase in the space for the animals can still add considerably to the revenue of the enterprise by rearing more cows. Thus, per unit increase in the land acreage would increase the objective function by 8003.55 whereas per unit increase in the fencing space would increase the objective function by 13107.385 with all other coefficients in the problem remaining the same.

On the other hand, a small increase in the right-hand-side of an abundant resource constraint will only change the amount of slacks or surplus and will not affect the value of the objective function. Thus the shadow price for any non-binding constraint is zero.

The other constraints, for example constraint (4) and constraints (6) to (30) except constraints (27) and (29), are not so binding in our case since an increase in their availability will leave more of them unused and add nothing to the revenue and, as such, their dual values are zero.

Constraints (1), (2), (3), (5), (27), and (29) are binding in our case and as such their dual values are positive. They are therefore scarce resources. If we go back to the primal problem we will see that these aforementioned constraints have all their slack values equal to zero and their corresponding dual variables are positive.

Variable	Dual solution or shadow price	Constraint
S_1	8003.55	(1)
S_2	13107.385	(2)
S_3	.121	(3)
S_4	0	(4)
S_5	7.632	(5)
S_6	0	(6)
S_7	0	(7)
S_8	0	(8)
S_9	0	(9)
S_{10}	0	(10)
S_{11}	0	(11)
S_{12}	0	(12)
S_{13}	0	(13)
S_{14}	0	(14)
S_{15}	0	(15)
S_{16}	0	(16)
S_{17}	0	(17)
S_{18}	0	(18)
S_{19}	0	(19)
S_{20}	0	(20)
S_{21}	0	(21)
S_{22}	0	(22)
S_{23}	0	(23)
S_{24}	0	(24)
S_{25}	0	(25)
S_{26}	0	(26)
S_{27}	.146	(27)
S_{28}	0	(28)
S_{29}	2.851	(29)
S_{30}	0	(30)

Table 14. Dual values for the recommended programme (model three whereby grown crops and concentrates are included).

The scarce resources in our model are therefore land, fencing space, energy supply from January to February, protein supply from January to February, energy supply from September to November and protein supply from September to November. Energy and protein supplies are scarce from September to November because these are dry months in Mbeya Region and as such food is scarce during this period. Similarly the supply of food from January to February is not adequate in Mbeya Region.

As for the abundant resources these have dual values equal zero in their constraints. An abundant resource worth mentioning is energy supply from March to April. The slack of this constraint S_{11} has the value 7179873.953 in the primal. This slack is an indication of surplus food available during rainy season in Mbeya Region which is mostly pronounced in March and April.

4.1.2. The objective function coefficients

It is important for us to know, for example, for what ranges of prices of the inputs in the objective function is the solution still optimal. To do this we assume the coefficient matrix A and the right hand side constraints b are unchanged but the profits vector c is changed to c+λc, where λ is any constant. The results are presented in Table 4.2.

Coefficient of Variables	Lower Limit	Original Value	Upper Limit
X_1	-9043.741	-1048	20358.836
X_2	NO LIMIT	-2553	6015.1
X_3	NO LIMIT	-2032	5040.77
X_4	NO LIMIT	-2566	-343.01
X_5	NO LIMIT	-3544	8003.55
X_6	-5666.778	-1653	672.409
X_7	NO LIMIT	-3661	8003.55
X_8	NO LIMIT	-3530	3401.85
X_9	-9099.061	-5820	2958.965
X_{10}	NO LIMIT	-10000	-4756.94
X_{11}	NO LIMIT	-3530	3843.18
X_{12}	NO LIMIT	-3048	5637.02
X_{13}	NO LIMIT	-3840	543.8
X_{14}	NO LIMIT	-5316	8003.55
Y_{19}	NO LIMIT	-1200	-5373.61
Y_{20}	NO LIMIT	-8000	-6662.69
Y_{21}	NO LIMIT	-6000	-4141.11
Y_{22}	-6529.272	-6000	-4058.013
Y_{23}	NO LIMIT	-6000	-4486.36
Y_{24}	NO LIMIT	-6000	-5622.5
Y_{18}	-5124.335	-4000	-2448.285
X_{15}	NO LIMIT	-2480	-1680.90
Z_{16}	2642.615	15750	NO LIMIT

Table 15. Sensitivity Analysis of Objective Function Coefficients

Of interest are the coefficients of the variables X_1, X_6, X_9, Y_{18}, Y_{22} and Z_{16}. the lower and upper limits within which the solution is still optimal are shown in Table 4.2.

For example, the solution is still optimal so long as -9043.741 < C_1< 20358.836 and so on. The cost (C_1) of natural pasture in the objective function is -1048 per hectare. As long as this cost lies between -9043-741 and 20358..836 the solution is still optimal so long as the other costs C_1's remain as they were in the primal.

4.2. The right-hand-side ranges

The right-hand-side ranges provide limits within which the objective coefficients of the dual problem are allowed to change without changing the solution. For changes outside the range the problem must be resolved to find the new optimal solution and the new dual price. We call the range over which the dual price is applicable the range of feasibility.

Assuming A and c are unchanged, b changes to b+χb where χ is any constant, the right-hand side ranges within which the objective function remains optimal are presented in table 4.3.

Constraint	Lower Limit	Original Value	Upper Limit
1	174.02	790	983.61
2	3212.67	4000	18158.99
3	-3640461.74	0	5212071.15
4	-12468000	0	NO LIMIT
5	-84149.1	0	21767.12
6	121920	0	NO LIMIT
7	43718442.08	0	NO LIMIT
8	4905442.08	0	NO LIMIT
9	328539.77	0	NO LIMIT
10	-389219.77	0	NO LIMIT
11	-7179873.95	0	NO LIMIT
12	-21391873.95	0	NO LIMIT
13	-150004.18	0	NO LIMIT
14	-288044.18	0	NO LIMIT
15	-39922442.08	0	NO LIMIT
16	-49054442.08	0	NO LIMIT
17	-300499.77	0	NO LIMIT
18	-389219.77	0	NO LIMIT
19	-23088000	0	NO LIMIT
20	-30776000	0	NO LIMIT

Constraint	Lower Limit	Original Value	Upper Limit
21	-176440	0	NO LIMIT
22	-251120	0	NO LIMIT
23	-11624000	0	NO LIMIT
24	-24864000	0	NO LIMIT
25	-89000	0	NO LIMIT
26	-217640	0	NO LIMIT
27	-25622891.49	0	NO LIMIT
28	-18952000	0	NO LIMIT
29	-281115.74	0	NO LIMIT
30	-184160	0	NO LIMIT

Table 16. Sensitivity Analysis of Right Hand Ranges

Of interest are constraints (1), (2), (3), (5), (27) and (29) i.e., land for cultivation, fencing space, energy and protein supply from January to February and energy and protein supply from September to November constraints. These are the binding constraints in our model. As shown in Table 4.3, the ranges of constraints (1) and (2) are all positive i.e. 174.04 < Land size < 983.61 and 3212.67 < fencing space < 18158.99 and so on. For example, land size could be increased up to 983.61 hectares so long as the A matrix and the objective function vector are unchanged. The solution would still be optimal. An increase of one hectare of land would increase the objective function by 8003.55 as provided for by the dual.

Changes in the right-hand side of the constraints show how the optimal solution and net profit would change if we could obtain additional land or fencing space.

5. Conclusion

The model has managed to ascertain the profitability of a dairy farm. Indeed, this form of argument can be useful in the management of dairy farms of similar traits elsewhere. The assumption here is that the herd size was kept constant throughout the year. Perhaps this is an oversimplification but it provides a starting point. There is a need of formulating Operational Research models for which the need for having a fixed herd size can be relaxed. As can be seen from the input parameters of the L.P., the values are probably not in line with dynamics of time and technological advancement of raring /keeping dairy cattle. Perhaps there is a need of updating the input parameters so that they can match with time from farm to farm.

Author details

Rocky R.J. Akarro
University of Dar es Salaam, Tanzania

6. References

Akarro, R.R.J. (1995).. An Operational Study of Dairy Farming in Tanzania. *Unpublished Ph.D. thesis, University of Dar es Salaam.*

Akarro, R.R.J. (2009). Milk Breed Selection in Selected Farms in Tanzania by the Use of Simulation Techniques. *European Journal of Scientific Research,* Volume 34 Issue 1, July, 2009.

Annual livestock report of 1984-85 of UAC. *Unpublished Report, UAC.* Mbeya, Tanzania.

Armstrong, D.G.(1964). Evaluation of artificially dried grass as a source of energy for sheep.2. *J. Agric. Sci. Camb.* 62: 399 -417.

Balch, C.C.(1976). Ruminant digestion and nutritive value pages 214 – 218 in P.V.Fonnesbeck, L.E.Harris & L.C.Kearl, eds. *Proc.of the 1st Int. .Symp. on Feed composition. Animal Nutrient Requirements and Computerization of diets.* Utah Agr. Exp. Sta., Utah State University, Logan.

Daellenbach, H.G. & George, J.A (1978). *Introduction to Operations Research Techniques.* Allyn & Bacon, Inc.

Dennis,T.L. & Dennis, L.B .(1991). *Decision Support Software/ Management Science. Linear Programming.* Copyright 1991, West Publishing Co.

Blaxter,K.L.(1969). The efficiency of energy transformation in ruminants. In K.L. Blaxter, J.Klelanowski & G.Thorbek eds. *Energy metabolism of farm animals.* London, Routledge and Regan Paul, 21-28.

Blaxter,K.L.(1973). Metabolizable energy and feeding system for ruminants. In H. swan & D. Lewis, eds. *Proc. Of Nutrition Conf. for Feed manufacturers No 7.* London, Butterworths.

Bredon, R.M. (1963). Feeding livestock in Africa, quoted by Kurwijila, R.L. (1991). *Dairy Production and management Vol. I. Basic husbandry and management practices under Tropical Environment.* Lecture notes compendium for B.Sc. courses at SUA.

Gohl, BO. (1981) *Tropical Feeds,* FAO, Rome.

Hunt, L.A.(1966). Ash and energy content of material from several forage grasses. *Crop Science,* 6:507-509.

ILCA (1978) (International Livestock Centre for Africa). *Mathematical Modelling of Livestock Production Systems. Application of the Texas A & M. University beef cattle production model to Botswana.* System Studies No 1: Addis Ababa.

Kay, M. (1976).. Meeting the energy and protein requirements of the growing animal. In H. Swan & W.H.Broster, eds. *Principles of Cattle Production.* London, Butterworths.

Kearl, L.C. (1982). Nutrient requirements of Ruminants in Developing Countries. *International Feed Stuffs Institute.* Utah Agric. Exptal. Stat; Utah State University; Logan, Utah.

Kifaro, G.C. & Akarro, F.M.N.(1987). Livestock Research in Southern Highland of Tanzania. Past constraints and future prospects, *Paper presented at the National Workshop on Agricultural Research in Tanzania,* 1987. Arusha, Tanzania.

Konandreas P.A. & Anderson, F.M. (1982). Cattle herd dynamics: an integer and stochastic model for evaluating production alternatives. *ILCA (International Livestock Centre for Africa)* . Research Report No 2. Addis Ababa, Ethiopia.

Kurwijilla, R.L.(1991). Dairy Production and Management. Vol I. *Basic Husbandry and management practices under tropical environment.* Lecture notes compendium for B.Sc. courses at SUA.

MAFF (1975). *Energy allowances and feeding system for ruminants.* Ministry of Agriculture, Fisheries and Food, United Kingdom. Technical Bulletin No. 33, London.

MAFF (Ministry of Agriculture, Fisheries and Food LGR 21, 1979). Nutrient allowances and composition of feed stuffs for ruminants. Ministry of Agriculture, Fisheries and Food, United Kingdom.

Mbwile, R.P; Kanyawara, K.S. & Wiktorsson,H. (1981). Digestibility and dry matter intake of Desmodium intortum/grass mixture and Rhodes grass fed with or without concentrate to dairy cow. Swedish University Agricultural Science Report No. 84, 1981.

Myoya, T. J. (1980).. Pasture Research. 1970/71 – 1977/1978. *Uyole Agricultural Centre, Mbeya, Tanzania.* Research Report No. 28, October 1980.

Nehring, K. (1970). Futtermitteltablellen –work. DT. Landus, Verlag, Berlin. Quoted by Kearl (1982).

NRC (1971). Nutrient requirements of Domestic Animals No 2. *Nutrient requirements of Dairy cattle* . Fourth Revised ed, Natl. Academy of Sciences, National Research Council, Washington D.C.

Orskov, E.R. (1976). Factors influencing protein and non protein nitrogen utilization in young ruminants p 456 -476 in D.J. Cole, K.N. Boorman, P.J. Buttery, D. Lewis, R.J. Neale & H. Swan eds. *Proc. Of the ist Int. Symp. On Protein Metabolism and Nutrition.* EAAP Pub. No 16, Butterworths, London.

Patle, B.A. & Mudgal, V.D. (1976). Protein requirements of crossbred lactating cows. Indian J. Dairy Sci. 29: 247.

Pidgen, W.J; Balch, C.C. & Graham, M. (1979) eds. Standardization of analytical methodology for feeds. *Proc. Of a workshop.* Ottawa Int. Dev. Res. Centre (IDRC).

Poppe, S. S. & Gabel, M. (1977). Views on the requirements of Beef Cattle (including fattening cattle) for Protein, essential Amino Acids and Non Protein Nitrogen and new sources of these nutrients suitable for use in the feeding of beef cattle.

Ranjhan, S.; Mohan, D.V. & Singh, R. (1977). Energy and protein requirement of Holsten – Friesian and Holstein –Friesian X Hariana crosses for maintenance and milk production. *Indian J. Anim. Sci.* 45: 717.

Sen, K.C; Ray, S.N. & Ranjhan, S.K (1978). Nutrient value of Indian feeds and the feeding animals. *Indian Council Agr. Res. Bull.* 25.

Van Es, A.J.H.(1976). Factors influencing the efficiency of energy utilization by beef and dairy cattle. In H. Swan & W.H.Broster, eds. *Principles of Cattle Production.* London, Butterworths, p 237-253.

Webster, A.J.F.(1978). Prediction of the energy requirements for growth in beef cattle. *Wld. Rev. Nutr. Diet.* 30: 189-226.

Chapter 12

Breeding, Management and Environmental Issues at Peri-Urban Dairy Farms

M. Subhan Qureshi

Additional information is available at the end of the chapter

1. Introduction

Dairy animals have been companions of human beings since the time immemorial. Cattle, buffaloes, sheep and goats are being reared to meet demands for human food, clothing and industrial needs. The river banks have been the seats of civilization and provided an opportunity for dairy farming, especially buffaloes in the South East Asia. However, this farming has just occurred haphazardly, without any scientific, development or business support, in most of the cases.

In the maritimes these farms are facing huge economic losses, due to under managed health, fertility and productivity and a very hostile marketing system. Still the livestock holders survive due to lack of any alternate source of livelihood and a huge investment by the forefathers of these poor people (opportunity cost). Opportunity cost remains the main support for such farmers and new investment is usually avoided. As a resultant such farms can neither provide an appropriate return to the farmers, nor a cheaper food of acceptable quality to the consumers.

The huge investment made by the ancestors of the farming family and the rising levels of unemployment compel them to stick to the business, willingly or unwillingly. The farming family tries to continue the business without considering the financial inputs and products, and the products have been reported to recover only 75% of the cost of productivity. Under such type of income levels the dairy farmers possess no capital to invest in strengthening their dairy production operations.

The peri-urban dairy farming systems in the South East Asia have been reviewed, with a special focus on Pakistan as described earlier (Qureshi, 2008). As a part of the agricultural production system, dairy farming is a prerequisite to alleviation of poverty. It supplements other income generating activities to eradicate poverty and creates adequate opportunities for enhanced rural and peri-urban employment, income generation and economical access to food. The horizontal expansion in dairy farming is still in progress. The increasing human population of the urban areas, the rising income levels and the awareness about need of animal proteins in human diet, has resulted in increasing demand for milk and meat. This demand for food items and the rising levels of prices, calls for expansion of dairy and livestock industry.

Dairy farms provide a unique environment for development of special social norms. The dairy farms are located in the peri-urban areas of the major cities to meet the demand for milk of the urban populations. The farms are established without scientific planning for construction of buildings, roads, water supply and drainage and other requirements of the people and the dairy operations. The farmers are taking care of 57 million dairy animals (cattle and buffaloes) in Pakistan, the approximate value being Pak Rs.1.5 trillion and contributing to the national economy to the tune of Pak Rs.1.2 trillion per annum (US$ 1 = Pak Rs.85). But they do not get the desired contribution from the society.

The living standards of the farmers are low due to low profitability of the farms. The high and non-regulated cost of inputs and state-controlled lower price of the products make the profit margin low. Lack of state-subsidy and hostile marketing system damage these enterprises. Under such circumstances the living standard of the dairy farmers is definitely deteriorated. The farmers have little chance to send their children to better educational institutions, which usually are expensive. The children discontinue their education after passing the primary schools. A so-called self employment is provided to the children by their parents at the dairy farms and their potential for better contribution to the society is wasted.

The prospects for local dairy production have recently become more favorable in the developing countries (FAO, 1995), following the reduction of milk production subsidies in western developed countries and the introduction of more realistic exchange rates under structural adjustment programs. These recent changes have provided many developing countries with the opportunity to develop their own milk industries, primarily through small-scale production, which will have a major impact on different levels of cash income. The document recommended greater attention for the provision of facilities and credit that benefit the small-scale producer, rather than major investments in institutions and facilities, such as big slaughterhouses, dairy plants and feed-mills, which are usually oversized, overstaffed and over-equipped.

Within the industry there are differing views on the way government policy should be used to assist the industry (Wynn, 2006). There is a conflict between the need to generate higher returns for milk producers and a desire to maintain low retail prices for milk that relates to alleviation of poverty for the urban poor. These pressures are evident in other industries (e.g. sugar, wheat) and have led to the implementation of support policies. Pakistan lacks appropriate resources and hence there is no domestic support polices that raise returns for domestic milk producers. However, in time political pressures may emerge to introduce price regulations in order to stimulate increased milk production. Policy makers do not appear to be aware of the implications of these sorts of policy developments for future industry development. The author (Wynn, 2006) emphasized that there would be some value in making key policy makers of Pakistan aware of the mistakes with previous Australian dairy policies.

2. The issues

The peri-urban dairy farms face several challenges impeding its productivity, profitability and sustainability. Resultantly, these farms are at risk of elimination as they can not compete with similar business activities in the peri-urban areas. Poor reproductive efficiency has been reported in buffaloes, associated with their inherent lower fertility, a smaller population of recruitable follicles and problematic estrus detection. Breeding efficiency of the animals is low and the physiological and management factors are responsible for lowered reproductive efficiency.

The dairy buffaloes maintained at the peri-urban dairy farms in the region are primarily kept for milk yield and the rebreeding finds a lower priority in management decisions. Rather the rebreeding and conception are considered undesirable in most of the cases. One of the factors behind the low priority for rebreeding practices has been the higher cost of milk, labor and land cost in the peri-urban regions. It leads to lack of availability of calves in appropriate age groups for replacement of lactating females and breeding sires. Therefore, breeding efficiency has been reviewed in the following paragraphs.

The feeding practice at the peri-urban dairy farms is haphazard, ignoring the nutrient requirements of the various animals' groups. Same scale feeding is practiced which is associated with overfeeding of the low-yielding animals and under-feeding of the higher-yielders. It may lead to malnutrition, affecting the body functions like milk synthesis and growth and may not support the good health conditions of the animals. The quality of feed ingredients is also low. Improper storage of feed items expose them to contamination with aflatoxins which may lead to poor productivity, fertility and health status. In most of the cases the feed stocks are adulterated with undesirable items. Reproductive cyclicity and a successful gestation requires an optimum status of nutrients availability. Any deviation from the optimum range will result in partial or complete cessation of reproductive cyclicity. Therefore, the reproduction-nutrition interaction has been reviewed in this chapter.

Nutritional status of the animal is assessed through the intake of nutrients and its utilization for various body functions. Intake and utilization of nutrients may be assessed through body condition score (BCS). BCS has been used extensively in research studies, since long. It may be recorded a bi-weekly intervals and the changes may be used for predicting a normal parturition, postpartum lactation and post-insemination conception of a dairy cow and buffalo. The author has investigated this parameter extensively for its association with reproductive cyclicity, milk yield and fatty acid profiles and the quality of drinking water.

Milk yield in the dairy buffaloes decline after getting pregnant; which has been a matter of concern for the peri-urban dairy farmers. This may be because of the already stated reason that the animals are provided same-scale feeding. Under such conditions the pregnant animals may get escaped from getting pregnancy supplement rations. After pregnancy the conception gets higher priority over lactation in buffalo and a feed deficiency may lead to decline in milk yield. In addition, the hormonal changes during pregnancy in dairy buffaloes may lead to intervention in the lactation process. The receptors for lactation hormones may intercept with the reproductive hormones, causing a decline in milk yield. These phenomenons were investigated by our group and the findings are reviewed in this chapter.

The labor cost at the peri-urban dairy farms is higher as compared to rural areas. Milkmen are hired for milking the lactating buffaloes, instead of using the milking machines. The milkmen are conscious about their time spent at the farm. They try to get rid of the milking duty as soon as possible. As a routine practice, the young calf is used for milk let down through suckling stimulus. However, as the calf keeping is considered expensive, such calves are disposed off at an earlier age. Under such conditions the milkmen use oxytocin for the purpose of milk let down. It poses a public health problem as well an undesirable offence over the reproductive endocrinology of the dairy buffaloes. This phenomenon has been reviewed in this chapter.

The milk composition, especially the fatty acids vary with the changing nutritional and physiological states of an animal. The human consumers expect healthy milk fatty acids from the dairy animals. Unsaturated fatty acids are considered cardio-protective in nature and such fatty acids may arise as a result of a specific set of management and physiological conditions. This phenomenon has been discussed in this chapter.

The quality of drinking water used at the peri-urban dairy farms may affect the productivity, fertility and milk quality of animals. The heavy metals and toxic materials present in the water supply network need appropriate monitoring to assess its fitness for dairy animals' use. The farms are supplied form the tube-well in these areas, where the water table is usually higher and is exposed to the pollutants from the waste depots. Heavy metal contents of the drinking water pass on to the human diet which may affect the health status of the consumers adversely.

3. Breeding efficiency and associated factors

The major causes associated with the under-developed buffalo farms have been identified as: i) calf losses, irregular breeding, imbalanced feeding; ii) unfavorable loans and; iii) a hostile marketing system. For commercial buffalo herds of Pakistan, those three causes lead to annual losses of US$ 18 billions (Qureshi, 2000). The normal breeding season in the Indo-Pakistan sub-continent begins in August and coincides with the feeding of non-leguminous fodders, such as sorghum and maize. Poor reproductive efficiency has been reported in buffaloes, associated with their inherent lower fertility, a smaller population of recruitable follicles at any given time than the ovary of the cow (89% fewer at birth), and a problematic estrus detection (Drost, 2007).

Breeding efficiency is the percentage performance based on the number of parturitions within the period from first to last lactation in relation to the standard and actual age at first calving. Breeding efficiency (BE) of dairy buffaloes and potential contributing factors were studied by our group (Sohail et al., 2009). A total of 5033 reproductive and productive records from the year 1985 through 2004 were utilized for this purpose. We reported BE (72.24%) which seemed to be sufficient under the management conditions of large sized state farms. However, this trait showed a persistent downward trend over the year (78.20 to 71.38%) during the period from 1991-92 to 2003-04. This may have occurred due to the effects of inbreeding or deteriorating or a deteriorating management conditions at these farms. A higher birth weight of a dairy buffalo was found to be the most significant contributing factor to BE and it also supported an earlier age at puberty (AAP), age at first calving (AFC) and a better lactation yield (LY).

Increase in BE was associated with an increase in lactation yield, confirming the earlier findings of Qureshi et al. (2007) who reported that high yielding buffaloes were also efficient in fertility. However, after a certain level BE was reduced with further increase in production, indicating high priority of nutrients partitioning towards production than reproduction.

An average breeding efficiency of 64.0 % was reported in Nili-Ravi buffaloes (Bashir et al. (2007). Herd and year were found as important source of variation for breeding efficiency while season of calving or age at first calving had no effect in the reported study. This indicated that all the herds used by Sohail et al. (2009) were better managed in terms of reproduction and the animals were fed properly. They also concluded that optimum age at first calving favored breeding efficiency because the reproductive organs and neuro-endocrine system developed sufficiently to support optimum reproductive cycle and conception. Further increase in age at first calving may be the effect of aging leading to a lower reproductive performance.

In a previous study Khan et al. (2008) reported that the decline in milk yield with the onset of pregnancy was prevented by an increase in maturity of dairy buffaloes. It was suggested that the increasing maturity up to some extent results in maintenance of better reproductive performance.

4. Reproduction-nutrition relationship

Under the conventional farming system in the region, diet is not formulated according to the requirements of individual animals, resulting in decreased production and poor health and reproduction (Qureshi, 1995; Qureshi et al., 1999, 2002). The lactating buffaloes are fed green fodders plus concentrate feeds but dry and pregnant buffaloes are considered uneconomical and are mostly fed only low quality green fodders Consequently, animals getting adequate nutrients have higher body condition scores, which enable them to produce higher quantities of milk and they also are bred earlier.

In the absence of any ration formulation practice under this production system, excessive or deficient intake of some nutrients may decrease reproductive performance. Few investigations have studied the association of intake of protein and energy and the resulting serum urea levels and body condition score with reproductive performance in Nili-Ravi buffaloes, under field conditions in the northern Pakistan.

The buffaloes calving during the normal breeding season (NBS, August to January) (p<0.01) had a significantly shorter postpartum estrus interval (55.9 vs 91.2 days) than those calving during the low breeding season (LBS, February to July, Qureshi et al., 1999). Milk progesterone levels (MPL) in the LBS remained lower than the NBS (p<0.01). Shortest postpartum ovulation interval was noted during autumn (August to October), followed by winter (November to January), summer (May to July) and spring (February to April). The incidence of silent ovulations was higher during LBS than NBS (70.6% versus 29.4%). In autumn there was minimum intake of crude protein (CPI) and maximum intake of metabolizable energy (MEI, p<0.01). Calcium intake was higher in NBS than LBS calving buffaloes (p<0.01). Phosphorus, copper and magnesium intake was lower (p<0.05) and zinc intake was higher (p<0.01) in autumn (August to October). It was concluded that onset of breeding season was associated with increasing MEI and decreasing CPI and minerals intake.

Qureshi et al. (2002) reported that crude protein intake (CPI) averaged 1.8 ± 0.5 kg/day, ranged from 0.95 to 2.6 kg/day, varied significantly between seasons, and was positively correlated with serum urea levels (r=0.22, p<0.01). Degradable protein intake (DPI) was 1.32±0.01 kg/day and was significantly affected by season (p<0.01) with summer > spring > autumn / winter. There was a positive correlation between CPI and duration of placenta expulsion (r=0.21, p<0.01), postpartum estrus interval (PEI) (r=0.08, p<0.05) and postpartum ovulation interval (POI) (r=0.21, p<0.01). Excess intake of crude protein (excess to requirements) was lower in animals which expressed oestrus than those which remained anoestrus (p<0.05). The difference was marked from one month pre-partum to four months postpartum. Excess CPI delayed the duration of placental expulsion (r=0.37, p<0.01). The dietary ratio of crude protein/metabolizable energy (g CP/MJ ME) consumed by the buffaloes during the pre-partum and postpartum periods shows that the animals resuming to oestrus had a narrow and almost constant CP/ME ratio (11.9 to 12.2 g/MJ), while the anoestrus animals had a widely fluctuating ratio, ranging from 10.7 to 13.1 g/MJ. CP/ME ratio was related positively with POI (r=0.15, p<0.01).

In the same study, energy intake showed an overall mean value of 174.5±1.1 MJ/day, ranging from 84.5 to 252.8 MJ/day. ME intake was lower in cows that calved in the NBS than those that calved in LBS (p<0.01). Intakes of ME were similar during winter and spring but lower than those in summer or autumn, which was highest (p<0.01). Increasing energy intake increased BCS (r=0.16) and duration of expulsion of placenta (r=0.19) and discharge of lochia (r=0.24) but decreased POI (r=-0.27, p<0.01). Prepartum ME intake was higher in animals observed in oestrous than those remaining anoestrous (177.2 vs 155.9 MJ/day, p<0.05). Prepartum metabolizable energy intake above requirement (MEAR) was also higher in animals returning to oestrus than the anoestrus ones (p<0.01). Higher MEAR during prepartum period was accompanied by a higher BCS in animals which came into oestrus. The animals coming into estrus within 75 days postpartum showed a moderate intake of ME as compared to those coming into estrus after 75 days postpartum, which showed either deficiency or excess of ME intake (p<0.01).

5. Body condition score

Body condition score (BCS) reflects the overall energy status of the body, depending upon the intake of nutrients and their utilization for milk yield, growth and maintenance. Osoro and Wright (1992) and De Rouen et al. (1994), concluded that BCS at calving significantly affected postpartum reproductive performance in cows. O'Rourke et al. (1991) reported that cows with BCS of ≥8 had a conception rate 33% higher than those with score ≤5 (scale 3-9). Buffaloes in poor BCS had inactive ovaries and long postpartum anoestrus periods (Jainudeen and Wahab, 1987). Bhalaru et al. (1987) reported that conception rates were significantly higher (88.3%) for buffaloes with moderate BCS (2.5 to 3.5) than for females scoring 1 to 2 (65.8%) or 4 to 5 (70.8%). The reason for low reproductive performance in the animals with low BCS was perhaps the non-availability of nutrients for reproduction, being the third candidate in partitioning of nutrients, after health and milk production. In the fat animals the low reproductive performance was perhaps due to abnormal physiological functions during the estrous cycle.

The BCS of the 51 buffaloes varied from 1.0 to 4.0 during the late prepartum and early postpartum periods (Qureshi, 2002). The buffaloes were grouped into three categories i.e. poor (BCS 1.0 to 2.0), moderate (BCS 2.5) and good (3.0 to 4.0). None of the buffaloes in the NBS calving group had poor BCS. Conversely, none of the LBS calving buffaloes had good BCS, which is consistent with the higher intake of metabolizable energy (p<0.01) during summer and autumn. BCS was significantly affected by the period of calving and season of the year. Animals calving during the NBS, had significantly higher BCS as compared to those calving during the LBS (2.82 vs 2.60), which is the probable cause of their better reproductive efficiency (animals coming into estrus within 45 days had higher BCS than those coming into estrus after 45 days). Body condition score was higher (2.97) prepartum than during the first two months postpartum (2.65, p<0.01). Placenta expulsion duration (r=-0.17, p<0.05) and PEI (r=-0.20) negatively correlated with BCS. In buffaloes resuming estrus, BCS was consistently higher than in those failing to resume oestrous activity.

6. Post-conception milk yield decline and progesterone stress

The decline in milk production of buffaloes after conception was investigated in a series of studies (Qureshi et al., 2007; Khan et al., 2009). The experimental buffaloes were selected in North-West Frontier Province (NWFP) of Pakistan. Complete milk yield records for 48 weeks of lactation were obtained for 465 pregnant and 179 non-pregnant buffaloes. Three different models were used to identify factors affecting milk yield reduction due to pregnancy.

Model-1, involved gestation stage in months was fitted using all the 30912 records. Then a reduced model-2 was fitted excluding gestation stage. The reduction in milk yield due to pregnancy was worked out relative to their non-pregnant counterparts. Only the data for lactation weeks after conception were analyzed to find out the milk reduction. Model 3 was used to analyze the factors affecting milk yield reduction due to pregnancy:

$$Y = L + P + LxP + LW + GM + E \qquad (1)$$

$$Y = L + P + LxP + LW + E \qquad (2)$$

$$RY = L + CS + P + LW + GM + E \qquad (3)$$

Where Y is milk yield, RY is the reduction in milk yield; L is location, P is parity, LW is lactation week, GM is gestation month, CS is conception season E is the residual term associated with the model. The milk records were divided in three subsets: lactation weeks 11-28 (early lactation); 29-36 (mid lactation); and 37-48 (late lactation) and analyzed separately to estimate the effect of pregnancy at different lactation stages.

Model 4 was used to modulate milk yield reduction with the onset of pregnancy at medium sized private farms comprising lactation records of 40 buffaloes. The data indicate that post-conception reduction in milk yield occurred earliest in those that conceived during 29-36 or 37-48 weeks of lactation, respectively. A noticeable reduction in milk yield was found during the 3rd, 5th or 6th month of pregnancy in the animals conceiving at earlier, mid or later stages of lactation. Initially the milk yield in pregnant animals increased up to 2 months post-conception and then decreased at an almost constant rate. The reduction was visible after 5th week post-conception. The decline in milk with advancing pregnancy was slight up to a point which we declared as joining point; thereafter the decline was much greater.

The onset of pregnancy may be associated with hormonal changes leading to the decline in milk yield of buffaloes. To investigate this, forty lactating buffaloes from 1st to 23rd weeks post-conception were selected in a study (Khan et al., 2009). The animals were assigned to three treatments: pregnant with traditional ration, pregnant with supplemented ration, non-pregnant with traditional ration and grouped according to milk yield: HMY, 66 to 75 l/week, n=12; MMY, 56 to 65 l/week, n=16; LMY, 46 to 55 l/week, n=12).

Milk samples (10 ml each) collected from the experimental animals were utilized for composition determination. Milk contents were determined through ultrasonic milk analyzer (model Ekomilk Total Ultrasonic Milk Analyser, Bullteh 2000, Stara Zagora, Bulgharia), using manufacturer's instruction, as already reported (Khan et al., 2007).

Milk progesterone concentrations were measured by enzyme-immunoassay (EIA). Group means were compared and correlation analysis was conducted. Progesterone concentrations increased in almost similar pattern with the advancing weeks post-conception. The high and low yielder showed greater progesterone concentrations in the supplemented than the animals on traditional ration (P<0.001) than the moderate yielders. Progesterone concentrations correlated positively with fat (%), negatively with milk yield, protein (%) and lactose (%) with milk fat content and negatively with protein content and lactose content.

The decline in milk yield became drastic when progesterone concentrations rose above 6.44 ng/ml. The pregnant animals on traditional ration exhibited a sharper decline in milk yield with the increasing progesterone concentrations as compared to pregnant animals with supplemented ration. It was concluded that concentrate supplementation induced an increase in progesterone levels. Progesterone concentrations and milk yield showed an inverse relationship.

7. Calf suckling and use of oxytocin

Calf suckling and oxytocin injections are commonly used for pre-milking stimulus in dairy buffaloes under field conditions. A study was conducted to investigate effect of these treatments on reproductive performance (Qureshi and Ahmad, 2008). We found a lower reproductive efficiency of dairy buffaloes under the peri-urban farming system reflected by ovarian cyclicity in 68.63% buffaloes within 150 days postpartum and silent estrus in 51.5% of the cases. Increasing suckling duration and use of oxytocin extended the postpartum ovulation interval (POI), however it was shortest in buffaloes suckled for one month. Fat-corrected milk (FCM) was significantly higher in estrus group as compared to anestrus one, during the first two months postpartum (15.09 versus 13.56 kg/day, $P < 0.01$). The moderate yielders had shortest postpartum uterine involution ($P < 0.01$) and estrus intervals (PEI, $P < 0.05$) and highest conception rate ($P < 0.01$). It was suggested that the high yielding buffaloes also manifested better reproductive cyclicity.

As the calves were allowed to suckle two times daily, it probably resulted in adverse effect on resumption of postpartum ovarian activity and increased PEI and POI. There was also a decrease in the duration of lochia discharge, which might have been due to sustained uterine contractions caused by oxytocin released in response to suckling. In agreement with this study, ovarian cyclicity was re-established earlier in non-suckled than suckled river, as well as swamp, buffaloes (El-Fouly et al., 1976; El-Fadaly, 1980; Jainudeen et al., 1984). In Nili-Ravi buffaloes, Usmani et al. (1985) reported that postpartum intervals to uterine involution, resumption of follicular development, first rise in milk progesterone, first palpable corpus luteum formation and first oestrus were longer for limited-suckled buffaloes than for non- suckled buffaloes.

For stimulation of milk letdown, oxytocin is released in response to tactile teat stimulation. The application of a fixed pre-stimulation of 30 to 60 s before milking has been recommended to ensure immediate and continuous milk flow after the start of milking (Rasmussen et al., 1992). However, recent investigations demonstrated the importance of the udder fill on the course of milk ejection (Dzidic et al., 2004). Therefore, pre-stimulation time according to the degree of udder fill in individual cows may improve the milking performance. According to previous results, no cisternal milk was available when milking started without pre-stimulation at a low udder fill (Bruckmaier and Hilger, 2001). At moderate udder fill, cisternal milk was immediately available for milking (Bruckmaier and Blum, 1996) and the alveolar milk ejection started about 70 s after the start of pre-stimulation, as indicated by the second rise of milk flow (Bruckmaier and Hilger, 2001). In full udders, the amount of cisternal milk was further enhanced (Pfeilsticker et al., 1996) and the lag time until the start of the alveolar milk ejection was further reduced (Bruckmaier and Hilger, 2001).

Milk yield in buffalo is lower than cattle because of the little progress made in its conversion to a specialized dairy animal. So the udder is not full and a stronger pre-milking stimulus is required for milk let down. In the present study, representative of the conventional buffalo farming, the suckling (twice a day for five minutes each time) was used or it was replaced by oxytocin in case of death of the calf. Qureshi and Ahmad (2008) suggested that increasing suckling duration and use of oxytocin delayed POI, however, POI was shortest in buffaloes suckled for one month.

8. Milk fatty acids

Buffaloes usually maintain higher body condition and do not produce milk at the cost of their own body reserves under tropical conditions. The mobilization of body reserves for fulfilling the demands of lactation has been extensively studied in dairy cows while limited work is available on this aspect in dairy buffaloes. Therefore, a study was conducted to examine variations in milk fatty acid profiles with body condition in Nili-Ravi buffaloes (Qureshi et al., 2010). We suggested that Nili-Ravi dairy buffaloes produce similar milk to dairy cows regarding content of cardioprotective fatty acids, with the highest concentration of C18:1 cis-9. Two HCFA (hyper-cholesterimic fatty acids, C12:0 and C14:0) were associated with higher body condition. Buffaloes with moderate body condition yielded milk containing healthier fatty acids (the unsaturated fatty acids).

The HCFA (C12:0, C14:0 and C16:0) found in this study on Nili-Ravi buffaloes were considerably lower and cardioprotective fatty acids (C18:1 and C18:2 and C18:3) level were higher than the Bulgharian Murrah buffaloes as reported by Mihaylova and Peeva (2007). They found that total amount of SFAs (saturated fatty acids) were 72.15% (varying from 64.92 to 77.60%), PUFA (poly unsaturated fatty acids) 3.15% and the HCFA were 43.62%. Our values were in close agreement with Fernandes et al. (2007) who reported that the total SFAs, MUFA (mono unsaturated fatty acids) and PUFA in Murrah buffaloes in Brazil were 65.04%, 31.68% and 3.28% respectively and the HCFA varied from 32.48 to 42.90%. Our values for dairy buffaloes were not much different from dairy cows where the SFAs varied from 60 to 65% and UFAs (unsaturatred fatty acids) 35 to 40% of the total fatty acids (Lock and Shinfield, 2004).

Talpur et al. (2008) compared milk fatty acid composition of Nili-Ravi and Kundi buffaloes in Sindh province of Pakistan. The average SFAs were; 66.96 g/100 g and 69.09 g/100 g; MUFA 27.62 and 25.20 g/100 g; PUFA 2.77 and 2.76 g/100 g and HCFA 42.8 and 46.54 g/100 g, of total fatty acids for Kundi and Nili-Ravi breed respectively. It appears that the cardioprotective quality of milk from Nili-Ravi buffaloes is almost similar to dairy cows and Brazilian and higher than in milk from Bulgharian Murrah buffaloes (Talpur et al., 2008).

The opposite pattern of BCS and UFAs concentration in milk fat (Qureshi, et al., 2010) in dairy buffaloes was probably due to lipolysis. In bovine adipose tissue, C18:1 cis-9, C16:0, and C18:0 account for nearly 90% of fatty acids in molar proportions (Christie, 1981) and body fat mobilization would probably increase direct accumulation of these fatty acids into milk fat. In addition, desaturation of stearic acid occurs in the intestinal epithelium and mammary tissues (Enoch et al., 1976). Some 40-50% of C18:1 cis-9 in milk fat is formed from C18:0 in the mammary gland via desaturase (Chilliard et al., 2000). The net outcome of all these processes is the higher level of UFA and more specifically the C18:1 concentration in milk fat.

9. Drinking water quality

The heavy metal content of human diets can adversely affect health status of the consumers. One source of animal feeds include drinking water. Free access to drinking water favorably affected fertility of buffaloes. A study was therefore conducted to investigate the mineral contents of milk (Qureshi and Khan, 2011). The study concluded that the drinking water was the major source of heavy metal contents (Cr, Cd, Pb) in milk produced in the peri-urban buffalo dairy farms in Peshawar. Peshawar was below the desirable limits of the beneficial inorganic minerals (Mg, Zn and Fe) whereas the toxic heavy metals (Cd, Cr and Pb) content of heavy metals were excessive in drinking water. Levels of the heavy metals Cd, Cr and Pb through milk alone was much more than the total daily intake of these heavy metals from all sources. Free access to drinking water effected milk yield, body condition and fertility favorably. The higher intake of lead was associated with depressed milk in addition to enhanced level of this element in the milk.

The study revealed that drinking water used at urban and peri urban dairy farms in Peshawar are below the maximum allowable intake (MAC) in essential minerals and the heavy metals are higher. The Ca and Mn concentration in drinking water were 63.6% and 40% above the MAC of Pakistan Council of Research on Water Resources (PCRWR). While other essential minerals Mg, Fe, Zn and Cu were 57%, 57%, 99.6% and 98% respectively below the MAC of PCRWR. The heavy metals Cd, Cr and Pb levels were above the MAC of PCRWR (700%, 1800% and 1240% respectively) and were also above the standards fixed for livestock drinking water by NRC (1500%, 850% and 4366.7% respectively). The concentrations of the three heavy metal were far above the standards of WHO (2566.6%, 1800% and 6600%).

Although, food is the major source of mineral nutrients in the diet, drinking water can contribute variable fractions of the total intake (WHO, 2004). The magnesium content of water is variable and depends on the region of its source and its manner of storage. 'Hard' water has a higher concentration of magnesium salts (COMA, 1991). The NIRS (National Inorganic Radionuclide Survey, WHO 2004) study provided data on many cationic inorganic ions in water including calcium and magnesium. They found that mean concentrations (49 mg/L for calcium and 16 mg/L for magnesium) were low compared to their dietary requirements. The 90th percentile values (97 mg/L for calcium and 36 mg/L for magnesium) would make more substantial contributions to dietary intake (WHO, 2004). The findings of present study showed a substantial contribution of Ca (327 mg/L) but a low contribution of Mg (24 mg/L) to dietary intake.

Somasundaram et al., (2005) found that technological progress, various industrial activities and increased roadway traffic have caused a significant increase in environmental contamination. Ubiquitous presence of some metal pollutants, especially cadmium (Cd), chromium (Cr) and lead (Pb), facilitates their entry into the animal food chain and thus increases the possibility of inducing toxic effects in humans and animals. Land application of sewage sludge, sewage water and industrial wastes gradually increases the toxic metals in the soil environment which are increasingly taken by plants and subsequently transferred into the food chain potentially causing severe damage to both animal and human health.

Due to lack of any strict legislation or its implementation for the proper disposal of industrial wastes in Pakistan these wastes become mixed with drinking water channels. In the present study the high level of heavy metals in the drinking water may be due to sewage water contamination and industrial pollution of livestock drinking water and the farmers pay no attention for the provision of clean drinking water to the livestock.

A comparison of the daily intake of minerals and heavy metals through milk to the maximum of allowable intake from all sources was done by the Expert group on vitamins and minerals (2003). Based upon the per capita milk consumption in the country calculated on the basis of Economic Survey (2009-10), the daily intakes of Ca, Mg, Fe, Zn, Cu, Mn, Cd, Cr and Pb through milk were 650.67, 78.27, 32.91, 10.42, 0.50, 0.46, 1.64, 30.84 and 9.27 (mg/day) respectively. The contributions of milk in the maximum daily intake of essential minerals were 14.45% (Ca), 5.5% (Mg), 74.79% (Fe), 13.53% (Zn), 4.54% (Cu) and 3.06% (Mn). The daily intake of toxic heavy metals Cd, Cr and Pb through milk were 1952.38%, 3896% and 1865.19% above the maximum allowable intake of heavy metals from all sources fixed by EVM (Expert Group On Vitamins And Minerals 2003). The milk produced in Peshawar contributes a slight amount of essential minerals to the total daily intake but a considerable amount of toxic heavy metals reflecting poor quality of milk.

In the present study iron significantly ($P<0.05$) affected milk yield, body condition score and services per conception. Iron in drinking water is probably the most frequent and important contaminant in dairy cattle. Whereas, iron deficiency in adult cattle is very rare because of abundant iron (Fe +3, ferric iron) in feedstuffs, excess total iron intake can be a problem; especially when drinking water contains high iron concentrations. Iron concentrations in drinking water of greater than 0.3 ppm are considered a risk for human health, and are a concern for dairy cattle health and performance. The first concern is that high iron in drinking water may reduce the palatability (acceptability) and therefore amount and rate of water intake. Also, formation of slime in plumbing by iron-loving bacteria may affect water intake and even the rate and volume of water flow through pipes.

The predominant chemical form of iron in drinking water is the ferric (Fe +3) form. The ferrous form is very soluble in water compared with the highly insoluble ferric (Fe) form present in feed sources. Highly soluble iron can interfere with the absorption of copper and zinc. The ferritin system in cells in the intestinal wall normally helps control the risk of iron toxicity in animals by controlling iron absorption. However, highly soluble ferrous iron can be readily absorbed by passing between cells; thus escaping the normal cellular regulation.

Once in the body, the transferrin and lactoferrin systems normally bind iron in blood and tissues to control its reactivity. These systems also help control risk of toxicity under normal conditions.

However, when excess, highly water-soluble iron in drinking water is absorbed there is an overload systemically within the animal and all can not be bound. Deleterious consequences of excess free iron include abundant and excessive amounts of reactive oxygen species (e.g., peroxides) that cause oxidative stress. Oxidative stress damages cell membrane structure, functions, and perturbs otherwise normal biochemical reactions. Consequences of iron toxicity and heightened oxidative stress that are magnified in transition and fresh cows include: compromised immune function, increased fresh cow mastitis and metritis, greater incidence of retained fetal membranes as well as diarrhea, sub-normal feed intake, decreased growth, and impaired milk yield. Excess iron (greater than 0.3 ppm) in drinking water is much more absorbable and available than iron from feedstuffs, and thus present a greater risk for causing iron toxicity (Beede 2006).

The number of services per conception (SPC) was increased with increasing Mn, showing its adverse effect. Expert Group on Vitamins and Minerals (2003) reported that manganese has low acute toxicity but has neurotoxic efffects on fertility. Amal (2003) concluded that long exposure of animals to lead affect the reproductive efficiency in the form of lower conception rates, as well as increased incidence of still births, SPC and abortions. The mothers exposed to Pb suffered reduced postnatal viability and lower birth weight.

10. Conclusion

Based upon the above review it may be concluded that buffaloes kept under the peri-urban dairy farming in Pakistan face huge challenges of survival due to poor physiological support to productivity and low socio-economic status of the farmers resulting in poor management. The breeding efficiency of buffaloes has been reported to show a persistent downward trend (78.20 to 71.38%) during the last decade which may have occurred due to the effects of inbreeding or a deteriorating management conditions at these farms.

Buffaloes show a seasonal breeding with lower milk progesterone levels and higher rates of silent ovulation during spring and summer. The breeding season commences during autumn with lowering intake of crude protein and increasing intake of metabolizable energy. Calcium and zinc intake is higher and phosphorus, copper and magnesium intake is lower during autumn. The onset of breeding season was found to be associated with increasing intake of metabolizable energy and decreasing intake of crude protein. Post-conception decline has been associated with onset of pregnancy and rising levels of milk progesterone beyond certain levels. However, the decline may be combated through feed supplementation. The traditional use of oxytocin was found to decrease reproductive efficiency of dairy buffaloes under the peri-urban farming system reflected by ovarian cyclicity in 68.63% buffaloes within 150 days postpartum and silent estrus in 51.5% of the cases. Increasing suckling duration and use of oxytocin extended the postpartum ovulation interval (POI), however it was shortest in buffaloes suckled for one month.

Nili-Ravi dairy buffaloes produce similar milk to dairy cows regarding content of cardioprotective fatty acids, with the highest concentration of C18:1 cis-9. Buffaloes with moderate body condition yielded milk containing healthier fatty acids (the unsaturated fatty acids). Drinking water used at urban and peri urban dairy farms in Peshawar are below the maximum allowable intake (MAC) in essential minerals and the heavy metals are higher. The higher intake of lead was associated with depressed milk in addition to enhanced level of this element in the milk.

Author details

M. Subhan Qureshi
Agricultural Unviersity, Peshawar, Pakistan

11. References

Amal, RAH, 2003. Clinicopathological studies on the effect of chronic exposure to lead on different reproductive phases in female baladi goats. Ph. D. Vet. Sci, Fac. Vet. Med. Cairo University, Egypt.

Bashir, M. K., M. S. Khan, S. A. Bhatti and A. Iqbal. 2007. Lifetime performance of Nili-Ravi buffaloes in Pakistan. Asian-Aust. J. Anim. Sci. 20(5):661-668.

Bhalaru, S. S., M. S. Tiwana and N. Singh. 1987. Effect of body condition at calving, on subsequent reproductive performance in buffaloes. Indian J. Anim. Sci. 57:33-36.

Bruckmaier, R.M., Blum, J.W., 1996. Simultaneous recording of oxytocin release, milk ejection and milk flow during milking of dairy cows with and without prestimulation. J. Dairy Res. 63, 201–208.

Bruckmaier, R.M., Hilger, M., 2001. Milk ejection in dairy cows at different degrees of udder filling. J. Dairy Res. 68, 369–376.

Chillard, Y., A. Ferlay, R. M. Mansbridge and M. Doreau. 2000. Ruminant milk fat plasticity: nutritional control of saturated, polyunsaturated, trans and conjugated fatty acids. Ann. Zootech. 49:181-205.

Christie, W. W. 1981. The composition, structure and function of lipids in the tissues of ruminant animals. Pages 95-191 in Lipid Metabolism in Ruminant Animals (Ed. W. W. Christie). Pergamon Press, Oxford, UK.

COMA 1991. Report on Dietary Reference Values for Food Energy and Nutrients for the United Kingdom; Committee on Medical Aspects of Food Policy (COMA). Department of Health and Social, Security, London.

Beede, D.K, 1993. Water Nutrition And Quality For Dairy Cattle. Western large herd management conference las vegas nevada.

DeRouen, S. M., D. E. Franke, D. G. Morrison, W. E. Yatt, D. F.Coombs, T. W. White, P. E. Humes and B. B. Greene. 1994.Prepartum body condition and weight influences on reproductive performance of first-calf beef cows. J. Anim. Sci. 72:1119-1125.

Drost, A., 2007. Advanced reproductive technology in the water buffalo. Theriogenology, 68 (3): 450-453.

Dzidic, A., Weiss, D., Bruckmaier, R.M., 2004. Oxytocin release, milk ejection and milking characteristics in a single stall automatic milking system. Livest. Prod. Sci. 86, 61–68.

Economic Survey of Pakistan, 2009-10. Economic Survey. Government of Pakistan, Economic Advisor's Wing, Islamabad.

El-Fadaly, M.A., 1980. Effect of suckling and milking on breeding efficiency of buffaloes. II. First postpartum estrus. Vet. Med. J. Egypt 28, 399–404.

El-Fouly, M.A., Kotby, E.A., El-Sobhy, A.E., 1976. Postpartum ovarian activity in suckled and milked buffaloes. Theriogenology 5, 69–79.

Enoch, H. G., A. Catala and P. Strittmatter. 1976. Mechanism of rat liver microsomal stearoyl-CoA desaturase. J. Biol. Chem. 251:5095-5103.

Expert Group on Vitamins and Minerals, 2003. Safe upper limits for vitamins and minerals. Crown copyright Published by Food Standards Agency. ISBN 1904026-11-7

FAO, 1995. Livestock - a driving force for food security and sustainable development. World Animal Review, Version 84/85. Viale delle Terme di Caracalla, 00100 Rome, Italy.

Fernandes, S. A. A., W. R. S. Mattos, S. V. Matarazzo, H. Tonhati, M. A. S. Gama and D. P. D. Lanna. 2007. Total fatty acids in Murrah buffaloes milk on commercial farms in Brazil. Ital. J. Anim. Sci. 6(Suppl.2):1063-1066.

Jainudeen, M. R. and S. Wahab. 1987. Postpartum anoestrus in dairy buffalo. Proc. Int. Symp. Milk Buffalo Reprod. Islamabad. pp. 69-77.

Jainudeen, M.R., Sharifuddin, W., Yap, K.C., Abu-Bakar, D., 1984. Postpartum Anestrus in the Suckled Swamp buffalo. In: "Use of Nuclear Techniques to Improve Domestic Buffalo Production in Asia". FAO/IAEA Division of Isotopes, Vienna, Austria.

Khan, S., M.S. Qureshi, N. Ahmad, M. Amjed and M. Younas, 2009. Feed Supplementation Prevents Decline in Milk Progesterone Levels associated with Post-conception Production Stress in Dairy Buffaloes. *Tropical Animal Health and Production*, 41: 1133-1142.

Khan, S., M.S. Qureshi, N. Ahmad, M. Amjed, F.R.Durrani and M. Younas, 2008. Effect of Pregnancy on Lactation Milk Value in Dairy Buffaloes. *Asian Aust. J. Anim. Sci.* 21(4): 523-531.

Lock, A. L. and K. J. Shingfield. 2004. Optimising milk composition. In: Dairying-Using Science to Meet Consumers' Needs (Ed. E. Kebreab, J. Mills and D. E. Beever). Occ. Pub. No. 29, Brit. Soc. Anim. Sci., Nottingham University Press, Loughborough, UK. pp. 107-188.

Mihaylova, G. and T. Peeva. 2007. Trans fatty acids and conjugated linoleic acid in the buffalo milk. Ital. J. Anim. Sci. 6(Suppl. 2):1056-1059.

O'Rourke, P. K., V. J. Doogan, T. H. McCosker and A. R. Eggington. 1991. Prediction of conception rate in extensive beef herds in north-western Australia. 1. Seasonal mating and improved management. Australian J. Exp. Agric. 31:1-7.

Osoro, K. and I. A. Wright. 1992. The effect of body condition, live weight, breed, age, calf performance and calving date on reproductive performance of spring-calving beef cows. J. Anim. Sci. 70:1661-1666.

Pfeilsticker, H.U., Bruckmaier, R.M., Blum, J.W., 1996. Cisternal milk in the dairy cowduring lactation and after preceding teat stimulation. J. Dairy Res. 63, 509–515.

Qureshi M.S., 2000. Productivity problems in livestock in NWFP and their sustainable solutions. *J. Rural Dev. Admin.*, 32 (2): 73-81.

Qureshi, M.S. and R. Khan, 2011. Heavy metals in drinking water of dairy buffaloes. VDM Verlag, Germany.

Qureshi, M.S., 1995. Conventional buffalo farming system in NWFP Pakistan. *Buffalo Bull.*, 14: (2) 38-31. (ISSN 0125-6726). Int. Buffalo Inf. Center, Bangkok.

Qureshi, M.S., 2008. Dairy Farms and farmers. Social norms and training needs. Pak J Agric Sci,45 (2):215-217.

Qureshi, M.S., G. Habib, H.A. Samad, M.M. Siddiqui, N.Ahmad and M. Syed, 2002. Reproduction-nutrition relationship in dairy buffaloes. I. Effect of intake of protein, energy and blood metabolites levels. *Asian-Aust. .J.Anim.Sci*, 15(3): 330-339.

Qureshi, M.S., N.Ahmad 2008. Interaction among calf suckling, use of oxytocin, milk production and reproduction in dairy buffaloes. Animal Reproduction Science, 106: 380-392.

Qureshi, M.S., S. Khan and N. Ahmad, 2007. Pregnancy depresses milk yield in dairy buffaloes. *Ital. J. Anim. Sci.*, 6 (Suppl. 2): 1290-1293.

Qureshi, M.S.; H.A. Samad, G. Habib., R.H. Usmani and M.M. Siddiqui, 1999. Study on factors leading to seasonality of reproduction in dairy buffaloes. I. Nutritional factors. *Asian-Aust. J. Anim. Sci*, 12 (7): 1019-1024.

Qureshi, MS, A Mushtaq, S Khan, G Habib, ZA Swati, 2010. Variation in milk fatty acids composition with body condition in dairy buffaloes *(Bubalus bubalis)*. Asian Aust J Anim Sci, 23: 340-345.

Rasmussen, M.D., Frimer, E.S., Galton, D.M., 1992. The influence of pre-milking teat preparation and attachment delay on milk-yield and milking. J. Dairy Sci. 75, 2131–2141.

Sohail, S.M., M.S. Qureshi, S Khan, G Habib, 2009. Relationship among various production and reproduction Contributors of Breeding Efficiency in Dairy Buffaloes of Pakistan. . Pak J Zool., Sup. No.9, 297-301.

Somasundaram, J, R. Krishnasamy, P Savithri. 2005. Biotransfer of heavy metals in Jersey cows. Indian J Ani Sci 75:1257–1260.

Talpur, F. N., M. I. Bhangera, A. A. Khooharob and G. Zuhra Memon. 2008. Seasonal variation in fatty acid composition of milk from ruminants reared under the traditional feeding system of Sindh, Pakistan. Livest. Sci. 118:166-172.

Usmani, R.H., Ahmad, A., Inskeep, E.K., Dailey, R.A., Levis, P.E., Lewis, G.S., 1985. Uterine involution and postpartum ovarian activity in Nili-Ravi buffaloes. Theriogenology 24, 435–448.

WHO, 2004. Rolling revision of World health organization guideline for drinking water quality, George Hallberg, USA.

Wynn, P., D Harris, R Moss, B Clem, R Sutton and P Doyle, 2006. Report On Dairy Mission To Pakistan. Mission carried out under the auspices of the Australia-Pakistan Agriculture Sector Linkages Program. Government of Punjab, Lahore.

Chapter 13

Effect of Environmental Conditions in Milk Production Under Small-Scale and Semi-Extensive Conditions in Kosovo

Hysen Bytyqi

Additional information is available at the end of the chapter

1. Introduction

During the war in 1998-99, farmers in Kosovo lost 200,000 cattle, or approximately half the national cattle population (Kodderitzsch & Veillerette, 1999). After the war the Food and Agriculture Organization (FAO) and the World Bank implemented a joint cattle-restocking project in Kosovo as part of the Emergency Farm Restocking (FAO/WB/EFR) project, to improve the nutrition and food security of poor households affected by the conflict. The project was started in October 2000 lasting till 2003, importing around 4500 cattle of the Simmental (S), Brown Swiss (BS) and Tyrol Grey (TG) breeds, in three project phases. The cattle were distributed to six municipalities (Deqan, Skenderaj, Gllogoc, Klina, Vushtrri, and Peja) that had suffered the greatest losses to their livestock. The cattle were given to the households who had lost all their animals during the war, such that these could restart livestock production activities. Moreover, the project contributed to upgrade the national cattle population in Kosovo.

1.1. Review of the literature

The literature review is made up of four parts. Part one offers a general description about the rural communities and agro-ecological factors in Kosovo, part two describes cattle production in Kosovo, part three provides brief information about the history of the three imported cattle breeds (S, BS, and TG), while part four describes the characteristics of analysed traits (i.e., milk production, growth rate, service period, non-return rate, body condition score, shape of lactation curve, and milk production efficiency), as well as giving an introduction to estimation of environmental sensitivity from variance components of the three breeds.

1.2. Rural communities and agro-ecological factors in Kosovo

Kosovo is located in southeast of Europe, with about 2 million inhabitants. The land area is 10,887 km², of which about 53% is cultivable. The climate in Kosovo is typically semi-continental, with average annual rainfall of 631 mm and average temperature of 11°C, for the last 20 years. At present, family sizes in rural Kosovo are large, with an average of 9.64 members. Farm sizes are small, about 55% ranging 1 – 3 Ha.

1.3. Cattle production in Kosovo

Cattle production is the most important segment of the economy in rural household in Kosovo. The cattle production includes various breeds and categories of cattle, in total around 320,000 heads. To date, cattle production in Kosovo can be clearly defined as an industry consisting of two sectors, the small-scale cattle farmers (about 95% of the national cattle population), mainly producing for home consumption (but also for market sale during some periods of the year), and commercial farmers (less than 5% of the national cattle population) producing solely for the market. The cattle production of Kosovo on small-scale farms has many characteristics typical for cattle production in developing countries, as it depends almost exclusively on the resources locally available on the farm and basically aims at fulfilling the households' own needs.

Normally, the cattle are kept in the barn from the second half of November until the end of April (winter period). Feeding with fresh-green feedstuffs usually starts in May. However, due to small size of land owned by the private farmers, cultivation of such feedstuffs is often fragmented in 4-5 plots, sometimes located at large walking distances from the house. The grazing period for cattle normally starts after the first harvest of grass, in June and lasts until November, i.e., for approximately six months (summer period). Small diversity of feed sources, limited resources for grazing, and other factors (i.e., small barns, poor hygiene, etc.) make intensive dairy production difficult and often unfavourable in such small-scale farms.

The cattle population in Kosovo mainly consists of dual-purpose breeds of widely different body sizes and production capacities (i.e., Busha, Simmental, Brown Swiss, Tyrol Grey, Holstein and their crosses). Through the aid operations foreign breeds (S, BS, and TG) were brought to Kosovo and allocated to small-scale farms.

1.4. History of imported cattle breeds (S, BS, and TG)

Although these breeds are characterized as dual-purpose, literature suggests clear differences (Simon & Buchenauer, 1993). The Simmental (Fleckvieh) is amongst the oldest and most widely distributed of all cattle breeds in the world. The origin of this breed is the Simme valley in the western part of Switzerland, from where the name derives. Today, this breed accounts for about 40–60 million cattle on all six continents and is known by a variety of names (i.e., "The Fleckvieh" in Germany, Austria and Switzerland; "Pie Rouge", "Montbeliard", and "Abondance" in France and "Pezzata Rossa" in Italy). The Simmental is known as a typical dual-purpose breed that combines characteristics of milk and beef, but with somewhat different emphasis on traits in different sub-populations. Based on data from Germany and Austria, average milk yield in lactation is about 6500 kg (Cattle breeding in Austria, 2003; Rinder production in der Bundesrepublik Deutschland, 2001). Average growth rate of steers/bulls can amount to averages of 1400 g/day. The coloration of Simmental varies from yellowish brown to straw color and dark red, with white markings on the head, brisket, belly and legs. No current information is available to establish whether there are significant genetic and phenotypic differences between Simmental and local red and white cattle in Kosovo).

The Brown Swiss originates from "Braunvieh" cattle in Switzerland, which was well known for the dual-propose characteristics. The Braunvieh cattle were brought to USA between 1869 and 1880, where the cattle-breeding program put more emphasis on milk (Zogg, 1997). Today, this breed has a variety of names (i.e., "Brown Swiss" in USA and Canada; "Braunvieh" in the German speaking countries; "Bruna Alpina" in Italy; "Brunedes Alpes" in France, and "Parda Suizo" in Spain and Latin America). Based on German and Austrian data average milk production in lactation is about 6700 kg (Cattle breeding in Austria, 2003; Rinder production in der Bundesrepublik Deutschland, 2001). Average growth rate of steers can amount to 1000–1200 g/day (Atlas der Nutztierrassen, 1994; Gruter, 1997). The coloration of Brown Swiss consists of various shades of brown ranging from light brown with grey to very dark brown (Gruter, 1997; Herzog, 1997).

Tyrol Grey are grey cattle originating from Tyrol-Austria, where they are used in typical mountain farming under rough conditions. The milk yield of the breed is around 4700 kg in lactation (Cattle Breeding in Austria, 2003). On low feedlevels, the Tyrol Grey steers achieve an average growth rate of 1100 g /day (Atlas der Nutztierrassen, 1994; Frickh, 1999). Tyrol Grey herds have a higher proportion of older cows (> 8 yrs), compared with Simmental and Brown Swiss, respectively (Wallnofer, 1999; Rinder production in der Bundesrepublik Deutschland, 2001), which may indicate good breed characteristics for functional traits.

1.5. Characteristics of analysed traits

After the importation of all three cattle breeds to Kosovo, there was both a need and a unique opportunity to compare the introduced breeds with respect to their suitability to the smallholder management system of Kosovo. Introduction of highly productive breeds of dairy cattle into an extensive environment will often lead to reduced milk production, as well as an increased risk of reproductive and metabolic disorders (Calus & Veerkamp, 2003; Cienfugos-Rivas et al., 1999; Horan et al., 2005). Thus, breeds well adapted to the environmental conditions in Western Europe may be poorly suited to the more extensive Kosovo environment.

One way to identify the most appropriate cattle breed for Kosovo would be to rely on a profit approach, requiring measurement of all traits affecting profit. Alternatively, one could choose to select the breed that is best fitted to the local environment as measured by some indicator traits (e.g., milk production, interval from calving to first insemination, body condition score, shape of lactation curve, estimated milk production efficiency, and environmental sensitivity). The traits that were included in this analysis are productive traits (milk yield and calf growth rate) with major influence on income (Haile-Mariam et al., 2003), fertility performance (i.e., interval from calving to first insemination and nonreturn rate) affecting costs of production (Stott et al., 1999), and body condition score, which is a useful tool for assessing energy status of the cow (Lowman et al., 1976; Edmonson et al., 1989). The breeds were also compared with respect to the shape of the lactation curves, efficiency of milk production, and environmental sensitivity, estimated through heterogeneity of variance components.

In dairy cattle breeding, the largest emphasis in selection of most breeds has been for increased production, because this improves feed efficiency, i.e., feed cost per unit of milk produced (e.g., Svendsen et al., 1994). The logic is that increased milk production will "dilute" feed requirements over more units of milk, primarily the maintenance requirement, and thus improve efficiency of production.

Selection for greater milk production will lead to an increasing nutritional demand, primarily energetic, that has to be met by: 1) increasing the feed intake, 2) by body tissue mobilisation or by 3) partitioning from other traits. It is generally accepted that the genetically correlated response in feed intake when selection is on production ($r_g = 0.46 - 0.65$; Veerkamp, 1998) is not large enough to cover the additional requirements (energy) due to increased production (Van Arendonk et al., 1989, 1991; Veerkamp & Thompson, 1999). This is also so as there is little evidence for genetic variance for the rate of efficiency at converting nutrients into milk (Blake & Custodio, 1984; Gibson, 1986; Svendsen et al., 1993; Veerkamp & Emmans, 1995; Zamani et al., 2011).

In consequence, selection for increased milk production will lead to a larger negative energy balance (Gallo et al., 1996; Veerkamp et al., 2000, 2001), especially when selecting for a peaked lactation curve in early lactation, increasing the level of non-esterified fatty acids, impairing the glucose synthesis (Overton et al., 1999; Rukkwamsuk, 1999), enhancing the risk of ketosis and for the fatty-liver syndrome (Baird, 1982; DeVries & Veerkamp, 2000; Loeffer et al., 1999). Further, the large negative energy balance will reduce fertility (Butler & Smith, 1989; Nebel & McGilliard, 1993; Senatore et al., 1996; Domecq et al., 1997; Rukkwamsuk, 1999; Buckley et al., 2003) and may increase the risk of mastitis and milk fever. Direct selection against a negative energy balance, being the difference between what is consumed and the requirements for yield and maintenance, relies on measuring feed intake, and is therefore difficult to select for (Collard et al., 2000). An alternative is to base selection on the body condition score, that can be used to monitor energy balance during the lactation (Wildman et al., 1982). Pryce et al. (2000, 2001), and Dechow et al. (2001, 2002) have reported a genetic relationship to fertility, e.g., that improved body condition scoring was genetically correlated with a shorter interval from calving to first insemination, i.e., the service period, being strongly determined by the energy balance (e.g., Van der Lende, (1998), cows coming into heat when the energy balance becomes positive. However, non-return rate is physiologically more strongly related to early embryonic death (Van der Lende, 1998).

From this, one should expect a negative genetic correlation between milk production and the interval from calving to first insemination, e.g. found by Andersen-Ranberg et al. (2005a), but also to other traits that depend on available resources, primarily energy, i.e., ketosis, other aspects of fertility as retained placenta, clinical mastitis, and also milk fever, as also demonstrated by Heringstad et al. (2005). These latter traits, together with service period, make up a group of traits that can be denoted as metabolic health. To this group a part of fertility aspects, also belong the fatty-liver syndrome, displaced abomasum, animal behaviour and disease resistance.

Despite the negative genetic correlation between milk production and metabolic health, Heringstad et al. (2005) also showed genetic progress for all the examined traits, resulting from field recording, on large daughter groups, and with considerable weighting of both trait groups to the breeding goal. This is a rather different result than what is expected from one-sided selection for increased milk production, indicating that different partitioning between traits results as a consequence of selection. Another way of demonstrating differences in partitioning would be through breed comparison, for which the comparison of amongst others the Norwegian Red with Holstein in Ireland can be used as an example, the latter considered one-sided selected for increased milk production. Results of Buckley et al. (2000, 2003) again indicate rather different partitioning between breeds for different traits, e.g. for body condition score and milk production. Hence, to improve the partitioning between traits in the Holstein, an alternative to this Nordic scheme would be to rely on selecting for an improved energy balance directly, as measured by the body condition score. Differences in genetics, management practice and environment cause variation in the shape of the lactation curves, both within and between cattle breeds (e.g. Grossman et al., 1986; Ray et al., 1992; Tekerli et al., 2000; Dillon et al., 2003). Some studies indicate that dairy cows having a flatter lactation curve tend to be more persistent than those with a steeper curve (Ferris et al., 1983; Grossman et al., 1986; Tekerli et al., 2000). Further, a flatter lactation curve may also reduce incidence rates of metabolic disorders and reproductive problems that originate from physiological stress due to high milk production (Pryce et al., 1997; Dekkers et al., 1998).

When comparing breeds for milk production efficiency, one way is to only consider feed requirements for maintenance against milk production, the former being closely associated with the body weight of the cow (W), through the metabolic body weight ($\hat{W}^{0.75}$) (McDonald et al., 1995). The considerable differences that are known to exist between the breeds for body size and those observed for milk production should therefore be taken into account. Lately, it was observed that the high producing breeds may be more sensitive to the variable environment, between e.g. from farm to farm or from day to day, introducing a genotype by environment interaction (Calus & Veerkamp, 2003; Dillon et al., 2003, Hayes et al., 2003), which also may affect the level of production during the course of lactation. The interaction can be tested by calculating variance components both between and within cows of each breed, on a test-day basis. The logic is that a larger variance component for a breed indicates larger environmental sensitivity, i.e., genotype by environment interaction (Lynch & Walsh, 1998).

1.6. Aim of the review

The overall objective of this review was to possibly identify the most appropriate cattle breed for Kosovo. As these breeds differ both with respect to breed characteristics and breeding goals (Dillon et al., 2003), the different breeds may respond differently to different environments (Falconer & Mackay, 1996; Lynch & Walsh, 1998; Bourdon, 2000). Three sub-goals were identified, first, to compare production, fertility and body condition score of the three imported breeds under the small-scale farming system in Kosovo. Secondly, the goal was to compare the three breeds for their shape of lactation curve and milk production efficiency, and thirdly, to compare variance components for daily milk yield both between and within cows of the different breeds, to possibly identify environmental sensitivity, i.e., genotype by environment interaction (Veerkamp et al., 1995).

2. Materials and methods

2.1. Description of project

The project was carried out from October 2000 till June 2003, and cattle were imported during the years 2000 – 2002; S and BS cows the first year, and S and TG the last two years.

Import phase	Simmental		Brown Swiss		Tyrol Grey	
	Heifers	Bulls	Heifers	Bulls	Heifers	Bulls
2000	1749	32	678	13	-	-
2001	1182	25	-	-	199	10
2002	532	10	-	-	60	2
Total	3463	67	678	13	259	12

Table 1. Number of imported heifers and bulls of Simmental, Brown Swiss and Tyrol Grey in three importation phases.

The total number of imports were 3463, 678, and 259 heifers of S, BS, and TG, respectively. S and BS were from Germany and Austria, while TG was imported from Austria. At importation, heifers were 4-7 months pregnant (Table 1).

The cows were donated to farms (one per farm) that had suffered the greatest losses to their livestock during the war. The farms were distributed in 228 villages, with an average of approximately 20 donated cows per village. For animals imported in the first year, calving was mainly from December till end of May (about 60% in January and February). In the third year, calving was between August and December (more than 50% in October and November). After calving the heifers were re-mated to a bull of the same breed, mainly by use of artificial insemination (A.I.) (56%), but also by natural mating to imported bulls (44%).

2.2. Data recording

Data were from the FAO/WB/EFR project in Kosovo. The cattle were monitored for a period of 14-16 months, and the database that was built holds information from several sources, i.e., farmer, contracted veterinarians and project staff. The data consisted of eartag number, breed, year of importation, village of donation, birth date, different events (i.e., mortality date and dates of different diseases (mastitis, metabolic disorders, ketosis, etc.), calving date, milk production (monthly test-day milk yield), calf data (i.e., calf sex, weaning date and

hearth-girth circumference at birth, 3 and 10 months of age), fertility information (insemination
dates till third mating, non-return, and whether mating was natural or by A.I. (mating type)),
body condition score (i.e., subjective, within one week after calving and within one week after
first insemination, respectively), as well as socio-economic data (i.e., household headed by a
female, size of land , size of the family, existence of members within the family older than 65
years of age, existence of members within the family younger than 12 years of age, whether the
family had cows before the donated one, and sex of beneficiary). The farmers recorded the
milk yield themselves and were trained for detection and recording of different events as well
as date of mating, (natural mating). The project staff recorded heart-girth data as well as body
condition score, and also socioeconomic data. The contracted veterinarians were responsible
for recording the information on fertility by A.I.

In some farms the data were partially or completely missing, which might be explained by
lack of recording practice. Unreliable data (e.g., daily milk production smaller than 3 kg or
larger than 50 kg, the first and last record for one cow being either observed earlier or later
than, respectively, 30 and 280 days in milk, 1st insemination before 20 days postpartum) and
data deriving from incomplete lactations (254) were excluded from the final dataset, which
was the basis for the statistical analyses included in this study.

2.3. A comparison of the productive, reproductive and body condition score traits

2.3.1. Milk production

Daily milk production was measured as the average of the yield in the morning and evening
on a monthly basis. From the first 10 test-day records, the average over 305 days in milk in
the first lactation was calculated, for analysis. Only cows having all 10 records were
included (Table 2).

Traits	Simmental			Brown Swiss			Tyrol Grey		
	N	X	SD	N	X	SD	N	X	SD
AMY305	1900	11.96	2.68	444	12.69	2.84	172	10.82	1.89
BWC	328	39.45	2.68	73	38.15	2.34	73	32.59	2.01
GRC3	281	0.86	0.16	62	0.74	0.10	68	0.65	0.08
GRC10	254	0.96	0.16	59	0.76	0.09	66	0.72	0.07
SP	3179	102.33	47.35	582	119.37	50.23	242	97.02	40.23
NRR	3261	0.46	0.52	623	0.33	0.53	244	0.52	0.57
BCSC	791	3.28	0.34	200	3.21	0.33	118	3.24	0.29
BCSS	791	2.52	0.34	200	2.35	0.34	118	2.62	0.30

[1]AMY305 = Average milk yield over first 305 days of first lactation (kg/day); BWC = Birth weight of calf (kg);
GRC3 = Growth rate of calf over first 3 months of age (kg/day); GRC10 = Growth rate of calf over first 10 months of age
(kg/day); SP = Service period (days); NRR = Non-return rate at first insemination (%); BCSC = Body condition score,
within one week after calving (1-5); BCSS = Body condition score, within one week after service (1-5).

Table 2. Number of records (N), mean (X) and standard deviation (SD) for each trait and breed.

2.3.2. Calf growth rate

Calf body weights were estimated based on hearth-girth circumference at birth, at 3 months (weaning), and at 10 months of age, and used for calculation of calf growth rates from birth until 3 and 10 months of age, respectively (Table 2). These two later traits were analyzed as well as birth weight of calf.

2.3.3. Fertility

The interval from calving to first insemination, i.e., the service period was analysed as well as the non-return rate at first insemination, coded 1 if a cow did not return to service after the first insemination and 0, otherwise (Table 2).

2.3.4. Body condition score

In Table 2, the body condition was scored at the loin, pelvis and tail head within one week after calving, and within one week after first insemination. The scoring was from 1 (very thin) to 5 (very fat) (Edmonson et al., 1989). The traits described above were exposed to an analysis of variance, using the PROC GLM procedure (SAS Institute Inc., 1999) of the SAS-package. Generally, in order to estimate a possible breed effect, on the different traits: 305 days milk yield, birth weight of calf, calf growth rates until 3 and 10 months of age, service period, non-return rate, and body condition scores, at calving and at mating), univariate fixed effect models were used. The final model was chosen using backward elimination by removing non-significant (P ≥ .05) explanatory variables from the model, one at a time. Several socio-economic indicators were recorded and tested as explanatory variables. These were; gender of household head, farmland area, number of family members, existence of family members above 65 years of age, existence of family member below 12 years of age, gender of the beneficiary, and whether or not the family had owned a cow before the donated one. For all traits, the effects of importation phase x village and month of calving were included. Age at first insemination and mating type were only considered for non-return rate, while sex of calf was included for weight and growth traits, respectively. To estimate for a possible breed x ration effect on the formerly described traits, information on feeding was recorded in 166 randomly chosen farms, and used in a separate analysis.

2.4. Lactation Curves and production efficiency

In this study, milk yield test-day records for cows in first lactation of the three breeds (S, BS and TG) were used in a statistical analysis with a linear model aimed at comparing lactation curves and milk production efficiency of the three breeds.

The applied model was used:

$$Y_{ijklmn} = BLM_i + PV_j + CM_k + HHF_l + c_m + e_{ijklmn}$$

where:

Y_{ijklmn} = milk yield record n of cow m in breed-lactation month class i, importation phase-
village class j, calving month k, and gender effect of owner class l;

BLM_i = fixed effect of first-lactation breed-lactation month class i, in 30 classes from 3 breeds
and 10 months;

PV_j = fixed effect of phase-village class j, in 176 classes, from 3 phases and 99 villages;

CM_k = fixed effect of calving month k, in 12 classes;

HHF_l = fixed effect of gender of owner l, in 2 classes;

c_m = random effect of cow m: $\sim \left(0, \sigma_c^2\right)$, σ_c^2 being the cow variance; and

e_{ijklmn} = random residual term: $\sim N\left(0, \sigma_e^2\right)$, with σ_e^2 denoting the residual variance.

As in previous study, data were restricted to cows having 10 monthly test-day milk yield
records in first lactation. In total, 25,160 records from 2516 cows were included in the
analysis (Table 3).

Breed	Cows in project	Cows in analysis	Test-day records
S	3463	1900	19.000
BS	678	444	4440
TG	259	172	1720
Total	4400	2516	25,160

Table 3. Number of cows and number of test-day records for milk yields over the first 10 months of
lactation (305 days) in cows of Simmental (S), Brown Swiss (BS) and Tyrol Grey (TG) breed.

To compare milk production efficiency of the three breeds, average body weight of each
breed was derived, using measure of hearth-girth circumference. Body weight records were
available for 102, 64 and 47 cows of S, BS and TG, respectively. The S cows weighed on
average 572 kg, while the BS and TG were on average 533 kg and 445 kg, respectively. Based
on the estimated average body weight (W) for each breed, average metabolic body weights
were estimated as $\hat{W}^{0.75}$ (McDonald et al., 1995).

The statistical analyses were conducted using the PROC MIXED procedure of the SAS
software package (SAS Institute Inc., 1999). All effects that had shown a significant effect ($P
< 0.05$) on 305 days milk yield were considered in the analysis: Breed, importation phase x
village, calving month, and whether the household was headed by a female or not. With the
aim of modelling lactation curves of the different breeds, a breed x lactation month effect
replaced the main effect of breed. As cows had repeated records during the lactation, a
random cow effect was included, while the other effects were considered as fixed. From the
monthly least-squares mean of daily milk yield (LSM$_{DMY}$), metabolic body weight per kg of
milk per month was calculated as $\hat{W}^{0.75}$ / LSM$_{DMY}$.

2.5. Estimation of environmental conditions on milk production for dairy breeds comparison using random regression models

In this study were utilized the same data as in previous one. Here, daily milk yield was analyzed with seven different models, consisting of both repeatability and random regression test-day models.

All models had the following general characteristics:

$$DMY_{ijklmn} = BLM_{ij} + PV_k + CM_l + HHF_m + \sum_{p=0}^{q} r_{pn} Z_p(j) + e_{ijklmn}$$

where:

DMY_{ijklmn} = daily milk yield of cow n, of breed i, in lactation month j, importation phase village class k, calving month l, and gender effect of household head class m;

BLM_{ij} = fixed effect of breed · lactation month class ij, in 30 classes (3 breeds and 10 lactation months);

PV_k = fixed effect of phase village class k, in 176 classes (3 phases and 99 villages);

CM_l = fixed effect of calving month l, in 12 classes;

HHF_m = fixed effect of gender of household head class m, in 2 classes;

$Z(j)_p$ = p^{th} order orthogonal polynomial of lactation month j;

r_{pn} = p^{th} order random regression coefficient of cow n; and

e_{ijklmn} = random residual.

The following models were specified:

REP1 = initial repeatability model with $q = 0$, assuming homogeneous cow and residual variances;

REP2 = extension of REP1, with heterogeneous residual variance for each month of lactation;

REP3 = extension of REP2, with heterogeneous cow variance per breed;

REP4 = extension of REP2, with heterogeneous residual variance per breed · lactation month class;

REP5 = combination of REP3 and REP4, with heterogeneous cow variance per breed, and heterogeneous residual variance per breed · lactation month class;

RR1 = extension of REP5 with $q = 1$ (1st order random regression of cow effects); and

RR2 = extension of RR1, with $q = 2$ (2nd order random regression of cow effects).

Initially, yield records (Table 3) were analyzed using a repeatability test-day model similar to the statistical model in the second study. Subsequently, this model was extended to allow heterogeneous cow and residual variances for the different breeds, and random regression models of varying orders were tested. The models were compared using a likelihood-ratio test statistics and Akaike information criterion (Akaike, 1973).

$$LR = 2[\ln RL(j) - \ln RL(i)] \sim \chi^2_{v_j - v_i} \qquad (1)$$

The likelihood-ratio test statistics (**LR**) for two models i and j, with the restricted model i nested within the model j, was presented in Equation 1: where $\ln RL(i)$ and $\ln RL(j)$ are the $\ln RL$ values of the models to be compared, and v_i and v_j are the corresponding number of (co)variance components in the models.

$$AIC = 2[\ln RL(i) - \ln RL(0) - (v_i - v_{0})] \tag{2}$$

Models were also compared on Akaike information criterion (AIC) (Akaike, 1973), favoring models with fewer parameters (Equation 2): where $\ln RL(0)$ and v_0 are, respectively, the ln restricted likelihood and number of (co)variance components of the base model (i.e., REP1). For all likelihood-based criteria, the model with the largest values was considered as having the best fit. The ASREML software (Gilmour et al., 1999) was used in all statistical analyses.

To determine whether breed differences in size of variance components could be attributed to scale effects (Falconer and Mackay, 1996), the coefficient of variation (**CV**) was calculated for each breed as follows:

$$CV = \frac{\sigma}{\overline{y}} \times 100 \tag{3}$$

where σ is the square root of the estimated variance component for a specific month of lactation in the preferred model, and $\overline{y}$ is the corresponding estimate for BLM_{ij} (Equation 3).

3. Main results

3.1. A comparison of the productive, reproductive and body condition score traits

The breeds differed significantly with respect to milk production (P < 0.0013; Table 4). The BS yielded the highest average daily milk production, followed by S and TG, the least-squares means being less with by 0.59 and 2.72, respectively (Table 6).

Month of calving had a clear significant effect on milk yield (P < 0. 0001; Table 4), with the highest milk yield obtained during the winter period.

Service period was affected by calving month (P < 0.0001; Table 4), favoring the cows that calved during spring period. Cows from households headed by a man produced more milk (0.42 kg) than cows in households headed by a female (P < 0 .0051; Table 4). Significant breed differences (P < 0.0001) were found for weight of calf at birth and growth rates until 3 and 10 months of age, respectively (Table 4). The S calves had both the highest birth weight and the highest growth rate, compared with BS and TG (Table 6).

Sex of the calf showed a significant effect on birth weight and growth rate traits (P < 0.0001), with males having the largest values for all breeds (Table 4).

In Table 5, significant breed differences were found for service period (P < 0.0001), the leastsquares mean being longest for S (125 days), followed by BS (114 days) and TG (97 days) presented in Table 6.

Traits	Breed	Importation phase × village	Calving month	Age - 1st insemination in 1st lactation	Sex of calf	Method of insemination	HHF	SL	SF	FM>65	FM<12	FCBD	SB
AMY305	<.0013	<.0001	<.0001	-	-	-	<.0051	NS	NS	NS	NS	NS	NS
BWC	<.0001	NS	NS	-	<.0001	-	NS	NS	NS	NS	NS	NS	NS
GRC3	<.0001	NS	NS	-	<.0001	-	NS	NS	NS	NS	NS	NS	NS
GRC10	<.0001	NS	NS	-	<.0001	-	NS	NS	NS	NS	NS	NS	NS
SP	<.0001	<.0001	<.0001	-	-	-	<.0366	NS	NS	NS	NS	NS	NS
NRR	<.0048	<.0001	NS	NS	-	<.0001	<.0191	NS	NS	NS	NS	NS	NS
BCSC	NS	<.0056	NS	-	-	-	NS	NS	NS	NS	NS	NS	NS
BCSS	<.0059	<.0129	NS	-	-	-	NS	NS	NS	NS	NS	NS	NS

[1]NS = Not significant, i.e. level of significance ≥ 0.05; [2]Breed = (Simmental, Brown Swiss, or Tyrol Grey); Importation phase x village = (1,2 or 3) x (1, . . . , 228)); Calving month = (1, . . . , 12); Age-1st insemination in 1st lactation = (months 22, . . . , 45); Sex of calf = (female or male); Method of insemination = (artificial or natural); HHF = (household headed by a female (yes or otherwise); SL = (Size of land (<1 Ha, 1 Ha $\leq$ - < 2.5 Ha, and $\geq$ 2.5 Ha); SF = (size of the family (< 7, 7 – 11 and $\geq$ 12); FM>65 = (members within the family older than 65 years of age (yes or otherwise); FM<12 = (members within the family younger than 12 years of age (yes or otherwise); FCBD = (family had cows before donated one (yes or otherwise); and SB = (sex of beneficiary (female or male). [3]AMY305 = Average milk yield over first 305 days of first lactation (kg/day); BWC = Birth weight of calf (kg); GRC3 = Growth rate of calf over first 3 months of age (kg/day); GRC10 = Growth rate of calf over first 10 months of age (kg/day); SP = Service period (days); NRR = Non-return rate at first insemination (%/100); BCSC = Body condition score within, one week after calving (1-5); BCSS = Body condition score, within one week after service (1-5).

Table 4. Level of significance for effects model to affect various trait in analyses of a breed effect.

The cows managed in households headed by a female had shorter service period (4 days) than cows in households headed by a man (P < 0.0366; Table 4).

For non-return rate, significant breed differences were estimated (P < 0.0048), from significant differences in leastsquares mean between S and BS (Table 6). S had the highest success rate on conceiving at first insemination (53%), followed by BS (44%) and TG (40%) (Table 6).

In Table 5, the method of insemination significantly affected non-return rate (P < 0.0001), with about 57% success for natural service compared with 34% in artificial insemination.

Whether the household was headed by a man or a woman also significantly affected the non-return rate (P < 0.0191), with female headed households being better than those headed by a man (Table 4).

No significant differences between breeds (P > 0.1701) were found for body condition score at calving (Table 4).

However, significant breed differences were found one week after insemination (P < 0.0059; Table 4), from significant least-squares mean differences between S and BS (Table 6). The TG cows showed the smallest reduction of least-squares mean for body condition score, compared to S and BS (Table 6).

Traits	Breed × Ration	Importation phase	Village	Calving month	Age- 1st insemination in 1st lactation	Method of service
AMY305	<.0750	<.0298	<.8570	NS	-	-
SP	<.0011	<.7863	<.3386	NS	-	-
NRR	<.5992	<.5611	<.3974	NS	NS	NS
BCSC	<.1329	<.2183	<.2741	NS	-	-
BCSS	<.0017	<.2497	<.2245	NS	-	-

[1]NS = Not significant, i.e. level of significance ≥ 0.05.
[2]Breed x ration = (Simmental, Brown Swiss or Tyrol Grey x forage and concentrate or only forage); Importation phase = (1,2, or 3); Village = (1, . . . , 64); Calving month = (1, . . . , 12); Age-1st insemination in 1st lactation = (months 22, . . . , 45); Method of service = (artificial or natural).
[3]AMY305 = Average milk yield over first 305 days of first lactation (kg/day); SP = Service period (days); NRR = Non-return rate at first insemination (%/100); BCSC = Body condition score, within one week after calving (1-5); BCSS = Body condition score, within one week after service (1-5).

Table 5. Level of significance for effects modeled to affect various trait) in analyses of a breed x ration effects.

Traits	Simmental LSM	SE	Brown Swiss LSM	SE	Tyrol Grey LSM	SE	S vs. BS	S vs. TG	BS vs. TG
AMY305	11.76	0.26	12.35	0.29	9.63	0.92	< .0037	< .0270	< .0058
BWC	39.24	0.17	37.96	0.31	32.39	0.31	< .0002	< .0001	< .0001
GRC3	0.87	0.01	0.74	0.02	0.65	0.02	< .0001	< .0001	< .0004
GRC10	0.96	0.01	0.78	0.02	0.72	0.02	< .0001	< .0001	< .0026
SP	125.29	4.13	113.79	4.67	97.07	9.88	< .0001	< .0035	< .2282
NRR	0.53	0.02	0.44	0.03	0.40	0.11	< .0023	< .2355	< .7374
BCSC	3.28	0.02	3.21	0.04	3.14	0.17	< .0611	< .4552	< .7415
BCSS	2.54	0.03	2.40	0.05	2.54	0.19	< .0014	< 1.0000	< .4937

Table 6. Estimates of least-squares mean (LSM), their standard error (SE) and level of significance on the test of differences in least-squares mean between Simmental (S), Brown Swiss (BS), and Tyrol Grey (TG), for various traits 1). 1) AMY305 = Average milk yield over first 305 days of first lactation (kg/day); BWC = Birth weight of calf (kg); GRC3 = Growth rate of calf over first 3 months of age (kg/day); GRC10 = Growth rate of calf over first 10 months of age (kg/day); SP = Service period (days); NRR = Non-return rate at first insemination (%/100); BCSC = Body condition score within one week after calving (1-5); BCSS = Body condition score within one week after service (1-5).

The breed x ration effect was significant (P < 0.0011 and P < 0.0017) for service period and for body condition scoring at insemination, respectively (Table 5). Within the same breed differences in least-squares means between rations, on body condition scoring at service, was reduced for both S and BS, on the ration without concentrate, and significantly (P < 0.0012 and P < 0.0444), respectively (Table 6).

Village effects, nested within importation year had a highly significant effect on milk yield, service period, non-return rate and body condition score, at calving and at service, respectively (Table 4).

3.2. Lactation curves and production efficiency

The results from the Figure 1 show that estimated cow variance was twice as high as the estimated residual variance. However, it also should consider that allocating one cow per farm, the estimate also contains the effect of farm.

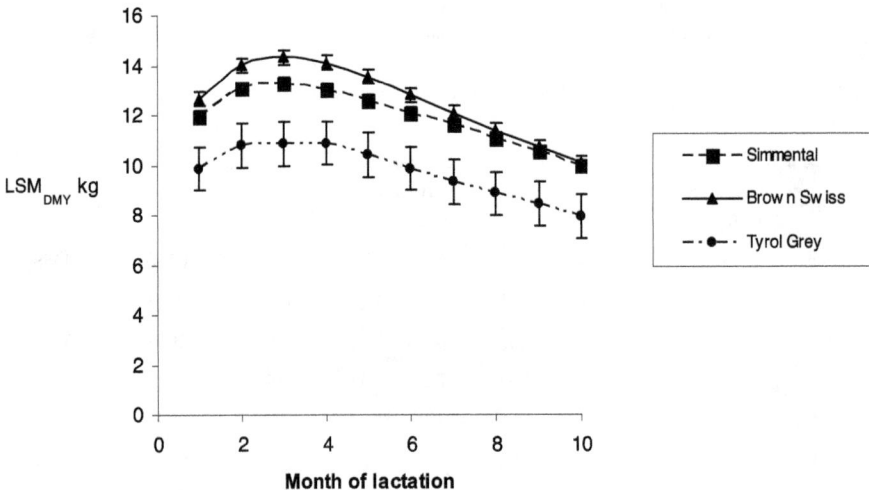

Figure 1. Least-squares mean of daily milk yield (LSM$_{DMY}$) by lactation month for Simmental, Brown Swiss and Tyrol Grey cows in Kosovo.

The lactation curve was consistently higher for BS than for S, the latter dominating the curve for TG (Figure 1). The milk production efficiency here is defined as milk yield per unit bodyweight. The lactation curve for TG cows tended to be less peaked than those for S and BS cows. BS cows tend to produce milk more efficiently throughout lactation compared to S and TG, the two latter being rather similar in this respect (Figure 2).

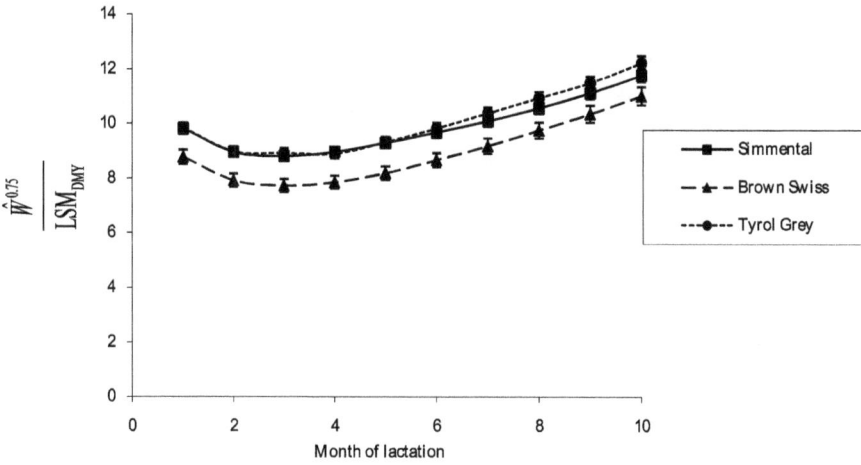

Figure 2. Ratio of metabolic body weight on daily milk yield $(\dfrac{\hat{W}^{0.75}}{LSM_{DMY}})$, LSM$_{DMY}$ being least-squares
mean of daily milk yield, by lactation month for Simmental, Brown Swiss and Tyrol Grey cows in
Kosovo

3.3. Estimation of environmental conditions on milk production for dairy breeds comparison using random regression models

In the Table 7 are presented the pepeatability (REP) and random regression (RR) test-day models and their estimates for ln of restricted likelihood (lnRL), Akaike information criterion (AIC), likelihood-ratio test statistics (LR), and level of significance (P), when comparing to the reduced model. In order to describe the statistical analyses of daily milk yield, seven models in total were developed. First model considered the homogenous cow and residual variances. Second model, was developed consisting of homogenous and heterogeneous cow and residual variances, respectively, the latter by month of lactation.

Full model	LnRL	AIC	Reduced model	LR	P
REP1	-26685	0	-	-	-
REP2	-24150	5053	REP1	5071	<0.01
REP3	-24129	5090	REP2	41	<0.01
REP4	-23553	6207	REP2	1194	<0.01
REP5	-23536	6236	REP4	32	<0.01
RR1	-19862	13572	REP5	7348	<0.01
RR2	-17415	18448	RR1	4894	<0.01

Table 7. Repeatability (REP) and random regression (RR) test-day models and their estimates for ln of restricted likelihood (lnRL), Akaike information criterion (AIC), likelihood-ratio test statistics (LR), and level of significance (P), when comparing to the reduced model.

Third model was based on heterogeneous cow and residual variances, by breed and month of lactation, respectively.

Fourth model described the homogenous and heterogeneous cow and residual variances, respectively, the latter by breed · lactation month. Model five showed the heterogeneous cow and residual variances, by breed and breed · lactation month, respectively.

Model six explains the first-order random regression of cow effects, assuming heterogeneous cow and residual variances, by breed and breed · lactation month, respectively, while the model seven considered the second-order random regression of cow effects, assuming heterogeneous cow and residual variances, by breed and breed · lactation month, respectively.

A second-order random regression model was preferred for statistical analysis of daily milk yield.

Generally, residual variances were largest in the first half of lactation and diminished towards the end, for all breeds (Table 7).

The rank order of breeds for the cow variance was observed for the residual variance (Figure 3). For BS the residual variance ranked from 1.62 in the beginning to 0.16 kg^2/d at the end of lactation, while for S and TG cows the residual variance was 1.04 and 0.38 kg^2/d milk in the beginning and 0.16 and 0.15 kg^2/d at the end of the lactation (Figure 4).

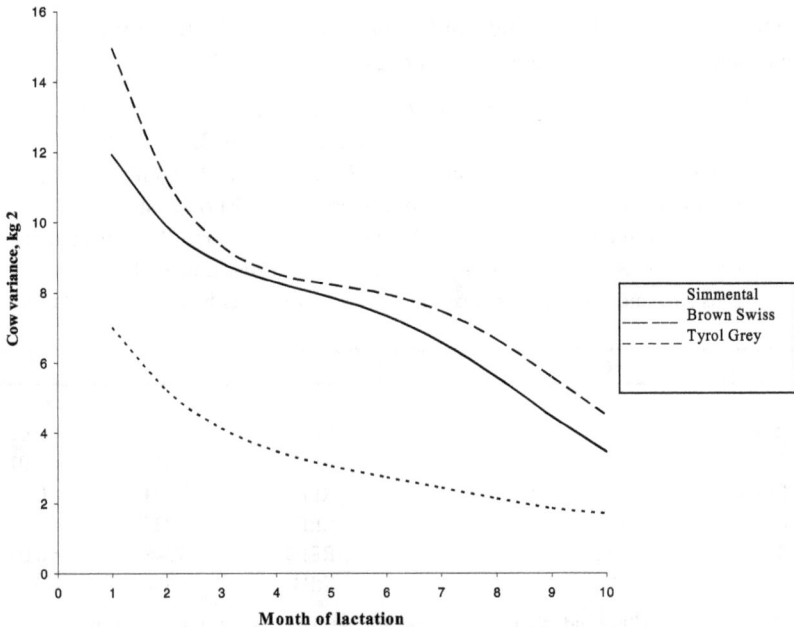

Figure 3. Trajectories of estimated cow variance by month of first lactation for Simmental, Brown Swiss and Tyrol Grey, using a second-order random regression model (RR2) for analysis.

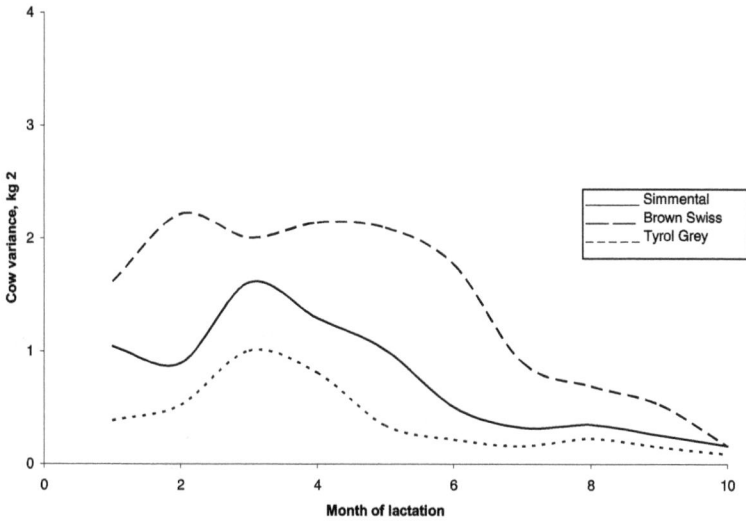

Figure 4. Trajectories of estimated residual variance by month of first lactation for Simmental, Brown Swiss and Tyrol Grey, using a second-order random regression model (RR2) for analysis.

Clear breed differences were observed also with respect to the coefficient of variation within breed (Figure 5), suggesting that scale effect alone might not explain the breed differences in size of the estimated variance components (Figure 6).

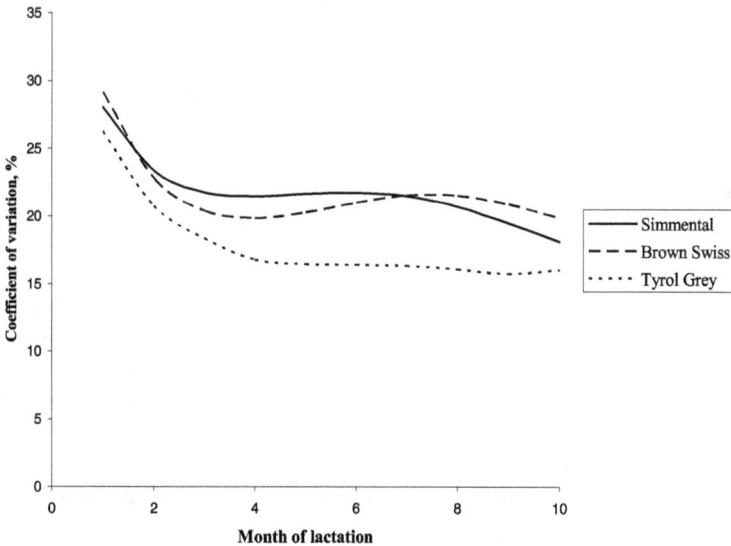

Figure 5. Trajectories of estimated coefficient of variation for cow effect by month of first lactation for Simmental, Brown Swiss and Tyrol Grey, using a second-order random regression model (RR2) for analysis.

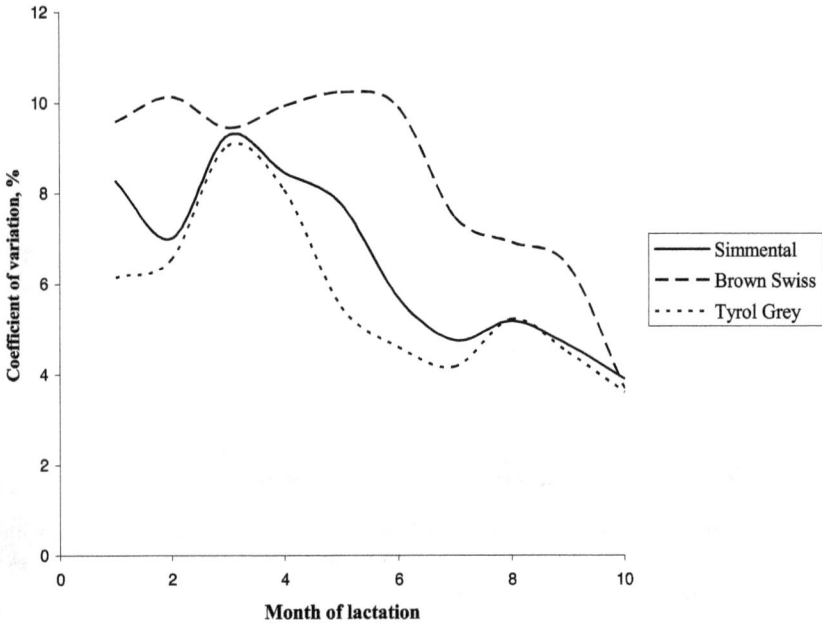

Figure 6. Trajectories of estimated coefficient of variation for residual effect by month of first lactation for Simmental, Brown Swiss and Tyrol Grey, using a second-order random regression model (RR2) for analysis.

4. General discussion

Cattle production in Kosovo is characterized by being predominately small-scale and semi-extensive. In this environment, focus in breed comparison should not only be on increased production (to improve feed efficiency), but also on how well the breeds are adapted to the local environment. An indication is given by studying the energy balance, here indirectly through measuring the body condition score and the length of the service period. It is also relevant to examine the pattern of the lactation curves for the breeds, for which a flatter curve should be favorable with respect to energy balance. The fit to the local environments was also studied through examination of genotype by environment interaction, e.g. through estimation of a breed x ration effect. Another approach was through testing for heterogeneity of variance components for milk production for the different breeds. In conclusion, the choice of traits to record for the breed comparison was in large sound generally appropriate.

The analyses showed that BS cows had higher yields compared with S and TG. The production levels of the different breeds in Kosovo were lower than in the countries of origin, with the most pronounced reduction for the BS and S breeds (Cattle breeding in Austria, 2003; Rinder production in der Bundesrepublik Deutschland, 2001; Tiroler Grauvieh, 1999). The TG breed is a smaller breed with lower milk production, and therefore also with lower nutritional requirements. This might explain why the Kosovo results for this breed seem to fit better with their milk production potential in the country of origin.

When comparing the efficiency of breeds only on the bases of production, the large differences in body weight are not taken into account. Hence, the efficiency, as the metabolic body weight per kg of milk, was calculated throughout lactation. This measure should be proportional to the expected maintenance requirement, given that weight records were representative for the entire lactation and that feed requirement per kg of milk was the same for all three breeds. After correction for weight, it was found that BS was the most efficient, while now TG and S breeds produced milk with similar efficiencies.

With regard to the service period and body condition score, results from this study showed that TG cows had shorter service periods and less negative energy balance during lactation (less reduction of body condition score), compared with the S and BS breeds. An important goal of dairy cattle breeding programs is to achieve an approximately 12-month calving interval (Schmidt, 1989; Schultz-Rajalla and Frazer, 2003), and in this context; the TG breed was closest to fit with this recommendation. However, it should be noted that an unfavorable genetic relationship exists between days open and non-return rate (e.g., Andersen-Ranberg et al., 2005b) such that cows coming into heat early often have high embryonic loss and reduced success of conceiving (Averdunk et al. 1995; Van der Lende, 1998), which might explain the lower non-return rate estimated for TG, compared with the other breeds.

The finding of less negative energy balance for TG relative to the other two breeds was also supported by the pattern of the lactation curves, tending to be less peaked for TG than for other two breeds.

In addition, the fit to the local environment was examined more directly than through indirect measures of the energy balance, by calculating the genotype by environment interaction directly, i.e., a breed x ration effect. Although the material was rather limited, the results for body condition score indicated that S and BS were more sensitive to an extensive environment than TG.

Existence of genotype by environment interaction was also examined by calculating variance components for milk production both between and within cows of the different breeds, with a random regression test-day model. The larger estimates obtained for BS than for S, again being larger than those for TG, indicate that performance of BS cows was more variable across the same range of physical environment than the two other breeds (breed and heifer were randomly allocated, one per farm), or from test-day to test-day. The smaller variance for TG implies reduced environmental sensitivity or greater phenotypic stability, relative to the two other breeds. In contract, the largest variances for BS imply greater phenotypic plasticity that might be desirable for herds with an improved environment. However, for the majority of herds, under small-scale and semi-extensive conditions, these latter results point to the choice of TG amongst the breeds compared.

Currently, beef cattle production is almost non-existing in Kosovo. Hence, dual-purpose cattle breeds (S and TG) should have preference over more specialised milk breeds (BS). A significantly higher growth rate was found for S compared with TG. However, breeds with high growth rate and larger mature weight normally reach maturity later, and also require more intensive feeding than smaller breeds (Geay & Robelin, 1979; Arango et al., 2002).

Finally, it should be mentioned that many farmers are used to dealing with local cattle and their crossbreeds, which are smaller in size and produce a lower amount of milk. Hence, to explore their potential they should have been included in the experiment that preferably also could have been more balanced for the number of animals of each breed.

5. Conclusion

In small-scale and semi-extensive management as in Kosovo, robust dual-purpose cattle breeds for production of both milk and beef should be preferred over more specialized dairy breeds highly adapted to intensive production systems, requiring intensive feeding and good management practices.

Substantial breed differences were found for the trajectories of cow and residual variances as well as for their coefficients of variation at different stages of lactation, indicating more environmental sensitivity in the larger and more productive breeds; S and BS compared to TG. Furthermore, TG had a shorter service period and thus a shorter calving interval than the other two breeds, less body reserve losses, in addition to a less peaked lactation curve and a satisfactory milk production, also relative to their metabolic body weight.

Author details

Hysen Bytyqi
Department of Livestock and Veterinary Sciences, Faculty Agriculture and Veterinary University of Prishtina; Str. "Lidhja e Pejes", Prishtina – Kosovo

6. References

Akaike, H. (1973). Information theory as an extension of the maximum likelihood principle. *Pages 267-281 in Petrov, B. N., and F. Csaki. (Editors), Second International Symposium on Informational Theory.* Akademia Kiado, Budapest.

Andersen-Ranbeg, M.; Heringstad, B.; Gianola, D.; Chang, M. & Klemetsdal, G. (2005a).
Comparison between bivariate models for 56-day nonreturn and interval from calving
to first insemination in Norwegian red. *J. Dairy. Sci.* 88:2190-2198.

Andersen-Ranbeg, M.; Klemetsdal, G.; Heringstad, B. & Stein, T. (2005b). Heritabilities,
genetic correlations, and genetic change for female fertility and protein yield in
Norwegian dairy cattle. *J. Dairy. Sci.* 88:348-355.

Arango, A.; Cundiff, V. & Van Vleck. D. (2002). Breed comparison of Angus, Charolais,
Hereford, Jersey, Limousin, Simmental, and South Devon for weight, weight adjusted
for body condition score, height, and body condition score of cow. *J. Anim. Sci.* 80:3123-
3132.

Atlas der Nutztierrassen. (1994). 250 Rassen in Wort und Bild / Hans Hinrich Sambras. – 4.,
erw. *Eugen Ulmer GmbH & Co. Wollgrasweg 41, 7059 Stuttgart (Hohenheim) Germany.*
ISBN 3-8001-7308-5, pp: 36–53.

Averdunk, G.; Aumann, J.; Thaler, G. & Gierdziewicz, M. (1995). Sire evaluation for
fertility and calving easy in Germany. *Interbull open meeting,* Prag, 07- 08.
September.

Baird, D. (1982). Primary ketosis in the high producing dairy cow: clinical and sub-clinical
disorders, treatment, prevention, and outlook. *J. Dairy Sci.* 65:1-10.

Blake, W. & Custodio, A. (1984). Feed efficiency: A composition trait of dairy cattle. *J. Dairy
Sci.* 67:2075-2083.

Bourdon, M. (2000). *Understand Animal Breeding.* - 2ⁿᵈ edition. Colorado State University.
Prentice Hall, Upper Saddle River, NJ 07458. ISBN 0-13-096449-2, pp: 3-17.

Buckley, F.; O'Sullivant, K.; Mee, F.; Evans, D. & Dillon, P. (2000). The relationship between
genetic merit for yield and live weight, condition score, and energy balance of spring
calving Holstein Friesian dairy cows on grass based systems of milk production. *J. Dairy
Sci.* 83:1878-1886.

Buckley, F.; O'Sullivant, K.; Mee, F.; Evans, D. & Dillon, P. (2003). Relationship among milk
yield, body condition, cow weight, and reproduction in spring-calved Holstein –
Friesians. *J. Dairy. Sci.* 86:2308-2317.

Butler, W. & Smith, D. (1989). Interrelationship between energy balance and postpartum
reproductive function in dairy cattle. *J. Dairy Sci.* 72:767-783.

Bytyqi. H., G. Klemetsdal, J. Ødegård, H. Mehmeti, and M. Vegara. (2005). A comparison of
the productive, reproductive and body condition score traits of the Simmental, Brown
Swiss and Tyrol Grey breeds in smallholder herds in Kosovo. Anim. Genet. Res. Inf. 37:
9-20.

Bytyqi. H., J. Ødegård, M. Vegara, H. Mehmeti, and G. Klemetsdal. (2006). Short
Communication: Lactation Curves and production efficiency for Simmental, Brown
Swiss and Tyrol Grey in Kosovo. Actae Scandinavica – Section – B, Volume 56, Issue 3
& 4, pages 161-164.

Bytyqi. H., J. Ødegård, H. Mehmeti, M. Vegara, and G. Klemetsdal. (2007). Environmental
Sensitivity of Milk Production in Extensive Environments: A comparison of Simmental,

Brown Swiss and Tyrol Grey using Random Regression Models. J. Dairy. Sci., 90: 3883-3888.

Calus, P. & Veerkamp, F. (2003). Estimation of environmental sensitivity of genetic merit for milk production traits using a random regression model. *J. Dairy. Sci.* 86:3756-3764.

Cattle breeding in Austria. 2003. *Zentrale Arbeitsgemeinschaft Osterrichischer Rinderzuchter.* Universumstrabe 33/8, 1200 Wien. Ausgabe, pp: 32-55.

Cienfugos-Rivas, G.; Oltenacu, A. Blake, W. Schaweger, J. Castillo-Juarez, H. & Ruiz, J. (1999). Interaction between milk yield of Holstein cows in Mexico and the United States. *J. Dairy. Sci.* 82:2218-2223.

Collard, L.; Poettcher, J. Dekkers, M. Petitclerc, D. & Schaeffer, R. (2000). Relationship between energy balance and health traits of dairy cattle in early lactation. *J. Dairy Sci.* 83:2683-2690.

DeVries, J. & Veerkamp, F. (2000). Energy balance of dairy cattle in relation to milk production variables and fertility. *J. Dairy Sci.* 83:62-69.

Dechow, D.; Rogers, W. & Clay, S. (2001). Heritabilities and correlation among body condition score, production traits, and reproductive performance. *J. Dairy Sci.* 84:266-275.

Dechow, D.; Rogers, W. & Clay, S. (2002). Heritability and correlations among body condition score loss, body condition score, production and reproductive performance. *J. Dairy Sci.* 85:3062:3070.

Dekkers, J.; Ten Haag, M. & Werrsink, H. (1998). Economic aspects of persistency of lactation in dairy cattle. *Livest. Prod. Sci.* 53:237-252.

Dillon, P.; Buckley, F.; O'Connor, P.; Hegarty, D. & Rath, M. (2003). A comparison of different cow breeds on a seasonal grass-based system of milk production. 1. Milk production, live weight, body condition score and DM intake. *Livest. Prod. Sci.* 83:21-33.

Domecq, J.; Skidmore, L. Lloyd, W. & Kaneene, B. (1997). Relationship between body condition scores and conception at first artificial insemination in a large dairy herd of high yielding Holstein cows. *J. Dairy Sci.* 80:113-120.

Drackley, K.; Overton, M. & Douglas, N. (2001). Adaptations of glucose and long-chain fatty acid metabolism in liver of dairy cows during the periparturient period. *J. Dairy Sci.* 84: (E.suppl.):E100-E112.

Edmonson, J.; Lean, J. Weaver, D. Farver, T. & Webster, G. (1989). A body Condition Scoring chart for Holstein dairy cows. *J. Dairy Sci.* 72:68-78.

Falconer, S. & Mackay, C. (1996). *Introduction to Quantitative Genetics. Fourth Edition (ed).* Longman Group, Essex, UK. ISBN 0582-24302-5..

Ferris, A.; Mao, L. & Anderson, R. (1983). Selecting for Lactation Curve and Milk Yield in Dairy Cattle. *J. Dairy Sci.* 68:1438-148.

Frickh, J. (1999). *Tiroler Grauvieh Juwel der Berge/ hrsg.vom Tyroler Grauviehzuchtverband. Insbruck: Ed.* Lowvenzahn, Austria. pp. 54-56. ISBN 3 – 7066-2199-1.

Gallo, L.; Carnier, P.; Mantovani, R.; Bailoni, L.; Cantiero, B. & Bittante, G. (1996). Change in
    body condition score of Holstein cows as affected by parity and mature milk yield. *J.
    Dairy Sci.* 79:1009-1015.

Geay, Y. & Robelin. J. (1979). Variation of meat production capacity in cattle due to
    Genotype and level of feeding: Genotype-Nutrition Interaction, *Livest. Prod. Sci.* 6:263-
    276.

Gibson, J. (1986). Efficiency and performance of genetically high and low milk-producing
    British Friesian and Jersey cattle. Animal Production, 42, pp 161-182.

Gilmour, R.; Bullis, R. Welham, J. & Thompson, R. (1999). *ASREML Reference Manual. NSW
    Agric. Biometric Bulletin. No. 3.* New South Wales Agriculture, Orange Agriculture
    Institute, Orange, NSW, Australia.

Grossman. M.; Kuck, L. & Norton, W. (1986). Lactation curves of purebred and crossbred
    dairy cattle. *J Dairy Sci* 69:195-203.

Gruter, O. (1997). The breeding Aim in the Course of Time. 100 Years Swiss Brown Cattle
    Breeder's Federation, pp 79-80.

Haile-Mariam, M.; Bowman, J. & Goddard, E. (2003). Genetic and environmental
    relationships among calving interval, survival, persistency of milk yield and somatic
    cell count in dairy cattle. *Livest. Production Sci.* 80:189-200.

Hayes, J.; Carrick, M. Bowman, P. & Goddard, E. (2003). Genotype X Environmental
    Interaction for Milk Production of Daughters of Australian Dairy Sires from Test-Day
    Records. *J. Dairy Sci.* 86:3736-3744.

Heringstad , B.; Chang, M. Gianola, D. & Klemetsdal, G. (2005). Genetic analyses of clinical
    mastitis, milk fever, ketosis, in three lactation of Norwegian Red cows. *J. Dairy Sci.*
    88:3273-3281.

Herzog, H. (1997). The breeding Aim in the Course of Time. 100 Years Swiss Brown Cattle
    Breeder's Federation, pp 55-75.

Horan, B.; Dillon, P. Faverdin, P. Delaby, L. Buckley, F. & Rath. M. (2005). The Interaction of
    Strain of Holstein-Friesian Cows and Pasture-Based Feed Systems on Milk Yield, Body
    Weight, and Body Condition Score. *J. Dairy Sci.* 88:1231-1243.

Kodderitzsch, S. & Veillerette, B. (1999). Kosovo: Re-launching the Rural Economy. *A
    medium term reconstruction and recovery program. ECSSD Environmentally and Socially
    sustainable development.* The World Bank working paper No. 19.

Loeffer, H.; De Vries, J. & Schukken, H. (1999). The effect of time of disease occurrence, milk
    yield, and body condition in fertility of dairy cows. *J. Dairy Sci.* 82:2589-2604.

Lowman, G.; Scott, N. & Somerwille, S. (1976). *Condition scoring of cattle. Bull. No. 6. East
    Scotland College of Agric.* Edinburg. Scotland.

Lynch, M., & Walsh, B. (1998). Genetics and Analyses of Quantitative Traits. *Sinauer,
    Sunderland,* MA. pp: 123-127.

McDonald, P.; Edwards, R. Greenhalgh, J. & Morgan, C. (1995). *Animal Nutrition.* Fifth
    edition. John Wiley & Sons, Inc., 605 Third Avenue, New York NY 10158.

Nebel, L. & McGilliard. L. (1993). Interaction of high milk yield and reproductive performance in dairy cows. *J. Dairy Sci.* 76:3257.

Overton, R.; Drackley, K. Ottemann-Abbamonte, J. Beaulieu, D. Emmert, S. & Clark, H. (1999). Substrate utilization for hepatic gluconeogenesis is altered by increased glucose demand in ruminants. *J. Anim. Sci.* 77: 1940–1951.

Pryce, E.; Veerkamp, F. Thompson, R. Hill, G. & Simm, G. (1997). Genetic aspect of common health disorders and measures of fertility on Holstein Friesian dairy cattle. *Anim. Sci.* 65:353-360.

Pryce, M.; Coffey, P. & Brotherstone, S. (2000). The genetic relationship between calving interval, body condition score and linear type and management traits in registered Holsteins. *J. Dairy Sci.* 83:2664-2671.

Pryce, M.; Coffey, P. & Simm, G. (2001). The relationship between body condition score and reproductive performance. *J. Dairy Sci.* 84:1508-1515.

Ray, E.; Halbach, J. & Armstrong, V. (1992). Season and lactation effect on milk production and reproduction of dairy cattle in Arizona. *J. Dairy Sci.* 75:2976-2983.

Rinder production in der Bundesrepublik Deutschland. (2001). (Ed). *Ausgabe 2002.* pp. 38-86. ISSN 1439-8745.

Robinson, J. (1986). Changes in body composition during pregnancy and lactation. *Proc. Nutr. Soc. UK.* 45:71-80.

Rukkwamsuk, T.; Kruip, T. A. M. & Wensing, T. (1999). Relationship between overfeeding and over conditioning in the dry period and the problems of high producing dairy cows during the post parturient period. *Veterinary Quarterly.* 21:71:77.

Santon, L.; Blake, W. Quaas, L. Van Vleck, D & Carabano, J. (1991). Genotype by environment interaction for Holstein in milk yield in Colombia, Mexico and Puerto Rico. *J. Dairy Sci.* 74, 1700-1714.

SAS Insitute Inc. (1999). The GLM Procedure. *In SAS/STAT User's Guide Version 8, Cary,* NC: SAS Institut Inc.

SAS Insitute Inc.(1999). The MIXED Procedure. *In SAS/STAT User's Guide Version 8, Cary,* NC: SAS Institut Inc.

Schmidt, H. (1989). Effect of length of calving on income over feed and variable cost. *J. Dairy Sci.* 72:1605-1611

Schultz-Rajalla, J. & Frazer, S. (2003). Reproductive Performance in Ohio dairy herds in the 1990s. *Animal Reproductive Science.* 76:127-142.

Senatore, M.; Butler, R. & Olenacu, A. (1996). Relationships between energy balance and post-partum ovarian activity and fertility in first lactation dairy cows. *J. Dairy Sci.* 62:17-23.

Simon, L. & Buchenauer, D. (1993). *Genetic Diversity of European Livestock breeds (EAAP publication No 66).* Wageningen Pres. Wageningen, Netherlands.

Sttot, W.; Veerkamp, F. & Wasessell, R. (1999). The economics of fertility in the dairy herd. *Anim. Sci.* 68:49-58.

Svendsen, M.; Skipenes, P. & Mao, L. (1993). Genetic correlation in the feed conversation complex of primiparous cows in the first two trimesters. *J. Animal Sci.* 71:1721-1729.

Svendsen, M.; Skipenes, P. & Mao, L. (1994). Genetic correlation in the feed conversation complex of primiparous cows at a recommended and reduced plane of nutrition. *J. Animal Sci.* 72:1441-1449.

Tekerli. M.; Akinci, Z. Dogan, I. & Ackan, A. (2000). Factors affecting the shape of lactation curves of Holstein cows from Belikesir province of Turkey. *J. Dairy Sci.* 83:1382-1386.

Tiroler Grauvieh- The Tyrol Grey: *Juwel der Berge / hrsg. Vom Tiroler Grauviehzuchtverband.* – Ed. Lowenzahn. (1999). pp: 38-56 ISBN 3-7066-2199-1.

Van Arendonk, M.; Hovenier, R. & De Boer, W. (1989). Phenotypic and genetic association between fertility and production in dairy cows. *Livest. Prod. Sci.* 21:1-12.

Van Arendonk, M.; Nieuwhof, J. Vos, H. & Korver, S. (1991). Genetic aspects of feed intake and efficiency in lactating dairy heifers. *Livest. Prod. Sci.* 29:263-275.

Van der Lende, T. (1998). Physiological aspects of reproduction and fertility in dairy cows. In: Proc. Int Worksh. Genet. Impr. Funct. Traits in Cattle: *Fertility and Reproduction Group, Germany, Interbull Bulletin 18,* 1998, 99 33-39.

Veerkamp, F. & Emmans. C. (1995). Sources of genetic variation in energetic efficiency of dairy cows; a review. *Livest. Prod. Sci.* 44:111-120.

Veerkamp, F.; Simm, G. & Oldham, D. (1995). Genotype by environment interaction – experience from Langhill. Pages 59-66 in Breeding and Feeding the high genetic merit cow. *Occas. Publ. 19. T.L.J. Lawrence, F. J. Gordon, and A. Carson. ed. Br. Soc. Anim. Sci., Edinburgh,* United Kingdom.

Veerkamp, F. (1998). Selection for economic efficiency of dairy cattle using information on live weight and feed intake. A review[1]. *J. Dairy Sci.* 81:1109-1119.

Veerkamp, F. & Thompson. R. (1999). A covariance function for feed intake, live weight and milk yield during lactation, estimated using a random regression model. *J. Dairy Sci.* 82:1565-1573.

Veerkamp, F.; Oldenbroek, K. Van Der Gaast, J. & Van der Werf, J. (2000). Genetic correlation between days until start of luteal activity and milk yield, energy balance, and live weights. *J. Dairy Sci.* 76:3410-3419.

Veerkamp, F.; Koenen, C. & De Jong, G. (2001). Genetic correlation among body condition score, yield, and fertility in first parity cows estimated by random regression models. *J. Dairy Sci.* 84:2327-2335.

Wallnofer, E. 1999. *Tiroler Grauvieh Juwel der Berge/ hrsg.vom Tyroler Grauviehzuchtverband. Insbruck: Ed.* Lowvenzahn, Austria. pp. 13-15. ISBN 3 – 7066-2199-1.

Wildman, E.; Jones, M.; Wagner, E.; Boman, L.; Troutt, F. & Lesch, N. (1982). A dairy cow body condition scoring system and its relationship to selected production characteristics. *J. Dairy Sci.* 65:495-501.

Zamani, P.; Miraei-Ashtiani, S. Alipour, D. Aliarabi, H. & Saki, A. (2011). Genetic parameters of protein efficiency and its relationships with yield traits in lactating dairy cows, *Livestock Science*, doi:10.1016/j.livsci.2011.01.006. In press (available online).

Zogg, M. (1997). The breeding program.*100 Years Swiss Brown Cattle Breeder's Federation*, pp: 43-53.

Chapter 14

Differential Characteristics of Milk Produced in Grazing Systems and Their Impact on Dairy Products

A.M. Descalzo, L.Rossetti, R. Páez, G. Grigioni,
P.T. García, L. Costabel, L. Negri, L. Antonacci, E. Salado,
G. Bretschneider, G. Gagliostro, E. Comerón and M.A. Taverna

Additional information is available at the end of the chapter

1. Introduction

Functional foods may exhibit health benefits beyond their nutritional value. Many traditional recommendations on food selection have included this view. In more recent years the interest in food with specific health benefits has greatly increased and stimulated the development of respective products for the food market. At the same time large efforts are made to substantiate health claims by validated experimental methods.

Dairy products have come to play an important role in this context. Regarding the nutritional and extra nutritional characteristics of milk, it can be considered a multifunctional food product. Milk is a source of proteins, lipids, vitamins and minerals, but also exerts other beneficial properties due to the presence of numerous bio-active molecules either through the direct consumption of milk or its derivatives (Guimont and others, 1997).

It is widely recognized that pasture, when efficiently grazed, is the cheapest feedstuff (Dillon, 2006) and that the cost of milk production is reduced as the proportion of grazed pasture in the diet of the cow increases (Dillon and others, 2008). Nevertheless, using pasture as the only source of food is inadequate for high milk producing cows. The lower intake of dry matter (DM) and energy would be the main cause of suboptimal production even under conditions of adequate quantity and quality of pasture (Kolver y Muller, 1998; Reis y Combs, 2000; Bargo and others, 2002).

Cows adapt to grazing through increased forage intake, but in order to achieve this, it is necessary to ensure around 60 kg DM/cow/day (Delagarde and others, 2004). Under these conditions, the maximum yield obtainable on high quality pastures seems to be at around 22-25 kg/cow/day in spring and 18 kg/cow/day in autumn. However, at high levels of energy supplementation, rumen pH, the diurnal pattern of volatile fatty acid concentration and fiber digestion may be altered because of changes in rumen fermentation (Tamminga, 1993). These changes may provoke lower consumption of forage and a drop in the fat content of milk. In this regard, studies carried out in EEA INTA Rafaela suggest that the optimal level of supplementation that could maximize milk production without affecting the rumen of dairy cows grazing high quality lucerne, would be 7 kg of concentrate containing approximately 70% corn (Salado and others, 2010 a and b).

On the other hand, it is also known that the amount of conjugated linoleic acid (CLA) in milk fat increases according to the level of pasture included in the diet, with the highest CLA concentrations found in cattle fed 100 % pasture (Dhiman and others, 1999).

Alike other country, Argentina is moving forward to more complex livestock production systems. In this regard, component feeding systems based on the use of grazed pasture are gradually being replaced by confining ones, in which conserved forages (hay, silage) and concentrates (cereal grains, agro-industrial byproducts) are fed to dairy cattle as a total mixed ration (TMR). This phenomenon could be explained by the fact that argentinean dairy farmers are looking for ways to release land for agriculture (mainly for the cultivation of soybean [Glycine max]), which is considered a more profitable and simpler activity than dairy farming.

Nowadays, it is supported that the extreme intensification of dairy production systems would lead to lose not only the advantage that grazed pastures have on the reduction of milk production costs (Garcia and Fulkerson, 2005; Dillon and others, 2008; Macdonald and others, 2008, McEvoy and others, 2008) but also the marketing niches that look for dairy products from pasture-fed cattle due to their high CLA content (White and others, 2002). In this respect, pasture-fed dairy cows produced a significantly higher concentration of CLA (83%) than TMR fed-confined dairy cows (0.66 vs. 0.36% of the total fatty acids) (White and others, 2001).

For that purpose, grass-based dairy production should consider some productive aspects such as high-quality pasture and livestock adapted to a high-forage diet.

Pasture is a highly organized structure which changes with season and management. These changes can affect photosynthesis and hence dry matter accumulation, the degree of utilization and the feeding value of herbage. At the same time, these changes in pasture can affect the quality of milk and thereby the quality of dairy products.

2. Functional compounds characterized in diets for dairy cows

2.1. Natural antioxidants: Lipid-soluble vitamins

Plant secondary metabolites are incorporated into the tissues through dietary delivery. Among these, alpha- tocopherol is the main active form of vitamin E in animal tissues. It is selectively incorporated among eight isomers (alpha, beta, gamma and delta tocopherol and tocotrienol respectively) that are naturally found in plants. Vitamin E is the main fat-soluble antioxidant incorporated into animal cell membranes and is also a regulator of cellular functions that are mediated by protein kinase C.

Carotenoids are lipophilic secondary plant and algae products consisting of eight isoprene units (C40). Among about 800 known carotenoids, up to ten of them have been determined in forages, namely oxygen-containing xanthophylls (lutein, epilutein, antheraxanthin, neoxanthin, violaxanthin and zeaxanthin), and carotenes (preferably β-carotene) of a hydrocarbon nature. Due to the high number of double bonds, carotenoids undergo oxidation and isomerisation (cis-isomers formation from all-trans-ones), in cut forage by the action of light. Carotenoids in cow's milk and consequently in milk products are important for human health and nutrition as natural antioxidants and some of them as precursors of vitamin A, among which all-trans-β-carotene is the main provitamin. Moreover, milk fat carotenoids cause the yellow color of butter and many cheeses, which is positively perceived by many consumers as "green image" because of its association with grazing animals. This vitamin is implicated in reproduction, immunity and normal function of retinal cells. Lutein is another carotenoid which is abundant in plants and can be used as a tracer of pasture feeding, as animals are not able to synthesize this molecule and it is stored in the animal's fat after absorption and thus found in milk and meat (Prache 2005).

These molecules are highly variable in forages, silage and hay (Müller 2007; Hidiroglou 1996), and depend on the fermentation form as well as on the conservation method. Praché and others (2009) reported values between 40 and 123 microg/g DM for beta carotene (60 microg/g DM for lucerne) and from 167 to 437 microg/g DM for lutein (142 for lucerne) in pasture collected between Spring (May) and Summer (July).

Similar variation in carotenoid levels in mountain grazing was found by Calderón and others (2006) with values that ranged between 18 and 60.4 microg/g DM for beta carotene and between 96 and 262 microg/g DM for lutein.

Data from four separated experiments, which were carried out in INTA, also show a high variability. As shown in table 1, fresh pasture showed the highest input of alpha- tocopherol and beta-carotene among feed components. The variation of its composition is high, depending on seasonal variations, phenological stage and conservation technology, among other factors. Other components of feed are also sources of dietary antioxidants, but they showed less variation and their contribution in vitamins was less significant compared with pasture.

2.2. Natural antioxidants: Polyphenols

Polyphenols are antioxidants that may prevent various pathologies (cancer, cardiovascular diseases, immune deficiencies, etc.). Particularly, isoflavones may have a preventive effect against breast and prostate cancers and, to a certain extent, against some disorders related to menopause or andropause. The active molecules are plant-based isoflavones (directly absorbed) as well as a transformation component in the large intestine, equol. The main source of isoflavones in human diet are soy and soy-derived food. Among forage plants, other legumes (certain types of clover and lucerne) contain high levels of isoflavones. They may constitute up to 2% of dry matter in purple clover and are well preserved after silaging. Of the active isoflavones in forage, daidzein and formononetin were transformed in the rumen into equol, which was secreted in the milk of goat and cow species in sufficient quantities to suggest they may exert a biological effect. Moreover, Besle and others (2010) found different phenolic compounds that can be used as tracers for the diet of dairy cows. Data shown in table 1, indicate that coumestrol was the main phytoestrogen found in lucerne (pasture and hay), corn silage, sunflower expeller and wheat bran. Concentrate mixture showed higher amounts of daizein, than other dietary compounds, probably derived from soy. In this experiment, the aglycone and glucoside forms were not differentiated. The extraction was performed in acid medium and therefore, the species were converted into the aglycone form.

Antioxidants (mg/Kg DM)[a]	Lucerne pasture	Oat pasture	Concentrate mixture	Soy expeller	Sunflower expeller
Gamma-tocopherol	19.679	9.10	67.231	101.538	3.519
Alfa-tocopherol	64.415	16.14	15.937	7.042	20.720
Beta-carotene	97.518	66.67	0.903	0.301	nd
Genistein	2.165	Not determined	1.110	Not determined	2.884
Lutein	481.92	1128.21	85.80	Not determined	Not determined
Daidzein	5.568	Not determined	14.061	Not determined	2.035
Coumestrol	25.821	Not determined	10.993	Not determined	11.943
% CV	61.98	17.15	34.51	nd	21.13
Antioxidants (mg/Kg DM)	Corn silage	Sunflower pellets	Wheat Bran	Lucerne Hay	
Gamma-tocopherol	20.111	1.521	48.809	3.345	
Alfa-tocopherol	17.702	2.445	27.533	24.612	
Beta-carotene	13.243	0.355	nd	5.493	
Genistein	1.593	Not determined	1.085	2.103	
Lutein	41.57	Not determined	Not determined	Not determined	
Daidzein	5.070	Not determined	1.665	1.962	
Coumestrol	22.731	Not determined	14.652	18.461	
% CV[b]	35.75	nd	9.95	38.68	

[a] For the determination of antioxidant vitamins, samples from individual components of the diet were extracted with hexane (after a saponification step) and quantified using high performance liquid chromatography. Carotenoids were monitored at 445 nm, tocopherols were detected by fluorescence at 296-330 nm (λ exc- λ em respectively) and retinol at 335 nm (Rossetti et al, 2010).
[b] Mean variation coefficient of all vitamins calculated for each dietary compound.

Table 1. Levels of fat-soluble antioxidants in dietary components used in INTA experiments with dairy cows.

2.3. Volatile compounds

Terpenes are direct biomarkers of animal diet as they are not modified by animal metabolism. In addition, the transfer of these compounds into milk fat is very fast and apparently exhibits no memory effect (Viallon and others, 2000). Therefore, their presence in milk and dairy products could be used to discriminate contrasting feeding conditions (Prache and others, 2005).

Lucerne pasture showed the highest volatile production (table 2). The most abundant terpene was d-limonene found in all dietary components. Only three minor compounds, nerol, menthol and linalyl acetate were found exclusively in lucerne. Terpinen-4-ol, was found in lucerne pasture and maize silage in a 2 to 1 proportion. In addition, ocimene was 4.5 times higher in lucerne pasture compared with the other components and linalool was 4 times higher than in maize silage and wheat bran. These results indicate that a group of terpenes, and their abundance and not a single compound could be an indicator of the different components of the diet.

Terpenoids (Relative Area Units)[a]	Lucerne pasture	Oat pasture	Concentrate mixture	Sunflower expeller	Corn silage	Wheat Bran
d-limonene	52.919	2.392	3.586	0.251	8.577	1.126
p-cymene	0.109	0.086	0.026	nd	0.009	nd
ocimene	0.164	nd	0.009	0.062	nd	0.077
linalool	0.985	nd	nd	0.072	0.116	0.136
menthol	nd	nd	nd	nd	nd	nd
terpinen-4-ol	nd	nd	nd	nd	0.002	nd
β-cyclocitral	0.412	0.051	0.004	nd	0.107	nd
nerol	0.052	nd	nd	nd	nd	nd
linalyl-acetate	0.021	nd	nd	nd	nd	nd
β-caryophyllene	0.061	nd	0.018	0.017	0.071	0.044
geranylacetone	0.089	nd	0.024	nd	0.122	0.028
%CV	36.10	12.01	14.29	17.45	18.07	8.83

[a] Volatile terpens were determined in individual feedstuff given to dairy cows in INTA experiments (2008-2009). For this assay, a fiber (CAR/PDMS/DVB 2cm-50/30mm) was used to extract/concentrate volatile compounds in the headspace (HS) of chopped samples. GC-FID was performed on a Shimadzu 14-B GC with a flame ionization detector (Negri et al., 2008).

[b] Mean variation coefficient of all volatiles calculated for each dietary compound.

Table 2. Volatile terpenoid compounds in dietary components used in INTA experiments with dairy cows.

Lucerne samples also presented a higher proportion of volatile aldehydes (table 3), at the expense of trans-2-hexenal, which was the most abundant volatile after d-limonene. Tava and Peccetti (1997) reported that trans-2-hexenal was the most frequent volatile in 13 genotypes of lucerne. This compound, originated in flowers, should have a physiological role for plant pollination by hymenopterans (Peccetti and others 2002).

Hexanal and trans-2 hexenal are end-products of the oxidation of C18:2 and C18:3 fatty acids respectively. In fact, fresh pasture shows the highest hexanal production, resulting from linoleic acid, whereas trans-2-hexenal indicates a high percentage of C18:3. These fatty acids are oxidized when fresh pasture is cut, producing the corresponding aldehydes. Both aldehydes and terpenes are responsible for the typical odour of different plant species and could be found in milk and dairy products.

Aldedhydes (Relative Area Units) [a]	Lucerne pasture	Oat pasture	Concentrate mixture	Sunflower expeller	Corn silage	Wheat Bran
3-methyl-butanal	0.161	0.022	0.026	nd	4.877	0.020
pentanal	0.363	0.016	0.008	nd	0.036	0.013
hexanal	0.889	0.360	0.130	0.192	0.129	0.190
trans-2-hexenal	6.689	1.911	nd	nd	0.302	0.030
heptanal	0.056	nd	0.018	nd	0.091	0.038
trans-2-heptenal	0.175	0.067	0.058	0.154	0.292	0.176
octanal	0.215	nd	0.041	nd	0.143	0.017
nonanal	0.235	0.039	nd	0.158	nd	0.205
trans-2-nonenal	0.292	0.019	nd	nd	0.022	0.081
%CV [b]	30.08	9.66	14.04	14.22	6.62	5.11

[a] Volatile aldehydes were determined as indicated in table 2.

[b] Mean variation coefficient of all volatiles calculated for each dietary compound.

Table 3. Volatile aldehydes in dietary components used in INTA experiments with dairy cows.

2.4. Fatty acids

Information available about the fatty acid composition of pasture lipids is scarce. Unsaturated fatty acids, particularly n-3 polyunsaturated fatty acids (PUFA) and conjugated linoleic acid (CLA) may be beneficial to human health. Therefore, maximizing C18:3 and CLA in milk and dairy products would benefit human health and nutrition. Milk and meat are the only significant source of CLA in the human diet and this appears to be related to the consumption by ruminant of fresh pastures (Elgersma and others, 2003).

Sources of variation in pasture lipid composition are plant species, growth stage, temperature and light intensity. Fresh grass contains a high proportion (50-75%) of its total fatty acid content in the form of n-3 linolenic acid. Levels of linolenic acid vary with plant factors such as stage of maturity and with light disponibility. Fatty acid profiles are distinctive to particular species, which confirm that fatty acid composition of grasses is under considerable genetic control (Dewhurst and others 2001). This offers the potential to select for grasses with higher concentrations or altered types of fatty acids.

Garcia et a. (2007) evaluated the range of fatty acid concentrations within various pastures and forage crops, at different cutting dates and season in Argentina for beef and milk production. Season affected ether extract (EE%) and fatty acid composition of lucerne cultivars (table 4). No interactions were detected between dormancy group and season (p<0.344). During the fall increased C18:2 and decreased C16:0, C16:1, C18:0. The differences between spring and summer were for C16:1, C18:0, C18:2 and C18:3. The EE% and contribution of C18:2 and C18:3 were higher in fall compared with the other seasons.

The fatty acid composition, percentage of ether extract and contribution (g/100g DM) of C18:2 and C18:3 of several cultivars are shown in tables 5 and 6. Cutting date has affected significantly these parameters.

SEASON	C16:0	C16:1	C18:0	C18:1	C18:2	18:3	EE%	18:3 mg	18:2 mg
Spring	29.4b	2.4c	5.6b	6.8	16.1a	39.4b	2.0a	0.80a	0.33a
Summer	31.0b	1.8b	6.6c	6.9	18.9b	33.9a	2.2a	0.75a	0.41b
Fall	20.0a	1.5a	3.2a	7.6	27.4c	40.8b	3.3b	1.33b	0.89c
SE	0.80	0.09	0.22	0.61	0.66	1.05	0.07	0.04	0.02

a b c Means in the same column with different letters differ (p<0.05)

Table 4. Fatty acid composition (%), ether extract (EE) and mg/g EE of C18:2 and C18:3 of different lucerne cultivars according to season.

		C16:0	C16:1	C18:0	C18:1	C18:2	C18:3	EE%
RGB	Cut 1	18.7c	2.0b	2.5c	4.5B	10.0 b CD	62.2a	23.3
	Cut 2	23.8ab	1.9bc	4.4ab	5.6B	11.8 a CD	52.4bc	5.5
RGF	Cut 1	19.6c	1.8bc	2.6c	4.7B	8.5b D	62.6a	22.7
	Cut 2	24.8ab	1.9bc	4.5ab	5.8B	10.9a D	51.6bcd	4.9
TC	Cut 1	21.3bc	2.0b	3.3abc	7.2AB	12.2b BC	53.8ab	22.4
	Cut 2	25.8a	1.8bc	4.9a	8.0AB	12.7a BC	46.1bcd	15.7
TG	Cut 1	24.7ab	2.5a	3.0bc	5.9AB	12.4b AB	51.6bcd	23.3
	Cut 2	25.4a	2.0b	4.0abc	8.6 AB	15.0a AB	45.3bcd	6.5
TDS	Cut 1	22.3abc	1.9b	3.5abc	7.3A	12.1b AB	52.8bc	21.9
	Cut 2	24.0ab	1.5cd	3.9abc	9.1A	15.7a AB	45.8bcd	16.5
CQ	Cut 1	25.4a	1.7bcd	4.8a	10.2A	13.4b A	43.1d	21.9
	Cut 2	25.9a	1.4d	3.4abc	8.3A	16.5a A	44.3cd	16.7
Pasture		***	***	NS	***	***	***	
Cut		***	***	***	NS	***	***	
Interaction		**	**	**	NS	NS	***	

Whole milk powder from pasture produced milk showed the same fatty acids profile as raw milk (table 5), and could constitute a natural source for n-3 fatty acids. In a similar way, antioxidants profile reproduced the composition of raw milk with estimated losses around 20 and 40% being higher for retinol and vitamin D3.

Higher natural antioxidants in whole milk powder lowered TBARS values in pasture (A70) samples compared to silage samples (A0) after 18-months shelf life assay (20°C in sealed bi- laminated plastic pouches under nitrogen atmosphere).

M: Raw Milk; WMP: Whole Milk Powder; CH: Cheese (regianitto).

a: Thiobarbituric Reactive Substances

b: Relative Area Units

c: Ferric Reducing antioxidant activity

d: A0, A35 and A70, feed treatments with 0, 35 and 70% Lucerne pasture

ND: not done; nd: not determined, NS: Not significant

* p<0.05 **1 p<0.01 ***p<0.001

Table 5. Fatty acid composition (%) and ether extract (EE) (g/kg DM) of several pastures and forage crops. RGB (Rye grass Bill), Rye grass Florida (RGF), wheat Charrua (WC), wheat Guapo (WG), triticale Don Santiago® (TDS) and Centeno Quehue(CQ).

		C18:2	C18:3	C18:2+ C18:3	18:3/18:2
RGB	Cut 1	0.23b CD	1.5a	1.7a	6.4a
	Cut 2	0.18a CD	0.8cd	1.0e	4.5bc
RGF	Cut 1	0.19b D	1.4a	1.6ab	6.4a
	Cut 2	0.16a D	0.8cd	0.9e	4.8b
TC	Cut 1	0.27b BC	1.2b	1.5bc	4.4bc
	Cut 2	0.20a BC	0.7d	0.9e	3.7bcde
TG	Cut 1	0.29bAB	1.2b	1.5bc	3.9bcde
	Cut 2	0.25a AB	0.8d	1.0e	3.0e
TDS	Cut 1	0.26b AB	1.2b	1.4c	4.4bcd
	Cut 2	0.26a AB	0.8d	1.0e	3.1de
CQ	Cut 1	0.29b A	0.9c	1.2d	3.4cde
	Cut 2	0.28a A	0.7d	1.0e	2.7e
Pasture		***	***	***	***
Cut		***	***	***	***
Interaction		NS	*	***	**

* p<0.05 **p<0.01 ***p<0.001, NS: not significant

Table 6. Contribution (g/100 g DM) of C18:2 n-6 and C18:3 n-3 of several pastures and forage crops

As expected, fatty acids profile was different among dietary components. As shown in table 7, the most abundant fatty acid in lucerne pasture was C18:3 and it can be the molecule that originates branched aldehydes, especially trans-2-hexenal, in this assay.

Fatty Acid (%)[a]	Lucerne pasture	Lucerne hay	Concentra te mixture	Sunflower expeller	Corn silage	Wheat Bran
C16:0	33.21	38.42	16.84	15.72	27.97	18.02
C16:1	5.14	3.79	0.39	0.48	1.67	0.58
C18:0	9.06	8.73	2.57	4.78	7.89	1.69
C18:1	17.20	14.56	27.31	30.01	16.74	21.23
C18:2	13.21	18.41	49.58	48.57	37.14	54.21
C18:3	22.06	15.89	3.24	nd	8.51	4.03
18:3/18:2	1.69	0.82	0.06	nd	0.23	0.07
%CV [b]	14.57	32.88	13.69	13.29	28.29	8.25

[a] mean of samples from years 2007 and 2008.

[b] mean variation coefficient of all fatty acids calculated for each dietary compound, nd: not detected.

Table 7. Fatty acids in dietary components used in INTA experiments with dairy cows.

3. Incorporation of antioxidant vitamins into milk

3.1. Properties of milk produced on pasture

The nature of cow's forage diet, i.e. botanical composition, maturity stage and preservation mode, strongly influences milk composition in fatty acids, vitamins and carotenoids (Chillard 2001, Hartman 1965). The contents of retinol, α-tocopherol and β-carotene, lutein, xanthophylls, saturated and polyunsaturated fatty acids profile in plasma, milk and milk fat are influenced by the diet. In addition, some of these parameters also differed according to sire and stage of lactation (Lucas 2006, Soren 1999).

Calderón and others (2007) reported that the incorporation of grass silage and lucerne protein (75:25), in an experimental diet designed for Montbéliarde cows in midlactation induced a rapid increase in plasma concentrations of beta carotene and vitamin E. The incorporation of these compounds varied in a linear form with the proportion of grass within the diet. Similar responses were observed in milk for vitamin E, whereas there was an apparent saturation in milk concentrations of beta carotene at high levels of carotenoid intake, i.e., when plasma beta carotene exceeded 5 microg/ml.

In an experiment conducted in INTA-Rafaela (Santa Fe, Argentina), two isoenergetic diets (1.55 Mcal EN/kgDM) of contrasting nature: 70% lucerne pasture (ALF) and grain sorghum silage (SS) were offered to Holstein cows on their second third of lactation. Milk and protein yields were higher (P < 0.01) in ALF than SS diet (31.78, 26.14 l/v/d and 1.019, 0.841 kg/v/d, respectively). Fat concentrations were lower on ALF than SS diet (P < 0.01). All remaining variables were not significantly different, with the only exception of total solid contents (P < 0.05). No significant differences were found on live weight variation (ALF= 0.084, SS= 0.271 kg/v/d).

As shown in figure 1, fat-soluble vitamins increased over the first 20 days after a dietary shift from silage to grazed pasture (70% lucerne), and the concentrations were stabilized at day 40 with persistence until day 60 (Rossetti and others, 2010). Similarly, Calderón and others (2007) observed a rapid increase during the first 14 days in antioxidant vitamins in plasma and milk of Montbéliarde dairy cows, after a dietary shift from a low-carotenoid diet based on hay and concentrates to a high-carotenoid diet based on grass silage that was prepared from perennial ryegrass (*Lolium perenne*). The concentration of α-tocopherol, 13-cis-β-carotene, all-trans-β-carotene and lutein reached a plateau between 21 and 28 days and persisted until day 42. The incorporation of zeaxanthine into plasma was reported but it was either not detected in milk (Calderón et al., 2007), or its concentration was around 1% of total carotenoids (Butler 2008).

3.2. Effect of increasing pasture content into a total mixed ration diet in relation to health promoting compounds

An increase of pasture in the diet, at the expense of a total mixed ration (TMR), has been shown to improve the incorporation of antioxidant vitamins into milk. Figure 2 shows the relationship between the percentage of pasture (grazed oat pasture) vs. α-tocopherol, retinol, all-trans-β-carotene and lutein in milk after 30-day treatments.

mg/g fat

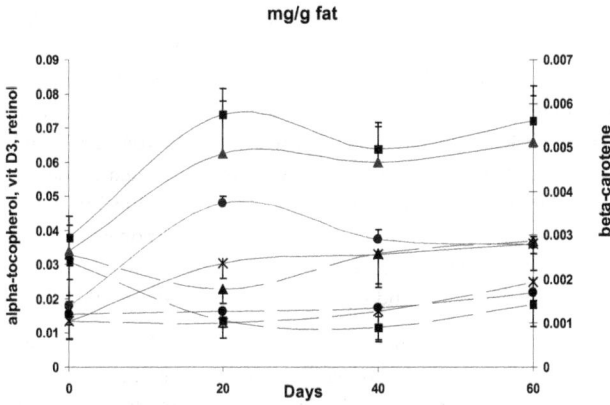

Figure 1. Lucerne or silage feeding. Evolution in α-tocopherol (●), all-*trans*-β-carotene (■), retinol (▲) and vitamin D3 (*) in milk of dairy cows fed diets differing in ALF (solid lines) or SS (dashed lines). All groups were fed sorghum silage during a pre-experimental period of 6 wk. Means ± SEM for 5 cows per group are presented.

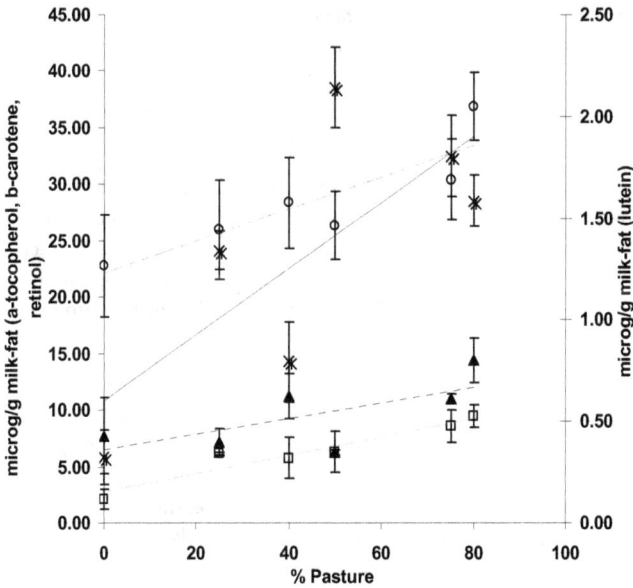

Figure 2. Oat pasture vs. TMR feeding. Concentration of α-tocopherol (○), all-*trans*-β-carotene (□), retinol (▲) and lutein (x) in milk-fat of dairy cows fed diets differing in the percentage of oat pasture. Linear tendency for α-tocopherol and retinol (dashed lines) and all-*trans*-β-carotene and lutein (solid lines) are shown. All groups were fed corn silage during a pre-experimental period of 6 wk. Each point represents eight cows per treatment. Data were compiled from two separated experiments with comparable diets.

Vitamin A in milk is incorporated under the form of retinol and carotenoids. Among these, all-trans-β-carotene was the mayor form recovered followed by lutein, which was consistent with what is generally reported. Although lutein was the major carotenoid found in pasture, its incorporation rate into milk was 10-fold lower than the incorporation of all-trans β-carotene. The same observation has been reported by Calderón (2007) and Butler (2008). Recovery of these species of vitamin A in milk was directly related to their concentration in feed. Pearson correlation coefficients for all-trans-β-carotene and lutein in feed and milk were R=0.8323 and R=0.7521 respectively. Concerning the lipoperoxyl radical-scavenging activity of all-trans retinol it behaves as a more effective antioxidant at a low partial oxygen pressure, low retinol concentration and high radical flux (Tesoriere 1997)

In a revision article, Nozière and others (2006), have discussed that the extent of carotenoid degradation by microorganisms in the rumen remains uncertain because of the wide range of results, mostly on β-carotene, from in vitro and in vivo studies. Whereas some authors reported no degradation (Dawson and Hemington, 1974; Cohen Fernandez and others, 1976a), others found moderate (10–25%; Davison and Seo, 1963; Potkanski and others,1974; Cohen Fernandez and others, 1976b; Mora and others, 1999) or higher β-carotene disappearance (40–55%; King and others, 1962). The carotenoid supplement form could explain discrepancies among experiments, because degradation rates were usually higher when carotenoids were supplied as purified products than when provided in forages. This hypothesis was recently confirmed in vitro at the INRA in a study where no apparent degradation was detectable for lutein provided in forage, whereas 50% of the initial amount disappeared when the same quantity of lutein was added as a pure commercially available source.

Vitamin E concentration ranged from 16.5 to 46 microg/g milk-fat for milk from silage or pasture diets respectively. Pasture values were even higher than concentrations reported by Al Mabruk (2004) for grass silage supplemented with vitamin E (22.26 microg/g milk fat). Calderón and others (2007) showed an increase from 8 to 14 microg/g milk fat depending on the proportions of experimental feed (75% grass silage + 25% lucerne protein concentrate) in the basal diet (from 0 to 66%).

Four different assays conducted in INTA (Argentina) indicate that after a shift from silage to pasture diet α-tocopherol augmented from 16.7 to 40.37 (70% grazed lucerne); from 11.59 to 26.05 and 28.53 (35 and 70% grazed lucerne pasture); 22.78 to 36.91 (80% grazed oat pasture) and 24.36 to 30.42 (75% oat pasture). In fact, Schingoethe and others (1978) reported a high variability in the concentrations of vitamin E in milk, depending on the grazing conditions. The concentration augmented from 20 to 50 microg/g milk fat at the time cows started grazing pasture. The same observation was reported by La Terra and others (2010) with an increase of 37% and 68% in α-tocopherol for cows fed 30:70 and 70:30 ratio for pasture:TMR, respectively in comparison to TMR diet. Therefore, results from INTA experiments showed the same tendency. However, one experiment conducted in autumn 2007, showed no differences between maize silage or pasture (lucerne) fed cows with an average of 27.5 microg/g milk fat, thus indicating that the quality of the diet and the pasture affects the incorporation of vitamin E into milk.

Vitamin D may be consumed in the diet as either ergocalciferol (D2) from plant sources or cholecalciferol (D3) from animal sources. With the exception of fatty fish, relatively few foods are naturally rich in vitamin D. In mammalians with ample sunlight exposure, the greatest source is endogenous vitamin D produced when 7-dehydrocholesterol in the epidermis and dermis of the skin is converted into vitamin D3 after exposure to ultraviolet B radiation. Vitamins D2 and D3 from dietary sources are transported to the liver, and from there to the tissues. Vitamin D3 from skin is transported through plasma. Therefore more evidence is needed to understand the factors that induce its variation in milk content. Consequently, for vitamin D, the dairy industry proposed technologically vitamin D-enriched dairy products to consumers (Graulet 2010). The vitamin D3 endocrine system has mostly been studied for its role in calcium and phosphorus metabolism and its possible role as an antioxidant has been neglected. Sardar and others (1996) demonstrated that D3 treatment brought about similar reduction in the extent of lipid peroxidation and induction in superoxide dismutase (SOD) activity, as with vitamin E supplementation in rats.

The concentration of vitamin D3 showed different tendencies among experiments. In one experiment, it ranged from 5.85 ± 1.21; 8.01 ± 2.13 and 10.29 ± 2.34 microg/g milk-fat for diets containing 0, 40 and 80% oat pasture respectively, showing a linear increase with the proportion of pasture in the diet. This observation agreed with the described in figure 1, although values were lower (average of 30 microg/g milk-fat, in figure 1 with 70% of lucerne pasture diet vs. 10 microg/g milk-fat, with an 80% oat pasture diet). In contrast, in a second experiment the concentrations were 0.80; 0.91; 1,98 and 1.34, without significant differences (P >0.05) for diets containing 0, 25, 50 and 75% of oat pasture. Therefore, a tendency for higher values of vitamin D3 with the increase of grazed pasture was observed but further assays are necessary to describe the productive parameters that may be used to improve the concentration of natural vitamin D3 in cow milk.

3.3. Effect of increasing concentrate content on grazed pasture.

The addition of concentrate on a pasture based feed was studied in two separated experiments: total feed assignment around 30 and 31.9 kg DM/head/day in experiments 1 and 2 respectively. Pasture consumption diminished with the increase of concentrate within the diet: 18.0; 16.1 and 14.2 kg DM/ head/day for 3.5; 7.0 and 10.5 kg of concentrate respectively in experiment 1 but not in experiment 2. In both cases, the addition of concentrate had no effect on the concentration of antioxidants in milk. The average concentrations were 5.34; 37.25; 1.09 and 6.51 microg/g milk fat for retinol, α-tocopherol, γ-tocopherol and β-carotene respectively. These values resulted similar to those obtained with pasture proportions above 40% in the diet (figures 1 and 2).

Calderón et. al, 2007 demonstrated that the transfer of β-carotene from plasma to milk reached a plateau when plasma levels exceeded 5 microg/ml and resulted a limiting factor in terms of the secretion of this antioxidant, that can be due to a limited uptake by the mammary gland or limited transport by binding to transport proteins like β-lactoglobulin (Dufuor and Hartle, 1991) and/ or to saturation of milk fat globules.

It can be concluded that good quality pasture favored a linear increase of antioxidant vitamins in milk, and that this increment was not affected by the addition of concentrate in the feed.

3.4. Functional fatty acids in milk

There is a growing interest in CLA, considered to be beneficial in prevention of carcinogenesis (Ip and others, 1999), the n-6 to n-3 polyunsaturated fatty acids (PUFA) relationship, and the low percentage of saturated fatty acids, as these conditions are beneficial for the prevention of cardiovascular diseases. Their percentages in milk products can be increased through a suitable dietary regimen. Changes in milk fatty acids profile are possible due to the plasticity of milk fat (reviewed in Chillard and others 2000).

Dietary changes after shifting from silage to lucerne pasture diet induced significant changes in some fatty acids. Saturated fatty acids showed higher values in lucerne than in silage produced milk (P<0.05): C12 (2.79 ± 0,123 vs. 2,45 ± 0,132), C14 (10,79 ± 0,344 vs. 9,30 ± 0,359); C16 (29,63 ±0,617 vs. 26,26 ± 0,650); C17 (0,72 ± 0,012 vs. 0,64 ± 0,013) and C18 (8,84 ± 0,504 vs. 12,22 ± 0,563) respectively throughout the experiment. Conversely, silage produced milk showed higher values for C18:1c (21,00 ± 0,823 vs. 25,22 ± 0,840); C18:2c (2,63 ± 0,085 vs. 2,98 ± 0,098) and C18:2t (0,40 ± 0,029 0,15 ± 0,031) than lucerne produced milk. In addition, pasture diet favored the incorporation of C18:3 (1,03 ± 0,051 vs. 0,54 ± 0,054) into milk. In this experiment, C18:1t (3.80 ± 0,269 vs. 3.51 ± 0.285) and CLA (1.52 ± 0.096 vs.1.25 ± 0.107) values for lucerne and silage milk respectively showed similar results (P>0.05). This result was not expected as Castillo and others (2006) reported a positive association between proportion of lucerne pasture in the diet and content of cis9 trans11 CLA and C18:1t in milk under similar experimental conditions within the same geographical region.

A second experiment was conducted in INTA with increasing percentages of lucerne pasture. Twenty four Holstein cows from the experimental herd (6 heifers and 18 multiparous) in mid lactation, were randomly assigned to three different treatments. During a first pre-experimental period of 7 days, considered as co-variable, all cows were fed the same diet, in a second period each group was assigned to their treatment during 21 days and in the third period of 28 days, each group received the corresponding diet, with 0 (A0 or maize silage control diet), 35 (A35) and 70 % (A70) lucerne pasture. Diets were isoenergetic and isoproteic. Lucern pasture was cut daily and offered to the cows. Rations were fed as TMR (20.34 kg DM/cow/day, in average).

Milk production, protein and urea content showed lower values for lucerne groups compared with milk from silage-fed cows (P<0.05). For other parameters, milk from the three treatments showed a similar overall quality.

As mentioned above, the proportion of pasture in the diet influenced the composition of saturated fatty acids (SFA) in milk (table 8). C4, C15, C16 and C17 FAs were higher in A70 milk compared with A35 and A0 treatments (P < 0.05). C12 FA was higher in A0 than in A35 and A70 milk (P < 0.05). Total C18:2 FA was higher in A0 milk, but differences were observed for individual isomers. Trans-9, trans-12 C18:2 and cis-9, trans-12 C18:2 showed similar percentages in all milk samples. However, trans-9, cis-12 C18:2 increased with the higher content of lucerne in the diet (A70 compared with A35 and A0), and conversely, cis-9, cis-12 C18:2 decreased with an increase of lucerne (P < 0.05). These results differed from Schroeder and others (2003) that reported higher levels of trans-vaccenic and c9 t 11 CLA in milk from pasture (Avena sativa L.) fed cows compared to milk from TMR fed cows. The mean percentage of CLA in argentine milk from the central Region was reported to be 1.2 (Castillo and others, 2006; Páez and others, 2007). The nature of the differences with previous reports could rely on the quality of pasture and its highly variable composition, as discussed in the section above. The mean C18:3 n-3 FA content in lucerne milk was approximately 1.7 and 2.9 times higher (A35 and A70 respectively) than in silage milk (P< 0.01). In addition, C 20:5 n-3 showed an lucerne-dependant increase (A70> A35 > A0). Therefore, the incorporation of lucerne pasture in the diet was associated with a higher proportion of milk n-3 FA compared with maize silage diet. Indeed, the plot of C 18:2 vs. C 18:3 for milk samples showed a differential distribution of milk according to the diet and could therefore be used to differentiate the incorporation of lucerne into the diet of dairy cows (figure 3).

This result is consistent with other results shown for grazed systems (Chillard and others 2001; Castillo and others 2006; La Terra and others 2010). The source of the variability remains to be further studied and strategies different that using pasture as the unique factor to improve trans- vaccenic acid and CLA in milk continue under experimentation.

Low percentages of CLA (around 0.34 %) was found to be associated to cows fed fermented roughage and concentrates (most intensive production farm) whereas high percentages (around 0.80 %) was found in the ecologically produced milk fat. The concentration of CLA correlated positively with and trans-vaccenic acid (Jahreis 1997). The high C18:3 content of young grass and its low fibre content interact to increase the production of CLA or its trans C18:1 precursors (Chillard and others 2000). The presence of lipid precursors is one of the factors described by Griinari and Bauman (1999), to improve CLA biosynthesis. The second factor are the changes associated with the microbial activity associated with ruminal biohydrogenation, which is incomplete and leads to an accumulation of trans vaccenic acid, the precursor of CLA biosynthesis. In addition, factors that regulate these patways as well as the activity of the Δ-9 desaturase in the mammary gland and its regulation are a matter of investigation.

CLA isomers exert different biological activities. Juárez and others (2010) have found that feeding 1068 IU vitamin E, reduced the total trans-18:1 content in backfat (P < 0.01), as well as the percentage of trans 10-18:1 (P < 0.001), which are related to an increased risk for cardiovascular diseases. On the other hand, trans 11-18:1 (vaccenic acid) the precursor for cis9, trans 11-18:2 (CLA), increased (P< 0.01). Vitamin E could, therefore, be used to decrease trans-18:1 in beef and improve its isomeric profile.

Fatty acids (FA) percentages				
Fatty Acid (FA)	A0	A35	A70	SEM [a]	P [b]
C4	3.27 a	3.43 ab	3.59 b	0.248	0.0137
C6	2.16	2.17	2.3	0.172	0.096
C8	1.34	1.26	1.29	0.099	0.178
C10	2.97	2.74	2.76	0.262	0.063
C10:1	0.24	0.22	0.22	0.027	0.086
C12	3.29 a	2.97 b	2.96 b	0.286	0.0108
C14	10.92	10.66	10.96	0.577	0.371
C14:1	0.79	0.71	0.7	0.103	0.065
C15	0.89 a	0.96 a	1.11 b	0.072	<0.0001
C16	25.15 a	26.35 a	28.32 b	1.899	0.0019
C16:1	1.38	1.42	1.51	0.198	0.241
C17	0.60	0.62	0.71	0.037	< 0.0001
C18	13.04	13.28	11.9	1.209	0.066
trans total C 18:1	3.17	2.94	2.93	0.426	0.398
cis9 C18:1	20.56 a	20.00 a	18.95 b	0.958	0.0016
trans9, trans12 C18:2	0.15	0.15	0.18	0.056	0.228
cis9, trans12 C18:2	0.07	0.16	0.09	0.096	0.105
trans9, cis12 C18:2	0.09 a	0.09 a	0.17 b	0.069	0.0080
cis9, cis12 C 18:2	3.85 a	3.27 b	2.37 c	0.475	<0.0001
cis9, cis12, cis15 C 18:3	0.31 a	0.53 b	0.92 c	0.180	<0.0001
cis9, trans11 18:2 (CLA)	0.87	0.9	0.86	0.076	0.512
C20:5 n3	0.03a	0.04 a	0.07 b	0.029	0.0060
C24	0.04 a	0.04 ab	0.07 b	0.026	0.0153
C22:4 n6	0.04	0.03	0.04	0.066	0.8467
C22:5 n3	0.04	0.06	0.06	0.057	0.6473
SFA	63.86 a	64.68 ab	66.17 b	1.50	0.0036
MUFA	25.90 a	25.07 ab	24.09 b	1.06	0.0015
PUFA	5.82 a	5.53 a	5.07 b	0.44	0.0017
Total Trans FA	3.59	3.62	3.62	0.61	0.9806

Table 8. Fatty acids profile in milk produced on 70 (A70), 35 (A35) and 0 (A0) percentage of Lucerne pasture.

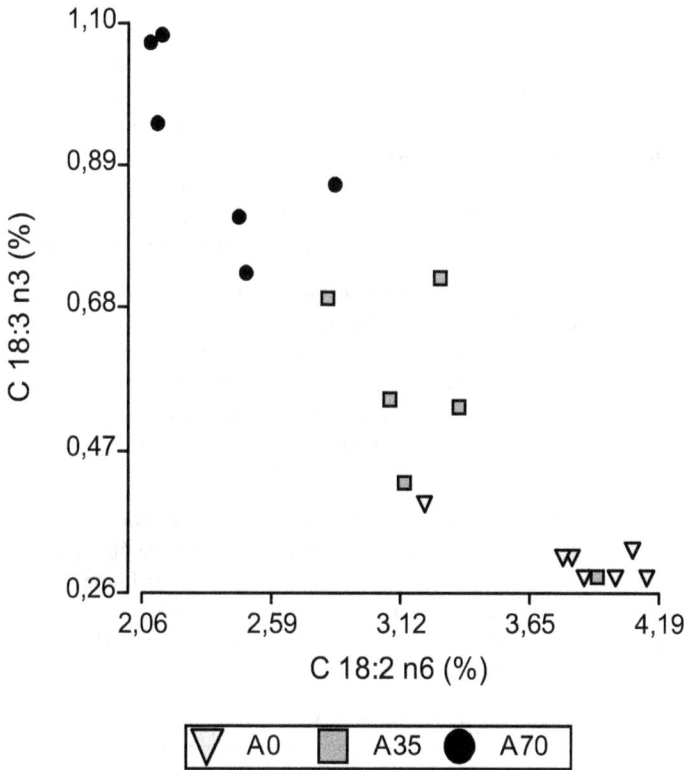

Figure 3. Relationship between linoleic acid (C18:2 n6) and linolenic acid (C18:3 n3) in milk from cows fed diets with different contents of lucerne (0, 35 and 70% for A0, A35 and A70 respectively) Different letters indicate significant differences (P<0.05).

4. Milk and dairy products

4.1. Oxidative status in milk and milk powder

Pasture feeding improved the antioxidant status in raw milk but also the peroxidative index at expenses of an increase of the n-3 fatty acids. However, feeding experiments carried out under the same production conditions showed quite dissimilar results depending on the seasonal variations.

Dry matter consumption differed between pasture and silage feeding regimes and also between Autumn and Spring. This feature is probably attributable to the differences between pasture qualities between assays. Total protein (TP) and butter fat (BF) were lower in pasture than in silage milk (3.38 vs. 3.71 % BF and 3.13 vs. 3.28 % TP respectively). α-tocopherol and β-carotene were significantly higher in pasture samples

(P<0.05), with a significant interaction between treatment and season for retinol and α-tocopherol, probably due to differences in pasture properties. β-carotene was highly correlated with the proportion of lucerne in the diet, as shown in figure 2 for oat pasture. As a consequence of the increase of cartonoids, milk, milk powder and cheese had higher b* values, which was traduced in a yellower fat colour that could differentiate dairy products from milk produced on pasture. Keenan and others., (1970) demonstrated that β-carotene was the major pigmented material encountered in bovine plasma membrane, and the plasma membrane origin of MFGM (milk globule fat membrane) would explain the origin of β-carotene in milk, which reflects the dietary origin of this provitamin.

Gamma-tocopherol was the minor isomer of vitamin E in milk and dairy products. Its incorporation into milk was not influenced either by seasonal variation or by dietary treatment. The significance of this minor form of vitamin E incorporated in milk should be further stated. Its concentration remains at least one order below α-tocopherol levels in vivo. However, its activity should be taken into account to prevent tissue damage. Particularly, γ-tocopherol has the ability to protect against nitrogen-based free radicals, whilst α-tocopherol cannot (Jiang and others, 2001).

Vitamin D3 was influenced by seasonal variation. The concentration in raw milk resulted 10-fold higher in spring than in autumn, probably due to degree of exposure to the sun which induces its biosynthesis in cows. Nevertheless, milk for human consumption is supplemented with 400 UI per quart (1 IU of vitamin D3 corresponds to 0.025 micrograms) to achieve the recommended mean intake of 5 micrograms per day (WHO). Natural vitamin D3 was not found in reggianito cheese.

Pasture feeding also modulated retinol content. Values were higher in pasture milk (p<0.05) with an interaction effect for seasonal variation. These differences were also found in milk powder and cheese, being enhanced in spring. . Therefore the susceptibility to peroxidation due to the number of double bonds in PUFAs in milk, increased with the proportion of pasture in the diet (table 9).

Oxidation in raw milk and milk powder was determined using the thiobarbituric acid reactive substances assay or TBARS, which is one of the oldest and most commonly used methods for assessing lipid oxidation in foods (Ladikos and others 1990).

This method is based on the spectrophotometric determination of extracted malonaldehyde, a minor product of oxidation, and can be performed either directly on the food product or on a steam distillate of the food. Due to the simple procedure and high correlation with sensory scores (Igene and others 1979), many studies have relied on TBARS for determination of oxidative status. The major disadvantage of the TBARS reaction is that it is not specific for malonaldehyde. No significant differences were found for TBARS (P>0.05), but this parameter showed consistently lower values in pasture milk, but was highly influenced by stationary variability.

Other method for the detection of intermediate peroxide compounds is the Peroxide Value. This parameter, as TBARS, was similar for all samples (data not shown).

Autumn	A0			A35			A70		
	M	MP	CH	M	MP	CH	M	MP	CH
L*	90.30	95.32	85.83	89.32	94.84	84.88	89.58	95.03	84.36
a*	-1.45	-1.24	3.46	-0.66	-0.77	5.20	-0.49	-0.90	5.48
b*	10.20	13.55	24.84	12.36	14.88	30.94	13.29	15.67	32.07
Beta carotene	0.08	0.72	0.41	0.25	1.56	0.78	0.27	1.58	0.74
Alfa tocoferol	1.13	6.68	5.44	0.97	8.07	6.54	1.06	7.04	5.93
Gamma tocoferol	0.06	0.30	0.24	0.02	0.19	0.15	0.02	0.09	0.04
Vit. D3	0.01	0.29	nd	0.02	0.36	nd	0.02	0.34	nd
Retinol	0.30	3.53	1.53	0.42	5.14	1.88	0.46	5.15	1.24
TBA (ppb)	44.01	1401.26		32.63	1165.21		47.19	1102.76	
Hexanal (RU/1000)	40.19	400		33.70	343		48.42	273	
FRAP (microM Fe+2)	355.1	291.43		374.5	319.85		390.2	310.22	
Spring	A0			A35			A70		
	M	MP	CH	M	MP	CH	M	MP	CH
L*	89.49	94.82	85.90	89.24	94.50	84.74	89.24	94.26	84.98
a*	-2.09	-1.52	4.11	-0.85	-0.92	6.43	-0.85	-0.95	5.97
b*	8.84	12.68	21.12	11.56	14.81	26.72	11.56	14.05	28.92
Beta carotene	0.02	0.23	0.17	0.19	0.78	0.51	0.19	0.87	0.84
Alfa tocoferol	0.35	2.82	3.23	0.75	3.81	5.48	0.75	4.56	5.88
Gamma tocoferol	0.06	0.63	0.65	0.05	0.38	0.46	0.05	0.39	0.41
Vitamin D3	0.11	0.90	nd	0.12	0.84	nd	0.12	0.90	nd
Retinol	0.47	6.96	1.88	1.47	14.28	3.84	1.47	13.85	2.76
TBARS (ppb) a	214.79	1073.54	ND	198.37	1132.32	ND	198.37	968.23	ND
Hexanal (RU/1000) b	102.45	629	ND	25.95	341	ND	27.45	305	ND
FRAP (microM Fe+2) c	308.36	166.43	ND	481.75	204.52	ND	481.75	163.67	ND

A0, A35 and A70: Diet with 0, 35 and 70% Lucerne pasture respectively.
M: Raw Milk; WMP: Whole Milk Powder; CH: Cheese (regianitto).
a: Thiobarbituric Reactive Substances
b: Relative Area Units
c: Ferric Reducing antioxidant activity
d: A0, A35 and A70, feed treatments with 0, 35 and 70% Lucerne pasture
ND: not done; nd: not determined
L*, a*, b*: CIELab Sysytem, D65 Illuminant and 10° geometry.

Table 9. Mean values of antioxidant vitamins, color parameters and oxidative stability indicators in two different seasons.

As indicated for feed components, numerous aldehydes are produced during oxidation, including octanal, nonanal, pentanal, and hexanal. Hexanal is the dominant aldehyde produced during oxidation (Dupuy and others 1987; Ajuyah and others 1993). It arises from both the 9 and 13 hydroperoxides of linoleate, and from other unsaturated aldehydes formed during the oxidation of linoleate (Shahidi and Pegg, 1994). It is a useful tool to assess lipid secondary oxidation products in milk (Erickson, 1999; Smet, 2009). This compound was higher in silage milk and could reflect a lower oxidative stability in milk and dairy products derived from silage-fed cows. Whole milk powder from pasture showed the same fatty acids profile as raw milk (table 9), and could constitute a natural source for n-3 fatty acids. In a similar way, antioxidants profile reproduced the composition of raw milk with estimated losses around 20 and 40% being higher for retinol and vitamin D3.

Higher natural antioxidants in whole milk powder lowered TBARS values in pasture (A70) samples compared to silage samples (A0) after 18-months shelf life assay ($20°C$ in sealed bi-laminated plastic pouches under nitrogen atmosphere).

The total antioxidant activity was measured using the Ferric reduction assay (FRAP). Antioxidant compounds such as α-tocopherol, trolox, vitamin C, uric acid and billirubin, among others, are able to reduce ferric- to ferrous-tripyridyltriazine which develops a blue colour (Benzie and Strain 1996) with an adsorption maximum at 593 nm. Smet and others. (2009), demonstrated that FRAP and DPPH assays provide useful information about the oxidation process, particularly about the very early changes in the oxidative stability of milk. FRAP values for A0, A35 and A70 milk, were 332, 428 and 448 micromolar equivalents of Fe+2 respectively ($P<0.05$) and served as an indicator of the antioxidant capacity of milk. In milk powder, no differences were detected, thus indicated that some thermo labile antioxidants could be destroyed due to the processing of milk.

It can be concluded that there is not a unique method to describe the antioxidant capacity of biological samples. The complexity and diversity of mechanisms that contribute to the onset of oxidation and the mechanisms that counteract oxidative reactions involve multiple pathways. Also α-tocopherol, β-carotene and retinol contents were higher in pasture samples throughout the shelf life experiment.

4.2. Instrumental sensory odour

The electronic nose (E-nose) has been successfully applied to distinguish seasonal variations in whole milk powder (Biolatto and others., 2005). Also its application permitted the differentiation of virgin olive oil that showed different volatile production and vitamin E degradation patterns during the frying process (Messina and others., 2009).

The application of e-nose measurement to milk samples permitted to demonstrate that a dietary shift from silage to Lucern pasture, induced changes in the odour profile of the raw milk.

As shown in figure 4, at the beginning of the experiment (time 0), all samples were grouped differently than after turn into pasture. Also changes in the profile were detected at time 60 days, thus indicated an interaction between diet and time in the experiment, probably due to changes in the nature of the volatile compounds produced by Lucern plants as the experiment proceeded.

Data were analyzed using a Discriminant Function Analysis (DFA) and were discriminated with 86% of recognition of each sample within each group for the original cases.

A positive correlation was found for e-nose measurements and the content of different antioxidant vitamins in milk. Rossetti and others., 2010, showed than individual e-nose sensors correlated with the concentration of antioxidant vitamins in milk. The sensor called LY2/gCTI correlated positively with α-tocopherol, β-carotene and retinol, and negatively with γ-tocopherol ($p<0.05$). The sensor P30/1 correlated positively with retinol and the sensor P10/1 showed a positive correlation with γ-tocopherol ($p<0.05$).

Odour components have shown also to be transferred from raw milk to whole milk powder from cows fed 0, 35 and 70% Lucern pasture. When the value of each sensor was compared in both products, a linear relationship was found with a correlation coefficient of 0.9851 (P<0.05). In reggianito cheese, and after six months maturation, this relation was different (figure 5b), probably due to the development of odour compounds during the fermentation and maturation processes.

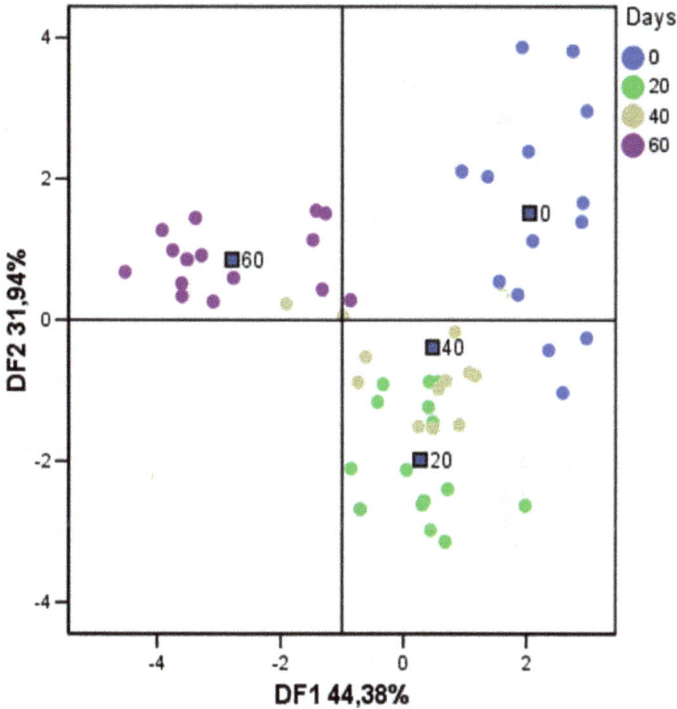

Figure 4. Odor profile of ALF (70% Lucerne pasture diet) milk at 0 (time 0), 20 (time 20), 40 (time 40) and 60 (time 60) days after the shift from silage to pasture diet. Odor profile was determined in the head space of milk samples (3 samples per cow per time) with 3 mL of sample in a vial, incubated at 50ºC for 10 min with an agitation speed of 500 rpm (Autosampler HS100, Alpha MOS). For electronic nose analysis (Alpha Fox 4000, Alpha MOS), 1 mL of headspace sample was injected and the acquisition was obtained either with 18- semi-conductor oxide metallic sensors (MOS) or 4 mL of headspace was injected in a quadrupole mass spectrometer (Alpha Kronos, Alpha MOS) with an electronic impact (70 eV) ion source (Rossetti et al., 2010).

Further research will allow explaining the biochemical changes that occur during the transformation of milk from different productive origin; in this concern the use of an e-nose approach represents an alternative powerful tool to traditional methods of odour measurements.

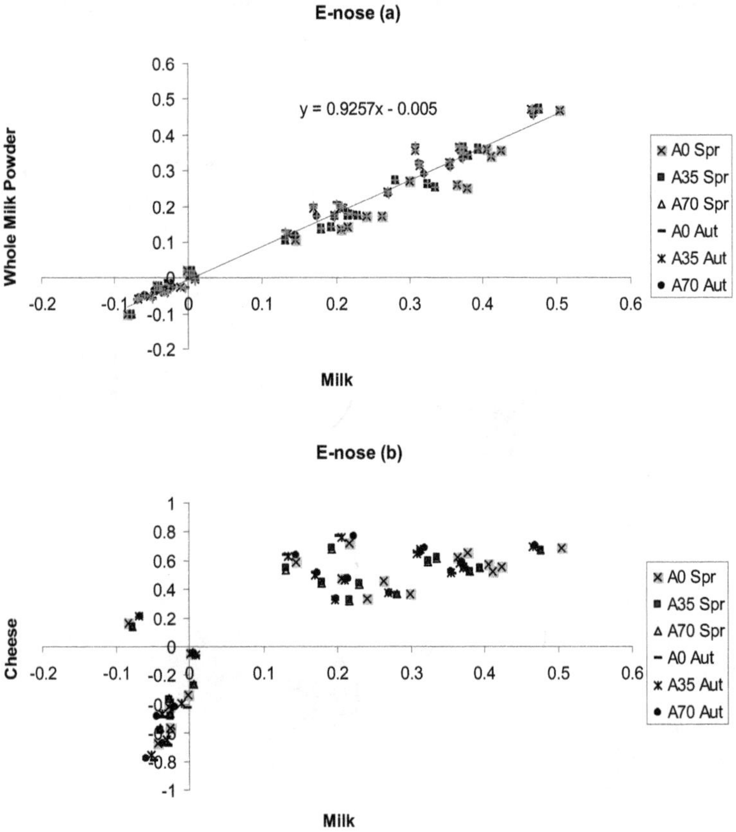

E-nose (a)

E-nose (b)

Figure 5. Relationship between e-nose sensors values in raw milk and its derived whole milk powder or reggianito cheese sample. Each point represents the value for each individual sensor applied to raw milk, and the whole milk powder or reggianito cheese that derived from the same alaboration batch. 0, 35 and 70 indicates Lucerne pasture percentage in the diet. Aut: autumn; Spr: spring.

4.3. Natural pigments in milk and cheese

Consumer's perception for natural and functional dairy products is an actual concern. In a survey over 179 consulted consumers, 87 % answered that would associate functional and natural concepts. In addition, 47% associated cheese with functional dairy products (other choices were yogurt, milk powder, ice-cream and dietary supplements).

In this concern, the concentration of carotenoids, especially β-carotene and lutein, could serve as natural pigment in milk fat and cheese.

Changes in beta carotene content were accompanied with changes in the yellow color in milk. As indicated in Calderon and others, (2007), the color index was a valuable tool to discriminate pasture from silage milk. The reflectance spectrum of raw milk samples was measured in the spectral region associated to light absorption by carotenoids, involving wavelengths from 450nm to 530nm (with a resolution of 10nm). In this range spectral data was translated to have reflectance values equal to zero at 530nm. On the translated spectrum, the integral value (IV) was calculated (Prache and others, 1999) as follows:

$$IV = \left(\frac{T450}{2} + T460 + T470 + T480 + T490 + T500 + T510 + T520 + \frac{T530}{2} \right) * 10$$

where T is the translated value corresponding to the reflectance intensity at each wavelength. Results in figure 6 indicate a linear relationship between IV and β-carotene. The β-carotene content could be predicted in function of IV by the following expression:

$$Z_{\beta-carotene} = -0.11 + 0.895\, Z_{IV}$$

Where Z is the standardized value of the variables β-carotene and IV ($Z_i = \frac{x_i - \overline{x}}{\sigma^2}$).

This result indicates that the transformation of spectral data in the range of carotenoids, may be used as a representative index of the feeding system.

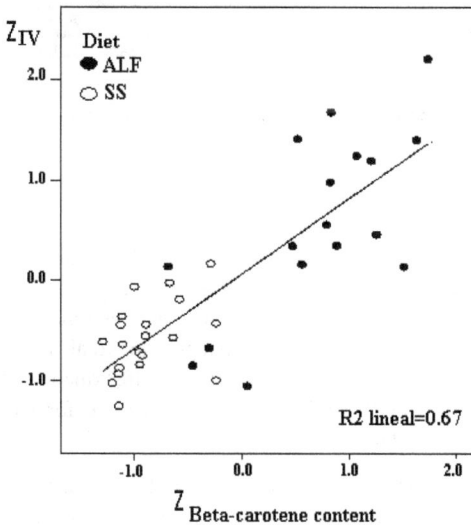

Figure 6. Correlation between IV and β-carotene content. ALF: 70% grazed Lucerne (•) values in the region of SS: Sorghum Silage diet (○) samples corresponded to milk at the beginning of the assay (time 0)

Experimental data obtained with reggianito cheese, which was elaborated within the experimental pilot plant of INTA are shown in table 5. Colour parameters indicate differences in cheese attributable to feeding management. Pasture cheese from A35 and A70 treatments presented higher b* values, which indicates a more yellow color, than A0. Differences were also distinguished by visual appreciation (figure 7). No significant differences were observed for L* parameter.

Figure 7. Aspect of reggianito cheese elaborated with milk from A0, A35 and A70 (feeding treatments containing 0, 35 and 70% Lucern pasture respectively) after six-month maturation. Left to right: A0, A35 and A70 treatments.

5. Conclusions

Productive strategies can be used to develop naturally added value dairy products. These strategies may be considered together with the consumer's perception about the role of functional and natural dairy products on human health. Preliminary studies on "pasture milk", clearly demonstrated the important changes in total lipids and fatty acid composition, especially in the contribution of C18:2 and C18:3 due to cultivar, cutting date and season. The importance of these fatty acids as substrate for CLA and n-3 PUFA concentration in milk lipids need to be considered.

The disadvantage of milk enriched with n-3 and CLA fatty acids is the possibility to suffer oxidation due to its high content of double-bonded molecules, which are prone to oxidation onset. This will not probably affect raw milk, but can bring inconvenient for the processing and commercialization of dairy product.

Therefore, pasture enhanced antioxidant status on raw milk, could be a strategy to solve part of the problem and consequently to avoid the development of rancidity and other off flavors and off odors in dairy products.

In addition, naturally pigmented cheese and milk products are positively accepted by consumers.

Pasture value added milk has an aspect which covers two purposes: preventing food spoilage and giving a fresh and healthier look to dairy products.

Author details

A.M. Descalzo, L.Rossetti, R. Páez, G. Grigioni, P.T. García, L. Costabel, L. Negri, L.Antonacci, E. Salado, G. Bretschneider, G. Gagliostro, E. Comerón and M.A. Taverna
Instituto Nacional de Tecnología Agropecuaria (INTA), Argentina

6. References

Ajuyah AO, Fenton TW, Hardin RT, Sim JS. 1993. Measuring lipid oxidation volatiles in meats. J Food Sci 58:270–3.

Al-Mabruk, R.M., Beck, N.F.G., Dewhurst, R.J.(2004). Effects of Silage Species and Supplemental Vitamin E on the Oxidative Stability of Milk. Journal of Dairy Science, 87(2): 406-412.

Bargo, F., L. D. Muller, J. E. Delahoy, and T. W. Cassidy. 2002. Performance of high producing dairy cows with three different feeding systems combining pasture and total mixed rations. J. Dairy Sci. 85:2948–2963.

Benzie, I.F.F. and Strain, J.J. (1996), The ferric reducing ability of plasma (FRAP) as a measure of antioxidant power: The FRAP assay. Anal. Biochem. 239, 70-76.

Besle, J.M., Viala, D., Martin, B., Pradel, P., Meunier, B., Berdague, J.L., Fraisse, D. Lamaison, J.L. Coulon, J.B. (2010). Ultraviolet-absorbing compounds in milk are related to forage polyphenols, Journal of Dairy Science, 93(7): 2846-2856.

Butler, G., Nielsen, J.H., Slots, T., Seal, C., Eyre, M.D., Sanderson, R., Leifert, C. (2008). Fatty acid and fat-soluble antioxidant concentrations in milk from high- and low-input conventional and organic systems: seasonal variation. J. Sci. Food Agric. 88:1431–1441.

Calderón, F., Chauveau-Duriot, B., Martin, B., Graulet, B., Doreau, M. and Nozière, P. (2007). Variations in Carotenoids, Vitamins A and E, and Color in Cow's Plasma and Milk During Late Pregnancy and the First Three Months of Lactation, Journal of Dairy Science, 90(5): 2335-2346.

Calderón, F., Tornambé, G., Martin, B., Pradel, P., Chauveau-Duriot, B., Nozière, P. (2006). Effects of mountain grassland maturity stage and grazing management on carotenoids in sward and cow's milk.Anim. Res. 55 (6) 533-544.

Castillo A ; Taverna M; Paez R. ; Cuatrin A ; Colombatto D; Bargo F ; Garcia M. ; Garcia P ; Chavez M ; Beaulieu A.; Drackley J (2006). Fatty acid composition of milk from dairy cows fed fresh lucerne based diets. Animal feed science and technology, 131: 241-254.

Chilliard Y., Ferlay A., Doreau M., (2001). Effect of different types of forages, animal fat or marine oils in cow's diet on milk fat secretion and composition, especially conjugated linoleic acid (CLA) and polyunsaturated fatty acids, Livest. Prod. Sci. 70 : 31–48.

Cohen Fernandez, S., Budowski, P., Ascarelli, I., Neumark, H., Bondi, A., 1976b. Pre-intestinal stability of betacarotene in ruminants. Int. J. Vit. Nutr. Res. 46, 439–445.

Cohen-Fernandez, S., Budowski, P., Ascarelli, I., Neumark, H., Bondi, A., 1976a. Low utilization of carotene by sheep. Int. J. Vit. Nutr. Res. 46, 446–453.

Comeron, E.A., Salado, E., Cuatrin, A., Paez, R., Riba, N., Molinelli, P., 2007. Modificación de la calidad de leche por la dieta ofrecida. 1.- Consumo, producción y composición de la leche. Rev. Arg. Prod. Animal, 27(1), 387.

Davison, K.L., Seo, J., 1963. Influence of nitrate upon carotene destruction during in vitro fermentation with rumen liquor. J. Dairy Sci. 46, 862–864.

Dawson, R.M., Hemington, N., (1974). Digestion of grass lipids and pigments in the sheep rumen. Br. J. Nutr. 32, 327–340.

Delagarde , R., Peyraud, J.L., Delaby, L. 2004. High grazing pressure in early-spring increases herbage intake of grazing dairy cows in late-spring. En : Land Use Systems in Grassland Dominated Regions. Proc. 20th General Meeting of EGF, 21-24 June, Luzerna, Suiza, 647-649.

Dewhurst, R.J., Scollan, N.D., Youell, J., Tweed, J.K.S. and Humpherys, M.O. (2001) Influence of species, cutting date and cutting interval on the fatty acid composition of grasses. Grass and Forage Science, 56:68-74

Dhiman, T. R., G. R. Anand, L. D. Satter, and M. W. Pariza. 1999. Conjugated linoleic acid content of milk from cows fed different diets. J. Dairy Sci. 82: 2146 – 2156.

Dillon, P., 2006. Chapter 1: Achieving high dry-matter intake from pasture with grazing dairy cows. In: Fresh herbage for dairy cattle, the key to a sustainable food chain. Eds. Elgersma, A., Dijkstra, J. & Tammingna, S.,Wageningen UR Frontis Series Volume 18, Springer, Dordrecht, Netherlands. pp 1-26.

Dillon, P., J. R. Roche, L. Shalloo, and B. Horan. 2005. Optimising financial return from grazing in temperate pastures. Pages 131–147 in XXth Int. Grassl. Congr., Cork, Ireland. Wageningen Academic Publishers, Wageningen, the Netherlands.

Dillon, P., T. Hennessy, L. Shalloo, F. Thorne, and B. Horan. 2008. Future outlook for the Irish dairy industry: A study of international competitiveness, influence of international trade reform and requirement for change: Review. Int. J. Dairy Technol. 61:16 - 29.

Dufour, E., and T. Haertlé. 1991. Binding of retinoids and β-carotene to β-lactoglobulin. Influence of protein modifications. Biochim. Biophys. Acta 1079:316–320.

Dupuy HP, Bailey ME, St. Angelo AJ, Vercellotti JR, Legendre MG. 1987. Instrumental analysis of volatiles related to warmed-over flavor of cooked meats. In: St. Angelo AJ,

Bailey ME, editors. Flavor quality of fresh meats: warmed over flavor. Orlando, Fla.: Academic Press. p 165–90.

Elsgerma, A. Ellen, G., Horst, H., Mouse, B.G., Boer, H and Tamminga, S. (2003) Influence of cultivar and cutting date on the fatty acid composition of perennial rygrass. Grass and Forage Science, 58:323-331.

Erickson, M. (1999). Chemical and microbial stability of fluid milk in response to packaging and dispensing. Intl. J. Dairy Technology, 50(3): 107-111.

Garcia, PT , Pordomingo,A, Perez CD, Rios,MD, Sancho, AM, Casal, JJ. (2007) Influence of cutting and season on the fatty acid composition of forage crops for forage crops beef production. In Proceedings of the 54 th ICoMST, pp:101-102.

Garcia, S. C and W. J. Fulkerson. 2005. Opportunities for future Australian dairy systems: a review. Aust. J. Exp. Agr. 45: 1041-1055.

Graulet B. (2010). Vitamins naturally occuring in bovine milk. In : Improving the safety and quality of milk, Volume 2 : Improving quality in milk products. Ed. M. Griffiths, Woodhead Publishing Ltd, Cambridge, UK.

Griinari J.M., Bauman D.E. (1999) Biosynthesis of conjugated linoleic acid and its incorporation into meat and milk in ruminants, in: Yurawecz, M.P., Mossoba M.M., Kramer J.K.G., Pariza, M.W., Nelson G.J. (Eds.), Advances in Conjugated Linoleic Acid Research, Vol. 1, Champaign, Illinois: AOCS Press, pp. 180–200.

Guimont, C., E. Marchall, J.M. Girardet , G. Linden. 1997. CRC in Food Science and Nutrition, 37:393-410.

Hartman A.M., Dryden L.P., Vitamins in milk and milk products, American Dairy Science Association, USA, 1965.

Hidiroglou, M., Batra, T.R., Roy, G.L. (1996). Changes in plasma [alpha]-tocopherol and selenium of gestating cows fed hay or silage. Journal of Dairy Science, 77(1): 190-195.

Igene JO, Pearson AM, Merkel RA, Coleman TH. (1979). Effect of frozen storage time, cooking and holding temperature upon extractable lipids and TBA values of beef and chicken. J Animal Sci 49(3):701–7.

Ip MM, Masso-welch PA, Shoemaker SF, Shea-eaton WK. (1999). Conjugated linoleic acid inhibits proliferation and induces apoptosis of normal rat mammary epithelial cells in primary culture. Exp Cell Res. 1999; 250:22-34.

Jahreis, G., Fritsche, J., Steinhart, H. (1997). Conjugated linoleic acid in milk fat: High variation depending on production system. Nutrition Research, 17(9):1479-1484.

Jiang, Q., Christen, S., Shigenaga, M. K., & Ames, B. N. (2001). Gamma tocopherol, the major form of vitamin E in the US diet, deserves more attention. American Journal of Clinical Nutrition, 74(6), 714–722.

Juárez, M, Dugan, M E.R., Aalhus, J L., Aldai, N., Basarab, J.A., Baron, V. S., McAllister, T.A. (2010). Dietary vitamin E inhibits the trans 10-18:1 shift in beef backfat. Canadian Journal of Animal Science. 90(1):9-12.

Keenan, T. W., Morre, D.J., Olson, D.E., Yunghans, W.N., Patton, S. (1970). Biochemical and morphological comparison of plasma membrane and milk fat globule membrane from bovine mammary gland. The journal of cell biology, 44: 80-93.

King, T.B., Lohman, T.G., Smith, G.S., 1962. Evidence of rumeno-reticular losses of Vitamin A and carotene. J. Anim. Sci. 21 (Suppl.), 1002 (abstr.).

Kolver, E. S., and L. D. Muller. 1998. Performance and nutrient intake of high producing Holstein cows consuming pasture or a total mixed ration. J. Dairy Sci. 81:1403–1411.

La Terra, S., Marino, V.M., Manenti, M., Licitra, G., Carpino, S. (2010). Increasing pasture intakes enhances polyunsaturated fatty acids and lipophilic antioxidants in plasma and milk of dairy cows fed total mix ration. Dairy Sci. Technol. 90: 687–698.

Ladikos D, Lougovois V. (1990). Lipid oxidation in muscle foods: a review. Food Chem 35:295–314.

Lucas, A., Rock, E., Chamba, J-F., Verdier-Metz, I, Brachet, P., Coulon, J-B (2006) Respective effects of milk composition and the cheese-making process on cheese compositional variability in components of nutritional interest Lait 86 (1) 21-41.

Macdonald, K. A., J. W. Penno, J. A. S. Lancaster, and J. R. Roche. 2008. Effect of Stocking Rate on Pasture Production, Milk Production, and Reproduction of Dairy Cows in Pasture-Based Systems. J. Dairy Sci. 91:2151-2163.

McEvoy, M., E. Kennedy, J. P. Murphy, T. M. Boland, L. Delaby, and M. O'Donovan. 2008. The Effect of Herbage Allowance and Concentrate Supplementation on Milk Production Performance and Dry Matter Intake of Spring-Calving Dairy Cows in Early Lactation. J. Dairy Sci. 91:1258-1269.

Messina, V., Biolatto, A., Descalzo, A., Sancho, A., Baby, R., Walsöe de Reca, N. (2009). Effect of pan-frying in extra-virgin olive oil on odour profile, volatile compounds and vitamins International Journal of Food Science and Technology, 44:552–559.

Mora, O., Romano, J.L., Gonzalez, E., Ruiz, F.J., Shimada, A., 1999. In vitro and in situ disappearance of betacarotene and lutein from lucerne (Medicago sativa) hay in bovine and caprine ruminal fluids. J. Sci. Food Agric. 79, 273–276.

Müller, C.E.; Möller, J.; Krogh Jensen S.; Udén, P. (2007) Tocopherol and carotenoid levels in baled silage and haylage in relation to horse requirements. Animal Feed Science and Technology 137: 182–197.

Negri, L, Sancho, A., Biolatto, A., Taverna, M.A. ; Descalzo, AM. (2008). HS-SPME-GC/FID: Application To The Study Of Volatile Components In Bovine Milk. Res Nº 456. COLACRO XII - Latin-American Congress of Chromatography and Related Techniques, October, 28-30, Florianopolis, Brasil.

Páez, R., Descalzo, A., Cuatrin, A, Comerón, E.A., Salado, E.E., Campos, S., Robledo, M. and Audero, G. (2007), Modificación de la calidad de leche por la dieta ofrecida. 2. Composición de ácidos grasos y estabilidad oxidativa. TPP 28. Revista Argentina de Producción Animal 27(1), 389-390.

Pecetti, L., Tava, A., Felicioli, A., Pinzauti, M. and Piano,E. (2002), Effect of three volatile compounds from lucerne flowers on their attractiveness towards pollinators. Bulletin of Insectology 55 (1-2), 21-27.

Potkanski, A.A., Tucker, R.E., Mitchell Jr., G.E., 1974. Pre-intestinal carotene losses in sheep fed lucerne or isolated carotene beadlets. Int. J. Vitam. Nutr. Res. 44, 3–7.

Prache, S., Cornu, A., Berdagué, J.L. and Priolo, A. (2005), Traceability of animal feeding diet in the meat and milk of small rumiants. Small Rumiant Research 59, 157-168.

Prache,S., Kondjoyan N., Delfosse,O., Chauveau-Duriot, B., Andueza, D., Cornu, A. (2009). Discrimination of pasture-fed lambs from lambs fed dehydrated lucerne indoors using different compounds measured in the fat, meat and plasma. Animal, 3:4; 598–605

Reis, R. B., and D. K. Combs. 2000. Effects of increasing levels of grain supplementation on rumen environment and lactation performance of dairy cows grazing grass–legume pasture. J. Dairy Sci. 83:2888–2898.

Rossetti, L., Langman, L., Grigioni, G.M., Biolatto, A., Sancho, A.M., Comeron, E. Descalzo. (2010) Antioxidant status and odor profile in milk from silage or lucerne-fed cows. Australian Journal of Dairy Technology. 65 (1):3-9.

Salado, E.E., Bretschneider, G., Cuatrin, A., Carassai, A., Espiñeira, M. y Haedo, E. 2010a. Suplementación de vacas lecheras en pastoreo con niveles crecientes de concentrado energético: 1. Producción y composición de leche. Rev. Arg. Prod. Anim. 30 (supl. 1): 520-521.

Salado, E.E., Bretschneider, G., Gaggiotti, M. y Cuatrin, A. 2010b. Suplementación de vacas lecheras en pastoreo con niveles crecientes de concentrado energético: 3. Ambiente y digestión ruminal. Rev. Arg. Prod. Anim. 30 (supl. 1): 523-525.

Sardar, S., Chakraborty, A., Chatterjee, M. (1996). Comparative effectiveness of vitamin D[3] and dietary vitamin E on peroxidation of lipids and enzymes of the hepatic antioxidant system in Sprague-Dawley rats. International Journal for Vitamin and Nutrition Research, 66(1): 39-45.

Schingoethe, D.J., Parsons, J.G., Ludens, F.C., Tucker, W.L., Shave, H.J. (1978). Vitamin E status of dairy cows fed stored feeds continuously or pastured during summer. J Dairy Sci 61:1582-1589.

Schroeder, G.F. , Delahoy, J.E. , Vidaurreta, I., Bargo, F., Gagliostro, G.A. and Muller, L.D. (2003). Milk Fatty Acid Composition of Cows Fed a Total Mixed Ration or Pasture Plus Concentrates Replacing Corn with Fat. Journal of Dairy Science, 86 (10), 3237-3248.

Shahidi F, Pegg RB. 1994. Hexanal as an indicator of the flavor deterioration of meat and meat products. In: St. Angelo AJ, editors. Lipid oxidation in food. ACS Symposium Series 500. Washington, D.C.: American Chemical Society. p 105–19.

Smet, K., De Block, J., De Campeneere, S., De Brabander, D., Herman, L., Raes, K., Dewettinck K., Coudijzer, K.(2009). Oxidative stability of UHT milk as influenced by fatty acid composition and packaging. International Dairy Journal 19: 372–379.

Søren Krogh Jensen, Anna Kirstin Bjørnbak Johannsen And John E. Hermansen (1999). Quantitative secretion and maximal secretion capacity of retinol, β-carotene and α-tocopherol into cows' milk. Journal of Dairy Research, 66, pp 511-522

Tamminga, S. 1993. Influence of feeding management on ruminant fiber digestibility. Page 571–602 in Forage Cell Wall Structure and Digestibility. D. R. Buxton, H.G. Jung, R. D. Hatfield, and J. Ralph, ed. ASA-CSSA-SSSA, Madison, WI.

Tava, A. and Pecetti, L. (1997), Volatiles from Medicago sativa complex flowers Phytochemistry 45 (6), 1145-1148.

Tesoriere, L., D'Arpa, D., Re, R., Livrea, M.A. (1997). Antioxidant reactions of all-trans retinol in phospholipid bilayers: effect of oxygen partial pressure, radical fluxes, and retinol concentration. Archives of Biochemistry and Biophysics, 343(1): 13-18.

Viallon, C., Martin, B., Verdier-Metz, I., Pradel, P., Garel, J.-P., Coulon, J.-B., & Berdagué, J.-L. (2000). Transfer of monoterpenes and sesquiterpenes from forages into milk fat. Lait, 80, 635–641.

White, S. L., G. A. Benson, S. P. Washburn, and J. T. Green Jr. 2002. Milk production and economic measures in confinement or pasture systems using seasonally calved Holstein and Jersey cows. J. Dairy Sci. 85:95–104.

White, S. L., J. A. Bertrand, M. R. Wade, S. P. Washburn, J. T. Green Jr., and T. C. Jenkins. 2001. Comparison of fatty acid content of milk from jersey and Holstein cows consuming pasture or a total mixed ration. J. Dairy Sci. 84:2295 –2301.

Chapter 15

Results of Automatic Milking System and Milk Performance on Selected Farms

Miroslav Maršálek, Jarmila Voříšková and Jana Zedníková

Additional information is available at the end of the chapter

1. Introduction

Milking times differ considerably in different countries based on the type of agriculture used. In US, where farms often hold large herds of dairy cows and the milking process is performed by specialised personnel a non-stop three-shift-milking production is not unusual. However, this type of production is economical only for large herds. In Western Europe, on the other hand, where small farms dominate and most work is performed by the farmer and his family members, the aim is to perform milking in as short time as possible (Sychra, 2001).

The first attempts at full automation of the milking process were made in the 1970's in several developed countries at the same time. The development was fastest in the Netherlands. The first fully automated milking system (Automatic Milking System- further AMS) was put into operation in 1992. Given the ever increasing average milk performance of cattle in the Czech Republic (6,870 l in 2009) and the increasing number of farms with average annual milk performance exceeding 10,000 litres of milk per cow, the first robotic milking machine in the country was installed in 2003. By 2009 the number had risen to 102. On 35 investigated farms Single Lely robots were the most frequent type used (Machálek, 2009).

Rotz et al. (2003) recommend the installation of milking robots on small family farms with the main motivation of achieved flexibility of working hours and therefore improved quality of life for the farmers, better working conditions and independence from outside-farm help. Rodenburg (2002) recommends machine milking for cattle herds from 60 to 200 cows based on lower costs of using milking robots as compared to using milking parlours.

Key factors for a successfully operating system in a dairy farm include breed, nutrition, and the environmental and human issues. When using AMS the management of dairy farming is transformed into so-called individual management. The flexibility and responsibility of such management as well as their ability to use PCs are important factors for the operation of the whole system (Havlík, 2007). Robotic milking requires no physical presence of an operator, thereby increasing productivity several times (Fleischmanová, 2005). It has been shown that robotic milking leads to 30–40% savings in physical labour as compared to conventional milking.

Most keepers are aware of the increased welfare after AMS has been installed. This effect can be further emphasised by installation of rubbing brushes in the stables. Although the acquisition cost is not insignificant, it is fast redeemed by the fact that the otherwise inevitable damage on fencing, gates, barriers and water troughs can be thus avoided, although the main factor for the acquisition of brushes is cleanness and health of the animals (Doležal et al., 2004). Also the absence of aggression from the care takers can often play an important part in the animal welfare and although the reaction of a cow toward an aggressive person is highly individual, experimental results show that milking production in older cows has decreased by 10 %, while the amount of residual milk has doubled as compared to the controlling sample cared for by a kind keeper. Therefore, it is clear that cows do recognise human character and their anxiety is mirrored not only in their behaviour but also in the internal reactions of their organisms, where the development of a stress load increases cortisol levels in the bloodstream thus decreasing milk production of the individual animal (Tančin & Tančinová, 2008).

Milk performance of dairy cows increases continuously from the first to the fifth lactation. This potential needs to be utilized by increasing the number of milking times using the milking automation system (Debrecéni et al., 1999). According to Přikryl (1997) and Stelwagen (2001) correct regulation of the milking process using the automation system enables milking several times per day which is especially significant in case of high performance animals. Doležal (2000) furthermore mentions a positive effect of multiple milking in cows with milking performance of over 35 kg of milk (increase by 18.9 %) in contrary to cows with milking performance below 25 kg of milk (increase by 1.4 %). This author therefore recommends utilisation of multiple milking for an average performance of cattle higher than 9.500 l of milk (Doležal, 2002). Kopeček and Machálek (2009) show an increase in milk performance after the introduction of AMS by up to 25%. By contrast, Kvapilík (2005) estimates the economic efficiency of the introduction of milking robots in the Czech Republic and calculates the expected increase in milk production per cow and year to be 8% on average as compared to the conventional parlour milking. Doležal (2002) and Kruip et al. (2002) also note that multiple milking results in worsened reproduction factors.

A voluntary visit to the milking robot box several times per day can only be achieved through correctly working motivation system. According to Šťastny (2010) such situation can be established by feeding each cow that has come for a correctly timed milking with granulated feed with higher (by 10%) energy value than what normal quantity feed contains. This makes the cows want to replenish their energy deficit in the AMS. Weisse et al. (2005) also state that offering concentrated feed represents the strongest motivation for a cow visiting the AMS.

Although the milking procedure in the AMS robot is very gentle (Debrecéni et al., 1999), a positive approach and no use of force toward the animal to be milked is still necessary during an adaptation period (about 15 days) after AMS installation. Furthermore, there are always a certain number of animals that will not adjust and voluntarily accept new behavioural routines. These cows will ignore the AMS completely and in such case it is necessary for the farmer to reserve time to manually lead these individuals to the milking boxes, or eventually exclude them from the milking system. The number of cows that must be led to the box usually increases with the number of individuals designated for each AMS station. There is also a risk of such cows getting used to the procedure and further expect such service. Another cause of cows getting excluded from the AMS system is their increased aggressively or anxiety (kicking, ducking away from the robot). Kic and Nehasilová (1997) state that up to 15% dairy cows must be excluded from the AMS system based on the above mentioned causes. The behaviour of animals, their quiet and balanced temperament, thus becomes one of the selection criteria when specialised herds suitable for the AMS system are being put together. Wagner-Storch and Palmer (2003) followed the activity of cows milked by the AMS system throughout the day and found that higher number of individuals is being milked between 8am to 1pm and 3pm to 7pm. On the other hand, the lowest frequency of milking was shown between midnight and 6am.

Another important decision factor for the introduction of AMS is economy. Building costs for a stable including milking parlour and milkhouse are according to Vegricht (2002) 7% higher than those for building a stable with AMS system. On the other hand, the cost for technical equipment of such stables is 185% higher as compared to that for stables with a milking parlour. The total investment costs for a stable with 300 dairy cows are 62 % higher than for stable with traditional milking technique. These investments are later incorporated into the total production costs of the stable and the total costs for the production of 1 litre of milk and thus the annual cost for 300 dairy cows being milked by the AMS system is 9.9% higher than for a similar farm using traditional parlour milking method. Maršálek et al. (2009) argue that although stables with AMS production report higher costs per feeding day they exhibit positive economic effect due to higher production and better quality of milk. At the same time cows in free stables using AMS milking also show positive results concerning reproduction. The increase in milk production through breeding can on the other hand induce the occurrence of many functional disorders, mostly of metabolic character or concerning milk secretion by decreasing the natural resistance in cows towards conditions of the outer environment (Hanuš et al., 2006).

On modern farms, the individual approach is to a great extent replaced by automatic data collection, radiotransmision of information and computer evaluation. Much of this received data can be used not only for production, reproduction and economy management but also as signalisation of possible health problems and thus for their early detection and treatment. It is important to state that this indirect individual care with the help of technical equipment for monitoring and evaluation does not rid the farmer of the obligation to secure at least one physical check of all animals daily (Doležal et al., 2004).

In the Czech Republic there are according to Kvapilík et al. (2011) 113 004 closed lactations for the Czech Fleckvieh dams and 157 634 closed lactations for Holstein dams in the production control procedure, where the milk productivity for these breeds is on average 6 472 kg and 8721 kg respectively. The Czech Fleckvieh cattle belongs to the world population of Fleckvieh Cattle with common phylogenetic origin, currently spread on all continents. The requirements for this breed are of dual performance with high milk production, medium to large body stature, excellent muscling and harmonic appearance (Bouška, 2006).

2. Results of the use of milking robots

The aim of this study was to evaluate the productivity of breeding cows kept on farms that use the AMS according to breed, number of robots used and the type of company ownership. Part of this study also considered the behaviour of the cows during the milking procedure, where thorough analyses of lactation curves for different productivity levels, the content of milk components during lactation process, correlation between milk productivity and fertility and between milking frequency and reproduction results were conducted.

Seven agricultural companies mostly from the Southern Bohemia region that use the progressive technology of automated milking system (AMS - Dutch made model Astronaut A3 from producer Lely) were included in this study. The chosen companies were divided according to the number of milking robots used into following categories: small (one robot – company nr. 1, 2 and 3), medium (two and three robots – farm nr. 4 and 5) and large (seven and eight robots – company nr. 6 and 7). According to the ownership the companies were designated as private (1 to 3) and cooperative (4 through 7). In companies 2, 3 and 7 the breed of dairy cows was the Czech Fleckvieh cattle (C), while the Holstein breed was kept in farms 1, 6 and 7 (H). Both breeds were kept in companies 4 and 5. These parameters were followed in the breeding cows: reproduction parameters (insemination interval, service period and meantime), the average amount of milk per cow and day (in kg), the average number of milking times per cow and day, the average number of refused milking times per cow and day, numbers of problematic cows (refusing being milked by the robot).

Basic data background was acquired from each company from program T4C (Time for cows), the steering unit of AMS that serves as the communication device and enables the back check of the milking data. The data set was consequently analysed by statistical programs Statistica.cz, Statsoft Co. and Statsoft CR Ltd. 200. Information for the analysis of problematic dairy cows that needed being lead to the robot was acquired directly from the keepers. Data evaluated concerned the period between 1.1.2009 and 30.11.2009 (334 days).

2.1. Results of herd reproduction and productivity

Table 1 shows an overview of breeds kept in individual companies including the average number of dairy cows, average number of cows per milking robot and chosen reproduction parameters. The Holstein breed was kept in companies 1 and 6. In company nr. 7 both breeds C and H were kept separately, while in companies 4 and 5 herds consisted of both breeds mixed together which prevented the evaluating of the followed parameters separately. The lowest number of cows was 51 in company 2, the highest 377 in company 6.

The number of robots varied from one in companies 1 and 3 to eight in company 7. The insemination interval was shown to vary greatly between 59 days (company 4) to 86 days (company 7), the service period varied from 89 (company 4) to 141 days (company 1). The length of the meantime was between 395 days (company 3 and 6) and 415 days (company 7). As a comparison Kvapilík et al. (2011) reported republic averages for dairy cows at productivity check for year 2009 to be 83.6 days for the insemination interval, 122.9 days as the length of the service period and 411 days as the length of the meantime.

Company number	Breed	Average number of dairy cows in herd (pc)	Number of milking robots	Insemination interval (days)	Service period (days)	Meantime (days)
1.	H	59	1	76	141	412
2.	C	51	1	62	97	398
3.	C	73	1	85	112	395
4.	H + C (30:70%)	97	2	59	89	401
5.	H + C (40:60%)	165	3	84	125	405
6.	H	377	7	75	118	395
7.	H	201	4	86	132	415
	C	160	4			

H - Holstein; C - Czech Fleckvieh

Table 1. Chosen reproduction factors in followed herds

The average daily production of milk achieved in individual farms is shown in Table 2. The highest productivity (40.43 kg) was measured for company 1. The second largest (30.16 kg) was found in company 6. On the other hand was the lowest productivity (21.04 kg of milk) measured for company 3 with the highest number of animals per robot (73 cows). Differences between the companies were highly statistically significant (P≤0.001). The obtained results father emphasise the premises of Doležal (2000) that higher milk production induces the higher effect of milking times. Compared to the average daily milk production for year 2009 in the Czech Republic (18.82 kg of milk - Kvapilík et al. 2011), the companies followed were able to show higher values.

Dividing the farms according to size corresponds well with the division according to ownership (Table 3). The highest daily productivity was achieved in small privately owned companies with one robot (28.79 kg of milk) as compared to medium and large cooperate companies. The data were despite large variation (s_x =8.54) and differences 3.57 resp. 1.45 kg of milk found statistically significant (P≤0.001). These results correspond to the conclusions made by Rotz et al. (2003) who recommend installation of milking robots on small family farms. According to Rodenburg (2002) is automated milking great advantage for dairy cattle herds with 60 to 200 heads, as the milking parlour technology is too expensive and automated milking represents more economical solution.

Company number	Number of cows per 1 robot	n	x̄	S_x	F test
1.	59	334	40.43	1.65	2745.0**
2.	51	334	24.83	1.71	1:2,3,4,5,6,7***
3.	73	334	21.04	1.40	2:3,4,5,6,7***
4.	49	668	22.92	3.00	3:4,5,6,7***
5.	55	1002	26.75	1.74	4:5,6,7***
6.	54	2338	30.16	2.97	5:6,7***
7.	45	2672	24.87	2.97	6:7***

Table 2. Average milk production in individual companies (in kg of milk)

Company size	Type of ownership	Number of cows per 1 robot	n	x̄	S_x	F test
Small (1)	private	61	1002	28.79	8.54	206.2**
Medium (2)	cooperative	53	1670	25.22	2.99	1:2***
Large (3)	cooperative	49	5010	27.34	4.00	1:3*** 2:3***

Table 3. Average milk production according to size and ownership of the company (kg of milk)

Table 4 shows the average productability with regard to the breeding type of the dairy cows. Highest values were obtained for the H breed, reaching 29.97 kg of milk. The difference compared to the C cattle was 1 7.17 kg (P≤0.001). Herds with both breeds have reached an average value of both (25.22 kg of milk). The number of cows per robot was identical for H breed and H and C mixed breed (53 heads), but lower for the C breed (47 heads).

Breed	Number of cows per 1 robot	n	x̄	S_{ix}	F test
Holstein Cattle (1)	53	4008	29.97	4.33	2903.0**
Czech Fleckvieh Cattle (2)	47	2004	22.80	2.26	1:2*** 1:3***
Both breeds (3)	53	1670	25.22	2.99	2:3***

Table 4. Average milk production according to breed (kg of milk)

2.2. Milking frequency using AMS

The average number of milking times per one dairy cow per day is an important parameter for the automated milking system. In Table 5 significant differences between the companies can be seen (P≤0.001), where company 1 achieved the highest frequency of 2.67. The productability of this company is thus markedly higher which corresponds to literature (Doležal, 2000, 2002, Stelwagen, 2001) regarding the positive influence of multiple milking times for high performance dairy cows. The lowest number of milking times, namely only 1.97, was found for company nr. 3, where the number of cows per robot was the highest, 73 heads. Lowering this number by 14 resulted in an increased number of milking times, up to 2.67. It is important to state here that both companies hold different breeds of cattle.

Company number	Number of cows per 1 robot	n	x̄	Sx	F test
1.	59	334	2.67	0.18	344.9**
2.	51	334	2.60	0.18	1:2,3,4,5,6,7***
3.	73	334	1.97	0.16	2:3,4,5,6,7***
4.	49	668	2.53	0.25	3:4,5,6,7***
5.	55	1002	2.44	0.14	4:5,6,7***
6.	54	2338	2.44	0.24	5:7***
7.	45	2672	2.41	0.21	6:7***

Table 5. Average number of milking per one dairy cow per day in individual companies

Significant differences (P≤0.001) in the number of milking times per head and day were also found for the companies when devided according to their size and ownership (Table 6) even though the differences were not large. The highest number of milking times was exhibited by medium and large companies (2.47 resp. 2.43), while the small companies put together obtained an average value of 2.34. Similar conclusions can be drawn when the ownership situation is considered simultaniously.

Company size	Type of ownership	Number of cows per 1 robot	n	x̄	Sx	F test
Small (1)	private	61	1002	2.34	0.33	96.9**
Medium (2)	cooperative	53	1670	2.47	0.20	1:2***
Large (3)	cooperative	49	5010	2.43	0.24	1:3*** 2:3***

Table 6. Average number of milking times per one dairy cow per day according to size and ownership of the company

Significant difference in milking times was also found when considering the different breeds of cattle (Table 7). The lower average of 2.32 milking times per head and day was found for the dual productivity cattle of breed C, while the breeding cows of the dairy cattle H were milked on average 2.45 times (P≤0.001). The correlation to the productivity average of the individual breeds is therefore clear.

Breed	Number of cows per 1 robot	n	x̄	S_x	F test
Holstein Cattle (1)	53	4008	2.45	0.23	245.2**
Czech Fleckvieh Cattle (2)	47	2004	2.32	0.26	1:2***
Both breeds (3)	53	1670	2.47	0.20	1:3** 2:3***

Table 7. Average number of milking times per cow and day according to breed

When a cow enters the robot before the right milking time has come, it is refused and released, which is defined as a refusal. Ideally the number of such refusals should be around 1.5 per day, which would mean one refusal for two concluded milking times (Šťastný, 2010). The differences found for the number of refusals per head and day were very significant (P≤0.001) for the different companies (Table 8). Clearly the highest number of refusals (3.66) was found for company 4, the second highest (2.63) for number 7. Both these companies have the lowest number of cows per milking robot (49 resp. 45). Extremely low values were then calculated for farms 2 (0.72) and 3 (0.88).

Company number	Average number of cows per 1 robot	n	x̄	S_x	F test
1.	59	334	1.74	0.51	524.1**
2.	51	334	0.72	0.27	1:2,3,4,5,7***
3.	73	334	0.88	0.28	2:3,4,5,6,7***
4.	49	668	3.66	2.07	3:4,5,6,7***
5.	55	1002	1.20	0.32	4:5,6,7***
6.	54	2338	1.93	1.15	5:6,7***
7.	45	2672	2.63	1.31	6:7*** 1:6**

Table 8. Average number of rejections per dairy cow per day in individual companies

The average number of refusals according to size and company ownership is shown in Table 9. Small privately owned companies with one robot showed low refusal count at about 1.11, as compared to medium and large cooperative companies with values around 2.18 resp. 2.30. Small and privately owned farms seem to place more emphasis on working relief supplied by the AMS than on milk production results. The differences are statistically highly significant for differences between companies 1 and 2 and 1 and 3 (P≤0.001). Milking robots are used more frequently in small private companies as there usually are more cows per robot (61) as compared to larger cooperatively owned companies with several AMS (53 resp. 49 heads/robot).

Table 10 shows that breeding cows of the Czech Fleckvieh cattle were refused by the robot 2.55 times per day on average, while the breeding cows of the Holstein breed only 1.88 times per day. These differences are highly statistically significant (P≤0.001). In herds with both breeds kept together the average number of refusals was calculated to 2.18.

Company size	Type of ownership	Number of cows per 1 robot	n	$\bar{x}$	S_x	F test
Small (1)	private	61	1002	1.11	0.58	231.1**
Medium (2)	cooperative	53	1670	2.18	1.80	1:2***
Large (3)	cooperative	49	5010	2.30	1.28	1:3***

Table 9. Average number of rejections per dairy cow per day according to company size and ownership

Breed	Number of cows per 1 robot	n	$\bar{x}$	S_x	F test
Holstein Cattle (1)	53	4008	1.88	1.01	154.5**
Czech Fleckvieh Cattle (2)	47	2004	2.55	1.59	1:2***
Both breeds (3)	53	1670	2.18	1.80	1:3** 2:3***

Table 10. Average number of rejections per one dairy cow per day according to breed

2.3. Problematic dairy cows

Figure 1 to 3 show percentage counts of problematic cows needed to be lead to the milking robot for a certain amount of time. The lowest number was found for company number 3 (8.2%), the highest for numbers 7 (25.2%) and 6 (15.1%). For the rest of the companies the values fluctuated around 10 % (Figure 1). When focussing on the companies according to size and ownership (Figure 2) the percentage of problematic animals obtained tends to be quite high (20.1%) in large companies as compared to medium large (even cooperative) and small farms where the numbers were kept just above 9%. Differences found between the breeds (Figure 3) point to least problematic individuals belonging to the Fleckvieh cattle (8.8%) as compared to the Holstein breed (14.4%). Veselovský (2005) explains that it is characteristic for classic motivation that the originally neutral stimulus (robot visit) eventually activates a certain behavioural pattern (milking) when induced by a reward here in the form of extra feed. The main target is to create association or connection between a certain activity and the reward. Šťastný (2010) says that the normal state is that 5 to 10% of the cows need being led to the robot box. According to Tančin and Tančinová (2008) thorough attention must here be turned to the quality of the person – keeper, as animals strongly feel the presence of an aggressive or unkind person and his presence tends to induce changes in behaviour, and consequently due to increased stress lead to lower productivity by as much as 10%. Weis et al. (2005) found based on own studies excellent capability of the cows to learn as after three days the majority were capable of entering the robot box without physical assistance from the keepers.

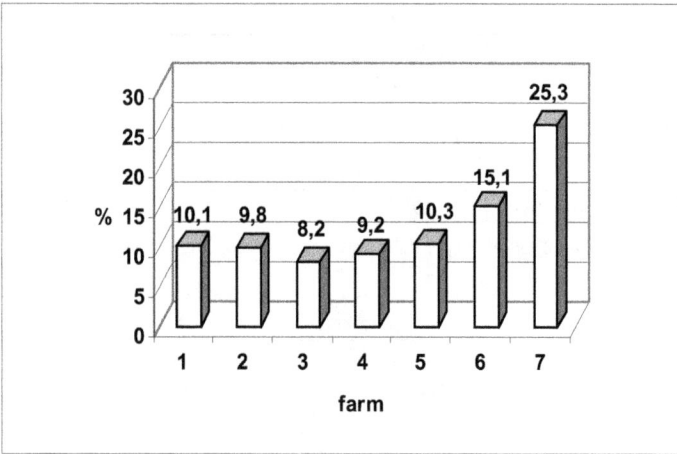

Figure 1. Number of leaded dairy cows to the milking robot in individual companies (in %)

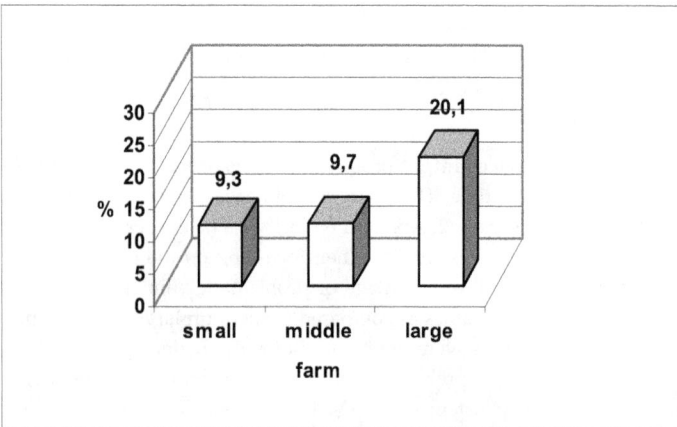

Figure 2. Number of leaded dairy cows to the milking robot according to company size (in %)

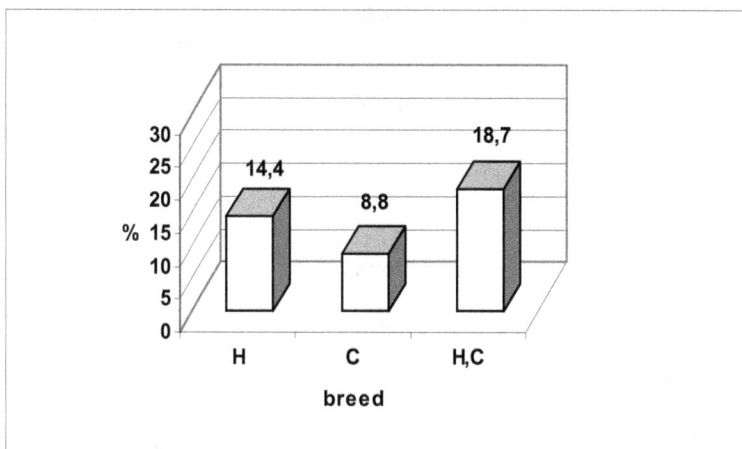

H – Holstein Cattle, C – Czech Fleckvieh Cattle

Figure 3. Number of leaded dairy cows to the milking robot according to breed (in %)

3. Analysis of milk productivity

An agricultural company with milking productivity corresponding to the average values found in the Czech Republic for the Czech Fleckvieh cattle was chosen for a thorough analysis of the milk productivity and the lactation process. Placed at an elevation of 500 to 550 m above sea-level and spread on 750 ha of agricultural land, with 600 ha being used as field lands, its production herd consists of 120 cows.

3.1. Progression of lactation curves

Lactation curves for first, second, third and following lactations are shown in Figure 4. Major differences in milking levels between the first and the remaining lactations are easily seen from the progression of the curves. The highest average milk production was found for the duration of the second lactation, about 10 kg of milk in its beginning as compared to the first lactation curve. During the third and the following lactations there is a slight decrease in milk production, although the progress of the curves resembles that of the second.

Figure 5 shows lactation curves with progression of lactation for cows with starting milk production up to 25 kg, between 25 and 30 kg and above 30 kg of milk. From the differences in the progression of the lactation curves it is possible to state that the initial milk production up to 25 kg has lactation curve very close to the optimal progression, the initial milking is higher by almost 4 kg of milk and this elevated value holds for four whole months. The lactation curve is very steep for high initial milk production where the initial milk increase is only 2.4 kg and keeps for two months which implies insufficient nutrition levels supplied by the company that does not cover the milk production in the initial lactation stage. For the group of cows with initial average milk production over 30 kg the lactation curve suggests no initial milk production stage (no production increase) and there is a continuous decrease of milk production from the beginning of lactation. This decrease is significant enough to level the milk production of this group with group 2 after just one month, thus implying that in given conditions of nutrition and care the cows with higher genetic potential towards milk production cannot make use of this potential and their production abilities are not fully used.

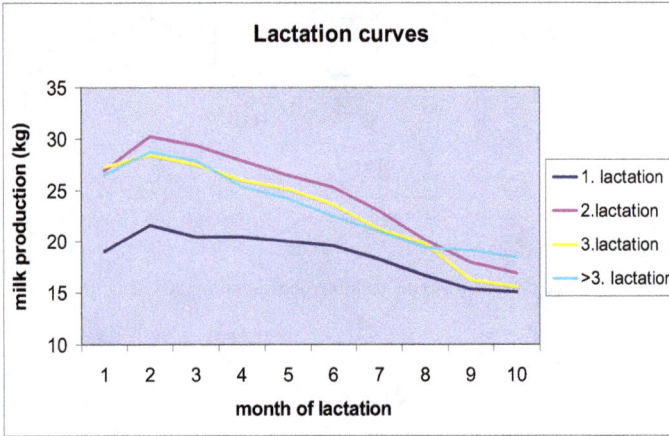

Figure 4. Progression of lactation curves according to lactation order

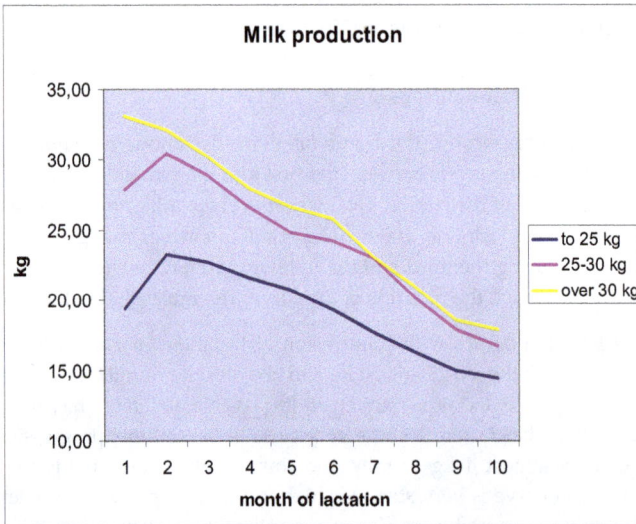

Figure 5. Progression of lactation curve for different milk production levels

Progression of lactation curves for the individual groups of cows (divided according to the initial milk production of up to 25, 25 to 30 and above 30 kg of milk) is shown individually in Figures 6 to 8. Expressing the progression of the curves linearly clarifies the differences in the slope of the individual lines and also in the fraction of the explained variability expressed by the coefficient of determination (R^2). While the fraction of variability for the lactation curve of cows with production up to 25 kg of milk in Figure 6 is 75.11%, the group depicted in Figure 7 with production between 25 and 30 kg exerts 92% and the cows with initial productivity over 30 kg reach 98.5%. This would then imply that we practically have not a lactation curve but lactation straight line. The straight-line parameters shown in Figure 8 ($y = -1.7735x + 35.451$) also show the relatively steep decrease of milk production for this group.

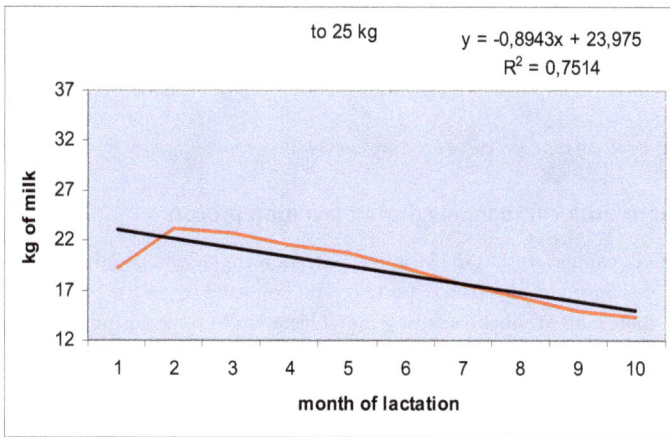

Figure 6. Progression of lactation curve for dairy cows with productivity up to 25 kg of milk

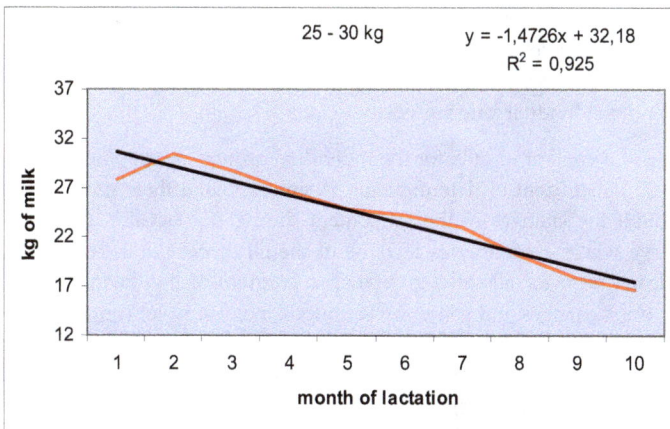

Figure 7. Progression of lactation curve for dairy cows with productivity from 25 to 30 kg of milk

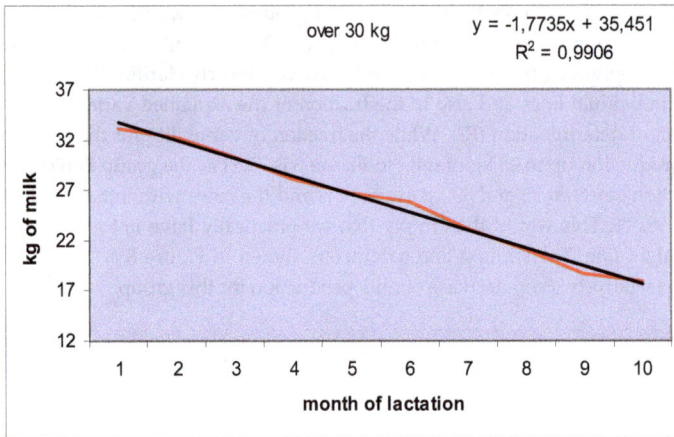

Figure 8. Progression of lactation curve for dairy cows with productivity over 30 kg of milk

3.2. Content of milk components during lactation process

Changes in milk content during the lactation process were judged based on analysis of milk samples taken during performance testing. The followed group of cows was divided according to their milk production during initial lactation into three groups with production up to 25 kg, from 25 to 30 kg and over 30 kg of milk and for each cow fat, protein and lactose content in the milk was established for a period of 10 months from initial lactation.

Values of fat content in milk samples show that there is no significant difference between the fat content for the three groups. The interesting feature are the maximum and minimum values measured for individual cows where individuals with very low fat content (2%) or quite high fat content (over 5% or even over 6%) appear in each of the three groups. These values suggest a possible use of the different fat content in milk as selection criteria for changed requirements on the quality of milk from the buyer or for changed financial evaluation for the individual milk ingredients.

The changes in fat content in milk for the individual groups are graphically represented in Figures 9 to 11. Coefficient of determination is similar for all three groups (R = 0.7623 to 0.7852) although the changes of the fat content during the lactation process resemble a parabolic curve where a continuous increase of the fat content at the end of the lactation process is clearly seen for all three groups. The progress of the theoretical curve for the individual groups changes and shows a continuous increase in fat content from the group with the lowest initial productivity ($y = 0.0196x^2 - 0.1464x + 3.9892$) to the group with the highest one ($y = 0.0085x^2 - 0.0571x + 3.8103$). The inflexion points that are characterized by the top of a parabola and represent the point when an increase in fat content appears after an initial decrease are shown to differ slightly for the three groups. The group with lowest productivity (up to 25 kg) exhibited the lowest fat content 3.71% after 3 months and 24 days of lactation, the second group showed a minimum of 3.70% after 3 months and 6 days and the third group with highest milk production (over 30 kg) reached the minimum after 3 months of lactation with 3.73% of fat.

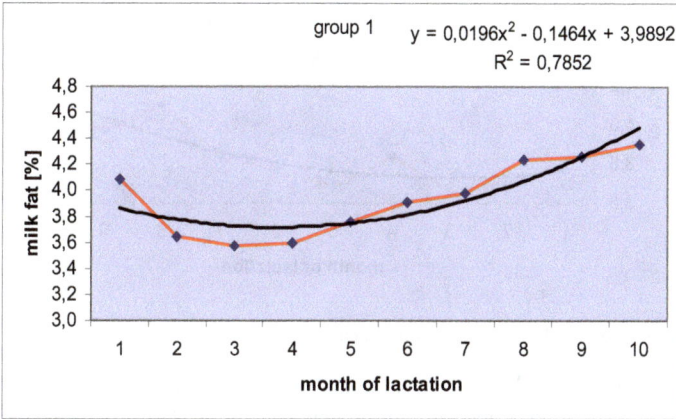

group 1 $y = 0{,}0196x^2 - 0{,}1464x + 3{,}9892$
$R^2 = 0{,}7852$

Figure 9. Milk fat during lactation in milk production up to 25 kg

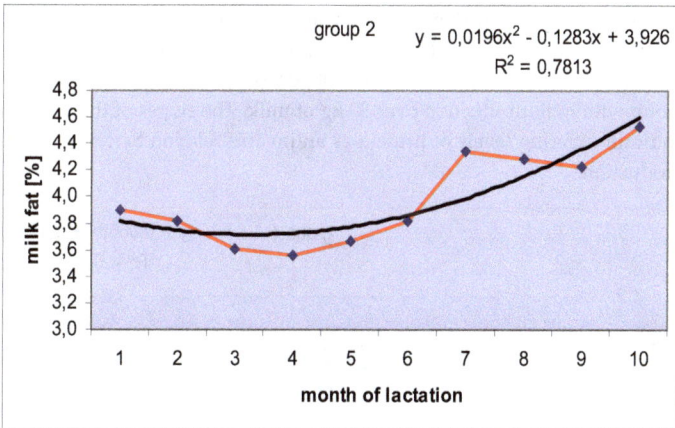

group 2 $y = 0{,}0196x^2 - 0{,}1283x + 3{,}926$
$R^2 = 0{,}7813$

Figure 10. Milk fat during lactation in milk production from 25 to 30 kg

Protein content was analysed for the same group of cows as above. Were founded, that not only lower average values of protein content as compared to fat, but also smaller variations of the values characterised by the spread and determinant deviation. Although the maximum values of protein content found in individual cows (in some cases high above 4%) suggest a possibility of selection towards attempts to increase protein content in milk or to increase the usability of selected dairy products.

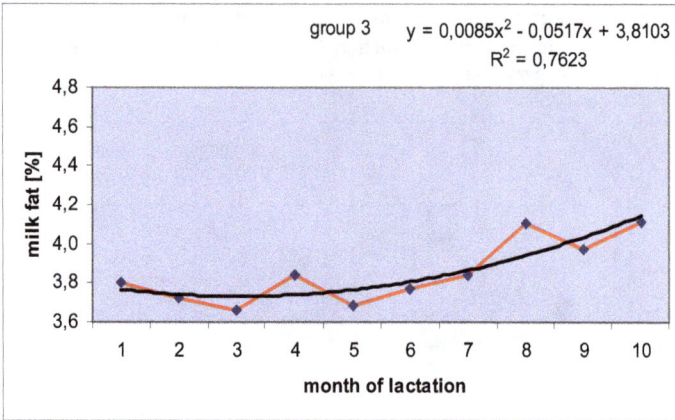

Figure 11. Milk fat during lactation in milk production over 30 kg

Graphic expression of the protein content in milk during the lactation process in Figures 12 to 14 shows a continuous increase in protein content in milk (of course under the presumption that the high protein percentage in colostrum during the first few days after calving is disregarded) with the tendency towards linearity. With increasing initial milk production after calving the real progression of the changes in protein content in milk approaches linearity also inferred by the determinant coefficient that was estimated to 0.8457 for the first group with up to 25 kg of milk, 0.884 for the second group and 0.9951 for the third group with milk production over 30 kg of milk. The slopes of the individual graph lines also indicate differing levels of protein as an inverse relation between protein content and milk production.

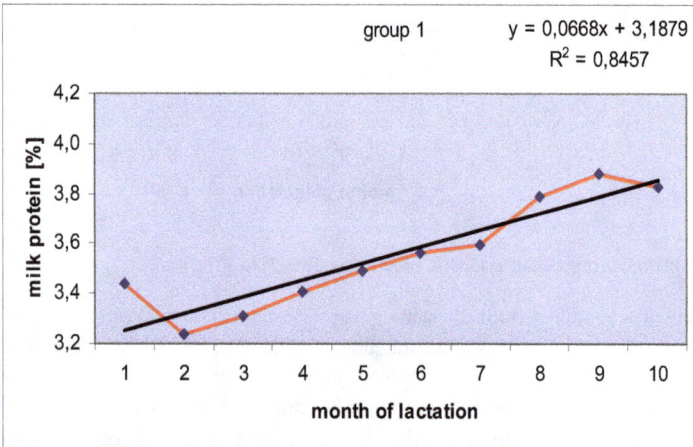

Figure 12. Protein content during lactation in milk production up to 25 kg

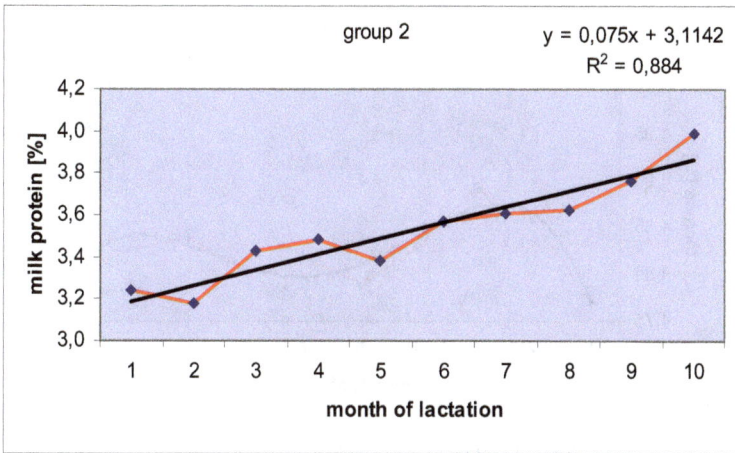

Figure 13. Protein content during lactation in milk production from 25 to 30 kg

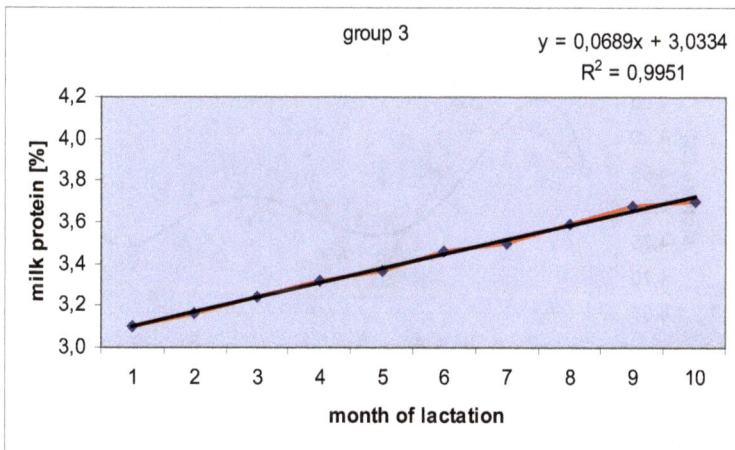

Figure 14. Protein content during lactation in milk production over 30 kg

Lactose content during lactation progress for the individual groups is shown in Figures 15 to 17. Average values of lactose content in the individual lactation months vary between 3.67 and 4.93 % and no trends among the groups or durations can be seen. The graphical expression especially shows that the lactation progression corresponds more to polynomic functions, although by different degrees and precisions (for example the group with milk production to 25 kg has determinant coefficient 0.8401). There is thus no correlation between lactose content and milk production and causes to changes in lactose levels need to be sought elsewhere.

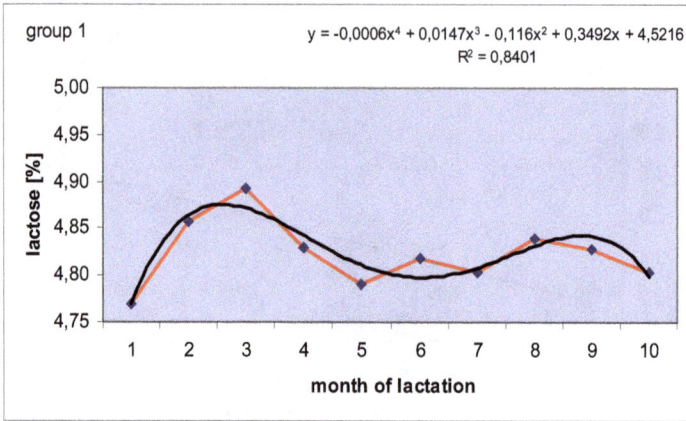

Figure 15. Lactose content during lactation in milk production up to 25 kg

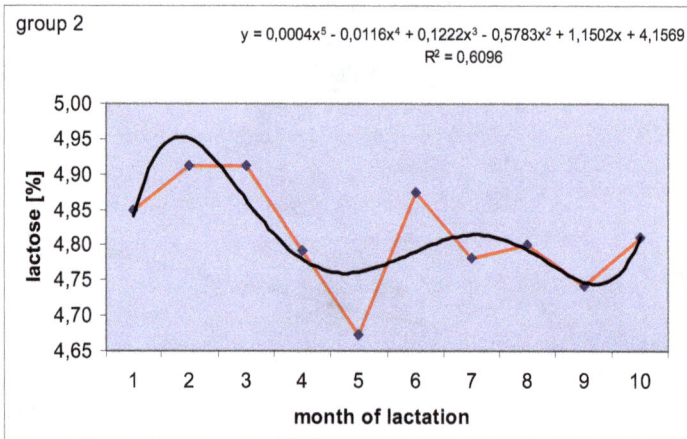

Figure 16. Lactose content during lactation in milk production from 25 to 30 kg

3.3. Correlation between reproduction parameters and milk content

Mutual dependencies between lactation order and parameters of milk productivity and fertility are shown in Table 11 expressed in terms of correlation coefficients. Statistically significant values are given in bold letters. Values in the first column show that satisfactory reproducibility was found for the experimental set (meantime of 375 days, insemination interval 64.4 days and service period of 111.94 days) and also that the milk content corresponds to the expected values. The average number of milking times per day is 2.62. It is clear from the obtained correlation coefficients that there is no connection between lactation order and the other followed parameters. Statistically significant is a positive but

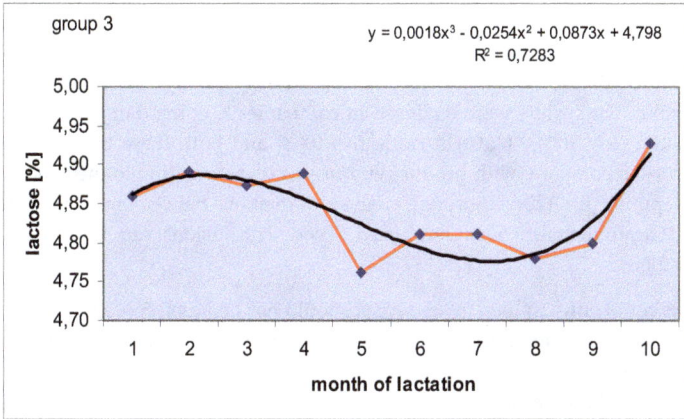

Figure 17. Lactose content during lactation in milk production over 30 kg

slight correlation between the length of insemination interval and the service period (r_{xy} = 0.309) which is quite logical. There is on the other hand a slightly negative statistically significant dependence between protein content in the milk and the length of the insemination interval (r_{xy} = -0.353) that would imply a connection to a deficit in cow nutrition after calving that is expressed by a lower protein content (also by less pronounced rutting period of the cows, their deteriorated health and therefore also in later release of the cows after calving). Low although statistically significant dependence was found between fat and protein content in milk (r_{xy} = 0.290). Increasing number of milking times is shown to have a negative impact on the fat content in milk (r_{xy} = -0.230).

Factor	Average	Standard Deviation	Lact.	Meantime	Ins.int.	SP	Fat	Protein	Number of cows
Lactation	2.50	0.796		0.056	0.039	-0.095	0.053	-0.143	-0.070
Meantime	375.78	48.669	0.056		-0.035	0.181	-0.068	**-0.264**	-0.001
Ins.int.	64.40	22.033	0.039	-0.035		**0.309**	0.131	**-0.353**	-0.080
SP	111.94	54.990	-0.095	0.0181	**0.309**		0.072	-0.214	-0.052
Fat	3.94	0.401	0.053	-0.068	0.131	0.072		**0.290**	**-0.230**
Protein	3.37	0.233	-0.143	**-0.264**	**-0.353**	-0.214	**0.290**		-0.055
Number of cows	2.62	0.305	-0.070	-0.001	-0.080	-0.052	**-0.230**	-0.055	

Table 11. Correlation between milking and reproduction parameters

4. Conclusion

Studies involving evaluation of results on milking with the help of automated milking systems (AMS) of the type Lely Astronaut A3 and finding differences that depend on size, levels of production and ownership were conducted in seven agricultural companies in the Czech Republic. The results were analysed in companies keeping dairy cows of the Czech Fleckvieh cattle (n = 2004), Holstein cattle (n= 4008) and both these breeds together (n = 1670). In a chosen company with productivity results corresponding to the average value of the Czech Republic (n = 120) a thorough analysis of relations between milk production, milk content and reproducibility was conducted. These conclusions can be drawn from the obtained results:

1. The number of cows milked by one robot should not by much exceed 60.
2. Significantly higher milk production was found in small privately owned companies as compared to large cooperative companies.
3. Disregarding the type of company where the cows are kept, the Holstein cattle breed shows significantly much higher milk production than the cows of the Czech Fleckvieh cattle.
4. Statistically significant differences in the average number of milking times per day were found between the individual farms. The frequency of milking time varied between 1.97 and 2.67 times per day.
5. Differences in the number of milking times according to the ownership are significant, though much smaller (2.34 to 2.47).
6. The Holstein cattle were statistically more often milked than the Czech Fleckvieh cattle (2.45 compared to 2.32).
7. The number of robot refusals is very individual for each stable and varies between 0.72 to 3.66 times per cow and day.
8. The refusals are much less frequent in small privately owned farms (1.11) as compared to medium (2.18) and large (2.30) cooperative companies.
9. Cows of the Czech Fleckvieh breed are significantly more often refused by the robot (2.55) in comparison to the Holstein dams.
10. The numbers of cows that need to be taken to the robot by a keeper vary between 8.2 to 25.3% in the different companies. It is much more frequent in large companies (20.1%) than in small (9.3%) and medium sized ones (9.7%).
11. The need to lead the cows to the robot is lower for the Czech Fleckvieh (8.8%) as compared to the Holstein breed (14.4%).
12. A significant difference was found in the progression height of the first lactation curve and the curves of the following lactations.
13. Only cows with initial milk production below 25 kg of milk exhibited an optimal progression of the lactation curve for an average productivity of the herd at 6450 kg of milk. Cows with milk production of 25 to 30 kg showed a steep lactation curve while the cows with milk production above 30 kg did not show any increasing milk flow and their lactation curve fell rapidly already from the beginning.

14. For high productivity cows (with over 30 kg as initial lactation) the progression of lactation resembled a straight line with a steep decreasing slope (R = 0.9875).

15. There are no significant differences found in milk fat content for the three groups of cows with different milk production (up to 25, 25 to 30 and above 30 kg). The differences in fat content in the milk of individual cows (from 2 to 6% of fat) would suggest some selective use.

16. The percentage content of fat in milk following the lactation curve resembles a parabola with the lowest point in the fourth month. This point occurs sooner (the third month after calving) for cows with higher milk production.

17. Protein content is less variable in comparison to the fat content and increases nearly linearly during the lactation progress, especially for high productivity cows (R= 0.9951).

18. Differences in lactose content during the progress of lactation are not dependent on the milk production levels and fluctuate within 0.2%.

a. Protein content in milk shows negative dependence on the meantime (r_{xy} = - 0.264) and the insemination interval (r_{xy} = -0.353).

b. Statistically significant inverse dependency was also found between the number of milking times per day and the fat content in milk (r_{xy} = -0.230).

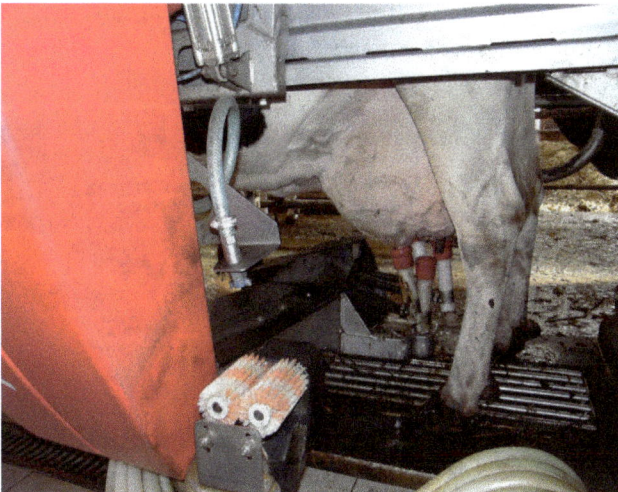

Figure 18. Milking by AMS (photo A. Váchová)

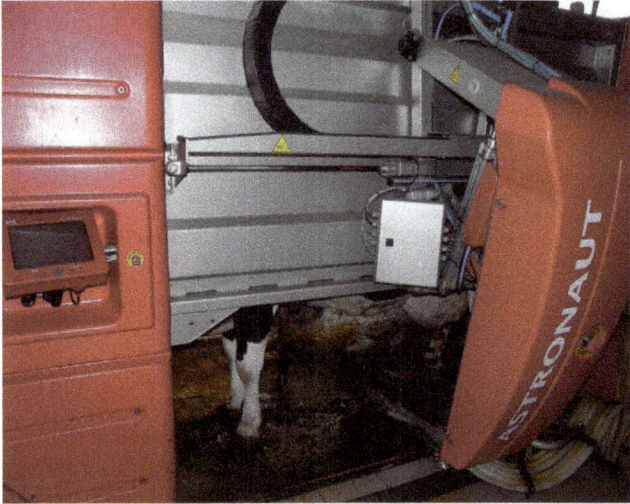

Figure 19. Computer install into AMS allowed getting the data during lactation (photo A. Váchová)

Figure 20. Cow get concentration feed in robot (photo A. Váchová)

Author details

Miroslav Maršálek, Jarmila Voříšková and Jana Zedníková
University of South Bohemia in Ceske Budejovice, Faculty of Agriculture,
Czech Republic

Acknowledgement

This study was supported by the Ministry of Agriculture of the Czech Republic, NAZV QH91260.

5. References

Bouška, J. (2006). *Chov dojeného skotu*. Profi Press, ISBN 80-86726-16-9, Praha

Debrecéni, O.; Bulla, J. & Vavrišínová, K. (1999). *Základy chovu kráv*. ÚVTIP, ISBN 80-85330-71-7, Nitra

Doležal, O. (2000). *Mléko, dojení, dojírny*. Agrospoj, Praha

Doležal, O. (2002). *Komfortní ustájení vysokoprodukčních dojnic*. VÚŽV Praha-Uhříněves, ÚVTIZ, ISBN 80-86454-23-1, Praha

Doležal, O.; Bílek, M. & Dolejš, J. (2004). *Zásady welfare a nové standardy EU v chovu skotu*. VÚŽV Uhříněves, ISBN 80-86454-51-7, Praha

Fleischmannová, H. (2005). Dojící roboty v podmínkách české prvovýroby mléka. *Náš chov*, Vol. LXV, No.1, pp.10-14, ISSN 0027-8068

Hanuš, O.; Jedelská, R.; Hering, P.; Klimeš, M.; Genčurová, V.; Janů, L. & Kopecký, J. (2006). Konstrukce algoritmu pro efektivní sofistikované grafické vyhodnocování výsledků složení a kvality bazénových vzorků mléka. *Výzkum v chovu skotu*. Vol.XLVIII, No.3, pp. 1 – 26, ISSN 0139-7265

Havlík, V. (2007). Dojící roboty Lely Astronaut ve světě a v České republice. *Náš chov*, Vol. LXVII, No.1, pp. 31-32, ISSN 0027-8068

Kic, P. & Nehasilová, D. (1997). *Dojící roboty a jejich vliv na zdravotní stav mléčné žlázy*. ÚZPI, ISBN 80-86153-32-0, Praha

Kopeček, P. & Machálek, A. (2009). *Ekonomická analýza výroby mléka na farmách s dojením roboty a v dojírnách*. Agritech Science (3), ISSN 1802-8942, pp. 1-8

Kruip, T. A. M.; Morice, H.; Robert, M. & Ouweltjes, W. (2002). Robotic Milking and Its Effect on Fertility and Cell Counts. *Journal of Dairy Science*, Vol. 85, No. 10, pp. 2576-2581, ISSN 0022-0302

Kvapilík, J. (2005). *Automatizované dojení krav*. ČMSCH Praha, ISBN 80-86454-58-4, 59 pp.

Kvapilík, J.; Růžička, Z. & Bucek, P. (2011). *Ročenka - Chov skotu v České republice*. ČMSCH, ISBN 978-80-904131-2-2, Praha

Machálek, A. (2009). Roboty na českých farmách. *Náš chov*, Vol. LXIX, No.12, pp. 13-14, ISSN 0027-8068

Maršálek, M.; Kottová, M.; Zedníková, J. & Voříšková, J. (2009). Ekonomický efekt zavedení dojících robotů v zemědělském podniku. *Agritech Science*, No. 2 , pp. 1-3, ISSN 1802-8942

Přikryl, M. (1997). *Technologická zařízení staveb živočišné výroby.* Tempo Press, ISBN 80-901052-0-3 Praha

Rodenburg, J. (2002). Robotic Milkers: What, Where…and How Much!? Dairy Management Conference, pp. 1-18, Ohio, December 16 -17, 2002

Rotz, C. A.; Coiner, C. U. & Soder, K. J. (2003). Automatic Milking System, Farm Size and Milk Production. *Journal of Dairy Science*, Vol. 86, No. 12, pp. 4167-4177, ISSN 0022-0302

Stelwagen, K. (2001). Effect of Milking Freguency on Mammary Functioning and Shape of the Lactation Curve. *Journal of Dairy Science*, Vol. 84, E. Suppl., pp. 204-211E, ISSN 0022-0302

Sychra, L. (2001).: *Nové trendy a poznatky v chovu skotu.* MZLU, 2001, ISBN 80-7157-541-0, Brno

Šťastný, V. (2010). Dojení robotem. 3.1.2010, Available from http://www.zootechnik.cz/zoodr3.php

Tančin, V. & Tančinová, D. (2008). *Strojové dojenie kráv a kvalita mlieka.* SPU, ISBN 978-80-88872-80-1, Nitra

Vegricht, J. (2002). Study of Using Automatic Milking Systems on Large Dairy Farms. *Research In Agricultural Engineering*, Vol. 48, No. 1, pp.1-6, ISSN1212-9151

Veselovský, Z. (2005). *Etologie-Biologie chování zvířat.* Academia, ISBN 80-200-1331-8, Praha

Wagner-Storch, A.M. & Palmer, R.V. (2003). Feeding Behavior, Milking Behaviourand Milk Yields of Cows Milked in a Parlor versus an Automatic Milking Sysstem. *Journal of Dairy Science*, Vol.86, No. 4, pp.1494-1502 ISSN 0022-0302

Weiss, D.; Moestl, E. & Bruckmaier, R. M. (2005). Physiological and Behavioural Effects of Changeover from Conventional to Automatic Milking in Dairy Cows with and without Previous Experience. *Veterinary Medicine*, Vol. 50, No.6, pp. 253-261, ISSN 0375-8427

Permissions

The contributors of this book come from diverse backgrounds, making this book a truly international effort. This book will bring forth new frontiers with its revolutionizing research information and detailed analysis of the nascent developments around the world.

We would like to thank Narongsak Chaiyabutr, for lending his expertise to make the book truly unique. He has played a crucial role in the development of this book. Without his invaluable contribution this book wouldn't have been possible. He has made vital efforts to compile up to date information on the varied aspects of this subject to make this book a valuable addition to the collection of many professionals and students.

This book was conceptualized with the vision of imparting up-to-date information and advanced data in this field. To ensure the same, a matchless editorial board was set up. Every individual on the board went through rigorous rounds of assessment to prove their worth. After which they invested a large part of their time researching and compiling the most relevant data for our readers. Conferences and sessions were held from time to time between the editorial board and the contributing authors to present the data in the most comprehensible form. The editorial team has worked tirelessly to provide valuable and valid information to help people across the globe.

Every chapter published in this book has been scrutinized by our experts. Their significance has been extensively debated. The topics covered herein carry significant findings which will fuel the growth of the discipline. They may even be implemented as practical applications or may be referred to as a beginning point for another development. Chapters in this book were first published by InTech; hereby published with permission under the Creative Commons Attribution License or equivalent.

The editorial board has been involved in producing this book since its inception. They have spent rigorous hours researching and exploring the diverse topics which have resulted in the successful publishing of this book. They have passed on their knowledge of decades through this book. To expedite this challenging task, the publisher supported the team at every step. A small team of assistant editors was also appointed to further simplify the editing procedure and attain best results for the readers.

Our editorial team has been hand-picked from every corner of the world. Their multi-ethnicity adds dynamic inputs to the discussions which result in innovative outcomes. These outcomes are then further discussed with the researchers and contributors who give their valuable feedback and opinion regarding the same. The feedback is then collaborated with the researches and they are edited in a comprehensive manner to aid the understanding of the subject.

Apart from the editorial board, the designing team has also invested a significant amount of their time in understanding the subject and creating the most relevant covers. They scrutinized every image to scout for the most suitable representation of the subject and create an appropriate cover for the book.

The publishing team has been involved in this book since its early stages. They were actively engaged in every process, be it collecting the data, connecting with the contributors or procuring relevant information. The team has been an ardent support to the editorial, designing and production team. Their endless efforts to recruit the best for this project, has resulted in the accomplishment of this book. They are a veteran in the field of academics and their pool of knowledge is as vast as their experience in printing. Their expertise and guidance has proved useful at every step. Their uncompromising quality standards have made this book an exceptional effort. Their encouragement from time to time has been an inspiration for everyone.

The publisher and the editorial board hope that this book will prove to be a valuable piece of knowledge for researchers, students, practitioners and scholars across the globe.

List of Contributors

Sammy K. Kiplagat and Moses K. Limo
Department of Biochemistry and Molecular Biology, Egerton University, Egerton, Kenya

Isaac S. Kosgey
Department of Animal Sciences, Egerton University, Egerton, Kenya & Laikipia University College, Nyahururu, Kenya

Yousef Mehmannavaz
Department of Animal Science, Maragheh Branch, Islamic Azad University, Maragheh, Iran

Abolfazl Gorbani
Department of Animal Science, Shabestar Branch, Islamic Azad University, Shabesta, Iran

Marcel Amills and Alí Zidi
Department of Animal Genetics, Center for Research in Agricultural Genomics (CSIC-IRTA-UAB-UB), Universitat Autònoma de Barcelona, Bellaterra, Spain

Jordi Jordana
Departament de Ciència Animal i dels Aliments, Universitat Autònoma de Barcelona, Bellaterra, Spain

Juan Manuel Serradilla
Departamento de Producción Animal, Campus de Rabanales, Universidad de Córdoba, Córdoba, Spain

Peter Hristov, Denitsa Teofanova and Georgi Radoslavov
Institute of Biodiversity and Ecosystem Research, Bulgarian Academy of Sciences, Bulgaria

Ivan Mehandzhiyski
Agricultural and Stockbreeding Experimental Station, Agricultural Academy, Bulgaria

Lyuben Zagorchev
Sofia University "St. Kliment Ohridski", Faculty of Biology, Bulgaria

Juan José Arranz and Beatriz Gutiérrez-Gil
Department of Animal Production, Faculty of Veterinary Sciences, University of León, Spain

Elli Pärna, Tanel Kaart, Heli Kiiman and Haldja Viinalass
Estonian University of Life Sciences, Tartu, Estonia
Bio-Competence Centre of Healthy Dairy Products, Tartu, Estonia

Tanel Bulitko
Animal Breeders´ Association of Estonia, Keava, Rapla County, Estonia

Narongsak Chaiyabutr
Department of Physiology, Faculty of Veterinary Science, Chulalongkorn University, Bangkok, Thailand

M.E. Caffaro, C. Suárez, D.L. Roldán, and M.A. Poli
Instituto de Genética, CICVyA-INTA, Castelar, Provincia de Buenos Aires, Argentina

Antonia Kécya França Moita
Department of Animal Science, Universidade Federal da Bahia, Brazil

Paulo Sávio Lopes, Robledo de Almeida Torres and Ricardo Frederico Euclydes
Department of Animal Science, Universidade Federal de Viçosa, Brazil

Humberto Tonhati
Department of Animal Science, Universidade Estadual Paulista, Brazil

Ary Ferreira de Freitas
Embrapa Gado de Leite, Brazil

Sezen Ocak and Sinan Ogun
Zirve University / Middle East Sustainable Livestock, Biotechnology and Agro-Ecology Research and Development Centre, Turkey

Rocky R.J. Akarro
University of Dar es Salaam, Tanzania

M. Subhan Qureshi
Agricultural Unviersity, Peshawar, Pakistan

Hysen Bytyqi
Department of Livestock and Veterinary Sciences, Faculty Agriculture and Veterinary, University of Prishtina; Str. "Lidhja e Pejes", Prishtina – Kosovo

A.M. Descalzo, L.Rossetti, R. Páez, G. Grigioni, P.T. García, L. Costabel, L. Negri, L.Antonacci, E. Salado, G. Bretschneider, G. Gagliostro, E. Comerón and M.A. Taverna
Instituto Nacional de Tecnología Agropecuaria (INTA), Argentina

Miroslav Maršálek, Jarmila Voříšková and Jana Zedníková
University of South Bohemia in Ceske Budejovice, Faculty of Agriculture, Czech Republic